History of
ORGAN AND CELL TRANSPLANTATION

History of ORGAN AND CELL TRANSPLANTATION

Editors

Nadey S Hakim

Vassilios E Papalois

St. Mary's Hospital, London

ICP

Imperial College Press

Published by

Imperial College Press
57 Shelton Street
Covent Garden
London WC2H 9HE

Distributed by

World Scientific Publishing Co. Pte. Ltd.
5 Toh Tuck Link, Singapore 596224
USA office: Suite 202, 1060 Main Street, River Edge, NJ 07661
UK office: 57 Shelton Street, Covent Garden, London WC2H 9HE

Library of Congress Cataloging-in-Publication Data
History of organ and cell transplantation / [edited by] Nadey S. Hakim & Vassilios E. Papalois.
p. ; cm.
Includes bibliographical references and index.
ISBN 1-86094-209-1 (alk. paper)
1. Transplantation of organs, tissues, etc.--History. I. Hakim, Nadey S., 1958- II. Papalois, Vassilios E.
[DNLM: 1. Organ Transplantation--history. 2. Cell Transplantation--history. 3. History of Medicine, 20th Cent. 4. Transplantation Immunology. WO 211.1 H673 2003]
RD120.7 .H576 2003
617.9'5'09--dc21 2002038719

British Library Cataloguing-in-Publication Data
A catalogue record for this book is available from the British Library.

Cover paintings by Professor Sir Roy Calne.
Top row: George Hutchings, Trudy Elion, Sir Peter Medawar
Centre row: Jean Dausset, Joseph Murray
Bottom row: Alexis Carell, The Organ Donor (sculpture)

Printed in Singapore by Mainland Press

DEDICATION

to my wife, Nicole, and children

Alexandra, David, Andrea and Gabriella

~NH

to my wife, Vana, and children

Timos, Sundy, Zoe and Pavlos

~VP

CONTENTS

CONTRIBUTORS

Abu-Elmagd, Kareem M., MD, PhD, FACS
Professor of Surgery
Director, Intestinal Rehabilitation and Transplant Centre
University of Pittsburgh School of Medicine
Thomas E. Starzl Transplantation Institute
Pittsburgh, USA

Bollinger, Randal R., MD, PhD
Professor and Chief, General Surgery
Duke University Medical Centre
Durham, USA

Calne, Sir Roy, FRS
Emeritus Professor of Surgery
Department of Surgery
Cambridge University
Cambridge, UK

Chavers, Blanche M., MD
Professor of Paediatrics
Division of Paediatric Nephrology
University of Minnesota
Minneapolis, USA

Dausset, Jean, MD
Professor of Medicine
Centre d'etude du Polymorphisme Humain (CEPH)
Paris, France

Gray, Derek D.W.R., DPhil, FRCP, FRCS
Professor and Consultant in Transplantation
The Nuffield Department of Surgery
John Radcliffe Hospital
Oxford, UK

Groth, Carl G., MD, PhD
Professor of Transplantation Surgery
Karolinska Institute
Huddinge Hospital
Huddinge, Sweden

Hakim, Nadey S., MD, PhD, FRCS, FACS, FICS
Consultant Transplant and General Surgeon and Surgical Director of the Transplant Unit
St Mary's Hospital
London, UK

Holzknecht Robert
Laboratory Research Specialist
University of North Carolina — Chapel Hill
Comprehensive Cancer Center
Chapel Hill, USA

Kamps, Melissa, PharmD
Clinical Pharmacist, Solid Organ Transplant
University of Minnesota
Minneapolis, USA

Kootstra, Gauke, MD, PhD
Professor of Surgery
Universiteit of Maastricht
Faculty of Medicine
Maastricht, The Netherlands

Kroshus, Timothy J., MD, PhD
Cardiac Surgical Associates
Minneapolis, USA

Kshettry, Vibhu R., MD, FRCS, FACS
Cardiac Surgical Associates
Minneapolis, USA

Madden, Brendan P., MD, MSc, FRCPI
Consultant Thoracic and Transplant Physician
St George's Hospital
London, UK

Matas, Arthur, MD
Professor of Surgery
Director, Renal Transplantation
Department of Surgery
University of Minnesota
Minneapolis, USA

McCann, Shaun R., MB, FRCPI, FRCPEdin, FRCPath
Professor of Haematology
St James' Hospital
Dublin, Republic of Ireland

Najarian, John S., MD
Emeritus Professor and Chairman of Surgery
Department of Surgery
University of Minnesota
Minneapolis, USA

Papalois, Vassilios E., MD, PhD, FICS
Consultant Transplant and General Surgeon
Transplant Unit
St Mary's Hospital
London, UK

Platt, Jeffrey L., MD
Professor of Surgery
Department of Surgery
Duke University Medical Centre
Durham, USA

Rapaport, Felix T., MD†
Distinguished Professor of Surgery
Director, Transplantation Service
Stonybrook Health Sciences Centre
State University of New York at Stony Brook
USA

Reinsmoen, Nancy L., PhD
Director, Clinical Transplantation Immunology Lab
Department of Pathology
Duke University Medical Centre
Durham, USA

Rostron, Chad K., FRCS
Consultant Surgeon
Corneal Transplant Unit
St George's Hospital
London, UK

Sugarman, Jeremy, MD, MPH, MA
Director
Center for the Study of Medical Ethics and Humanities
Duke University Medical Center
Durham, USA

Sutherland, David E.R., MD, PhD
Professor of Surgery
Head, Transplant Division
Department of Surgery
University of Minnesota
Minneapolis, USA

Ward, Frances E., PhD
Professor Emeritus
Department of Immunology
Duke University Medical Centre
Durham, USA

FOREWORD

Sir Peter Medawar was the first elected President of the Transplantation Society in 1968. For 50 years he was also my husband. You might think that during this time, enough of his encyclopaedic knowledge of transplantation would have rubbed off on me to enable me to write a foreword to this book with little effort and much pleasure. Unfortunately this was not so — I do not know nearly enough to do the book justice, but I did remember his excellent advice when starting to write anything on any subject: "Remember", he said, "the values of cogency, brevity and clarity — and clarity is the greatest of these three". I have had a wonderful time reading the 19 chapters which cover all aspects of the history of organ transplants since they became possible.

Apart from this knowledge, what the book has given me is enormous admiration for the skill, intelligence and patience of the research workers and the people with whom they worked. It is no exaggeration to claim that organ and cell transplantation has developed into the greatest therapeutic advancement of the second half of the 20th century. Transplantation has fascinated scientists and the general public for many reasons — almost all medical specialities are involved and no other discipline demands such an extensive co-operation between scientists, surgeons, physicians, and their patients. As for the patients, their appreciation of the new life that is given them is immeasurable, especially when contrasted with what they thought of the earlier treatments and their prognoses.

Lady Jean Medawar

FOREWORD

Sir Peter Medawar was the first elected President of the Transplantation Society in 1968. For 50 years he was also my husband. You might think that during this time enough of his encyclopaedic knowledge of transplantation would have rubbed off on me to enable me to write a Foreword to this book with little effort and much pleasure. Unfortunately this was not so — I do not know nearly enough to do the book justice, but I did remember his excellent advice when starting to write anything on any subject. "Remember", he said, "the values of economy, brevity and clarity — and clarity is the greatest of these three." I have had a wonderful time reading the 19 chapters which cover all aspects of the history of organ transplants since they became possible.

[illegible] the book has given me is enormous admiration for the [illegible] and the people [illegible] their [illegible]. It is no exaggeration [illegible] advances in [illegible] century. Transplantation has fascinated scientists and the general public for many decades — almost all [illegible] or other discipline demands such an extensive cooperation between scientists, surgeons, physicians and their patients. As for the patients, their appreciation of the new life that is given them is immeasurable, especially when contrasted with what they thought of the earlier treatments and their prognoses.

Lady Jean Medawar

PREFACE

Organ and cell transplantation is the greatest therapeutic advancement of the second half of the 20th century. Of all medical specialities, the pioneers of transplantation make up the largest number of experts awarded with, or nominated for, the Nobel Prize.

Over the years, transplantation has fascinated the scientific community as well as the general public for a variety of reasons. The development of transplantation involved almost all medical specialities. In the history of medicine, perhaps there is no other better example of so extensive a co-operation and exchange of knowledge and experience among basic scientists, surgeons and physicians in achieving a common goal. The rapid progress of transplantation has made it necessary for doctors to "re-write" the medical textbooks, since nothing was as before, and they had to deal with a great spectrum of post-transplantation issues, such as the physiology of transplanted organs, recurrence of initial disease in the transplanted organs, complications of immunosuppressive drugs, infectious diseases and cancer, maternity, child development and care for the elderly. Finally, the development of transplantation raised a great number of ethical issues regarding the acceptance of brain death, pre-death consent for organ donation, family consent for organ donation, live related donation, organ availability and sharing, selling of organs and, more recently, issues related to our right to use animal organs for transplantation and the possibility of transmitting infections from animals to humans.

The pioneers of transplantation proved in the modern world what Hippocrates said 2,500 years ago: "a great doctor has the value and the strength of many other people together". They were inspired and encouraged by their patients and their families (the "pioneers" that history will never fully be able to name and credit), and they tried what at times looked impossible, knowing well that, as Plato said: "what is of real value in life is not an effort within your grasp with secured success but an effort beyond your strength with a great risk of failure". They distinguish themselves for their scientific achievements as well as for their dedication, respect and humane behaviour towards their patients. Their pioneering clinical efforts were always characterised by the spirit so vividly expressed by Kant: "no matter how high are the ambitions and how great are the stakes, humanity should always be treated as an end only and never as a means".

In this book, we present in a systematic way the historic hallmarks of the amazing field of transplantation. We strongly believe that, concrete knowledge of the past is essential in order to understand the present and open the way to the future. We are truly grateful to all the contributors for their expertise in a sound, smooth and elegant way. History must be as objective as humanly possible, but at the same time it should also be something more than a thorough report of events. It must reflect the human efforts, thoughts and feelings related to those events. It is only this way that we can (rephrasing Eliot) advance information to knowledge and knowledge to wisdom. We would also like to express our sincere gratitude to Lady Jean Medawar for her inspiring foreword, and to Sir Roy Calne for his magnificent painting of the book cover.

We hope that, through the fascinating journey of the history of organ and cell transplantation, the readers of this book will discover the triumph of the human spirit. What perhaps is even more fascinating is that this journey is anything but over, as a matter of fact, it has just begun…

Nadey S. Hakim
Vassilios E. Papalois
London, January 2003

I

THE HISTORY OF HLA AND TRANSPLANTATION IMMUNOLOGY

Nancy L Reinsmoen
Frances E Ward

Introduction

Scientific advances in the fields of immunology and genetics provided a fertile environment for the discovery and characterization of the human major histocompatibility complex (HLA). By the middle of the 20th century, Landsteiner,[1] Levine[2] and Wiener[3] had applied principles of mendelian genetics to the study of human erythrocyte antigens; and Tyzzer[4] and Little[5] had demonstrated a genetic basis for susceptibility or rejection of tumor grafts. Snell[6] used the congenic strains of mice he had developed to identify several loci, designated histocompatibility or H genes, that were important in allograft rejection. At about the same time, Gorer[7] identified a genetic factor, called II, that was present on red cells and other tissues. If this inherited factor was present on donor tissue and absent in the recipient, it induced an immune response that destroyed the graft and could also induce an antibody reactive with antigen II, present on normal and tumor tissues. Together with Lyman,[8] Gorer and Snell showed that Gorer's antigen II segregated with Snell's strongest H antigen. It appeared they had recognized the same genetic factor and they called it H-2. We now know that H-2, the mouse major histocompatibility complex, is not a single genetic locus but rather a complex of linked genes.

The classic human skin graft experiments of Medawar and Gibson[9] in 1943 demonstrated the immunologic specificity of allograft rejection — donor-specific response, systemic immunity and immunologic memory. Shortly thereafter, Owen[10] showed that most dizygotic cattle twins were red cell chimeras; these chimeric animals were tolerant of skin grafts from each other but rejected grafts from a third party.[11] Bauer[12] similarly showed that skin grafts between human twins were accepted while those between other family members were rejected.

Concomitant with these advances in immunogenetics, vascular surgery had advanced to the point where it was technically feasible to perform renal transplants.[13] In 1954, clinical transplantation became a reality when Murray[14] performed the first successful kidney transplant between monozygotic twins. Defining and characterizing the human histocompatibility system and establishing its role in transplantation immunology became clinically relevant.

Serological Definition of the Human Major Histocompatibility Complex (MHC): Class I Antigens

In man, the major genetic system involved in transplantation rejection and antigen-specific immune responses is called HLA. Today, we know HLA is a gene complex found in a four-megabase region on the short arm of chromosome 6. Three groups of MHC loci have been described: Class I, Class II and Class III. The first alloantibodies to Class I HLA were identified by Dausset[15, 16] and Miescher and Fouconnet[17] in leukemia patients who had received blood transfusions. The first antigen described by Dausset was called MAC and has subsequently been shown to be approximately equivalent to the currently recognized HLA-A locus antigen, A2. Payne[18, 19] and van Rood[20] described additional antibodies found in sera obtained from multiparous women, enabling a more detailed study of the human MHC. Payne and Rolfs[18] showed that antibodies contained in maternal serum reacted with antigens on the leukocytes of their husbands and newborn infants, demonstrating that pregnancy as well as blood transfusions could induce anti-leukocyte antibodies. Payne[19] and van Rood[20] and their co-workers each described an allelic system, designated LA1, LA2 and LA3 (LA) and 4a and 4b (Four), respectively. Although the genetic basis for the leukocyte antigens that these investigators detected was unknown at the time, it was subsequently shown that these two allelic systems detected products of two distinct loci, HLA-A and HLA-B.

Initially, attempts to understand the genetic basis for human histocompatibility antigens were frustrating. Serologic studies of twins by Dausset and Brecy[21] and by Lalezari and Spaet[22] demonstrated that monozygotic twins reacted similarly while dizygotic pairs did not. Early family studies by Payne and Hackel[23] were confounded by the complex mixtures of antibodies in available sera. van Rood[24] also studied a number of families and reported evidence for apparent linkage and recombination. Based on population studies, Dausset *et al.*[25] proposed that the HLA system was a single locus analogous to the mouse H-2 system. A collaborative study of ten relatively large families by Bach and Amos,[26] using antibodies with more restricted specificities, demonstrated that the serologically defined leukocyte antigens *and* the mixed leukocyte response were inherited as a unit from parent to child. They later found a few exceptional siblings who were serologically identical but stimulated[27] in mixed leukocyte cultures. These were the first data showing that the serologically defined specificities and the cellularly defined mixed leukocyte reactively were products of distinct genes. Although the LA and Four series of antigens were indirect evidence for at least two loci within the human MHC, it was not formally proven until 1969, when Kissmeyer-Nielsen[28] and Ward *et al.*[29] published HLA recombinant families. Data obtained by Sandberg *et al.*[30] and Svejgaard *et al.*[31] suggested the existence of a third serologically defined locus (now called HLA-C), based on the reactivity of sera that gave odd patterns of reactivity and were associated with particular HLA-B specificities. A recombinant individual, found by Low *et al.*[32] permitted the new locus to be placed between HLA-A and HLA-B. Recombination between HLA-A and HLA-B occurs with a frequency of about 0.8%;[33,34] that between HLA-B and HLA-C is much less frequent and HLA-C maps closer to HLA-B than to HLA-A.

Although leukoagglutination was first used to identify human histocompatibility antigens, the complement dependent microlymphocytotoxicity assay[35, 36] soon became the serological method of choice due to its greater reproducibility and miniaturization. Rabbit serum as a source of complement,[37] platelet absorption of sera to improve specificity,[38] and introduction of a wash step[39] or antiglobulin reagent[40] to increase sensitivity, further improved the reliability of serological typing. Establishment of collaborative International Histocompatibility Workshops, in which antisera and other reagents were exchanged among investigators worldwide, contributed greatly to the rapid serological characterization of the HLA-A, -B and -C gene products.

Initial Use of Cell Culture Methodologies to Detect MHC Products

In 1964, Bain *et al.*[41] and Bach and Hirschhorn[42] reported that when mixtures of leukocytes from two unrelated individuals were cultured together for several days, large DNA synthesizing lymphoblasts appeared in the cultures. Bach and Hirschhorn[42] suggested quantitation of the degree of transformation in these cultures might measure the histocompatibility of recipient-donor combinations. These tests developed into what is known as the mixed lymphocyte culture (MLC). These early studies revealed that lymphocytes from monozygotic twins did not mutually stimulate in MLC while lymphocytes from HLA-A, -B, -C non-identical siblings and unrelated individuals usually stimulated each other in MLC. Based on these observations, the reactivity observed in the MLC was concluded to be linked to the HLA region. Subsequent studies of recombinant families revealed that strong mixed lymphocyte culture reaction (MLR) was controlled by a separate chromosomal region (now termed HLA-D) distinct from the genes encoding the Class I HLA-A, -B, -C antigens.[27, 43–45] In 1976, Bach and van Rood[46] summarized the HLA system by describing three types of antigens associated with the MHC: (1) SD, the classical serological-defined antigens, HLA-A, -B, -C; (2) Ia-like antigens detected by serological techniques on a limited cell population (B-cells and macrophages) analogous to the immune response (Ia) antigens in mice; and (3) LD antigens or lymphocyte-defined antigens which cause cellular proliferation in MLC.

Two approaches were used to type for the HLA-D determinants: (1) the use of cells homozygous for the HLA-D region determinants; i.e. homozygous typing cells (HTCs) as stimulators in the MLC, and (2) the use of B-cell alloantisera to define HLA-DR (for HLA-D related) serologically-defined specificities. Theoretically, cells from individuals who are homozygous for the HLA-D region determinants could be used as typing reagents/stimulator cells in the MLC to identify responder cells possessing the HLA-D specificity for which the HTC is homozygous. Lower MLC reactivity would be detected with responders who share HLA-D region determinants with the HTC than those responders who do not.[47–51] Most of the first studies involving HLA-D typing utilized lymphocytes from offspring of first cousin marriages who had inherited the same haplotype by descent.[48, 49] Subsequently, HTCs were identified in the random population. These HTCs were, also submitted to the International Histocompatibility Workshops that defined these "HLA-Dw" specificities.

The description of the mouse Ir genes[52] stimulated the search for the human equivalent Ir genes and a series of observations by several investigators resulted in the description of the HLA-DR (D-related) antigens. Several investigators observed that some HLA-A and HLA-B typing sera could inhibit the MLC reaction.[53–55] Walford *et al.*[56] detected extra reactions with certain HLA-A and -B typing sera when testing cells from patients with chronic lymphocytic leukemia (CLL), a B-lymphocyte tumor; this suggested that these typing reagents contained anti-B-cell antibodies in addition to anti-HLA-A and -B. van Leeuwen *et al.*[57] presented evidence that HLA-D related gene products could be recognized by serological methodologies utilizing sera which recognize determinants present on B lymphocytes and monocytes but absent on platelets. Using B-cell enriched lymphocyte suspensions and platelet absorbed antisera, this typing technique was improved. These antisera were submitted to International Histocompatibility Workshops and the first HLA-DR specificities-defined (DR1–8) were given numbers according to their association with the Dw specificities 1–8.

Studies from several investigators provided evidence for new serologically-defined B-cell loci associated with, but separate from the HLA-DR specificities. These new specificites were initially called MT, MB, DC, DS, and LB-E.[58–62] Through the collaborative studies of the 9th International Histocompatibility Workshop in 1984, it was commonly agreed that the supertypic specificities MB, DC, DS, and LB-E defined an allelic system encoded by loci now called HLA-DQ. The DQ allelic system is genetically linked to but separate from the HLA-DR loci. There was initial conflicting evidence concerning the molecular localization of MT2 and MT3[62–65] though it is clear that they differed from HLA-DQ. These specificities were designated DRw52 and DRw53 at the 9th International Histocompatibility Workshop in 1984. It is now clear that they are encoded by loci distinct from the locus encoding the HLA-DRB1 alleles. DR52 and DR53 are now known to be products of the HLA-DRB3 and HLA-DRB4 loci, respectively, and another serologically defined specificity, HLA-DR51, is the product of the HLA-DRB5 locus.

With the development of the primed lymphocyte typing (PLT) assay, characterization of the MHC antigens was facilitated by the generation of highly specific reagents. These primed cells, presumably memory cells, will respond in an accelerated, i.e. secondary response, when restimulated by cells from the original stimulator or other cells which share stimulatory determinants with the sensitizing stimulating cell. The PLT technique, first described by Sheehey *et al.*[66, 67] and Mawas *et al.*,[68] appeared to detect predominantly the HLA-D region determinants. Several investigators[69–72] generated highly discriminatory reagents against HLA-D region determinants by utilizing HTCs or heterozygous cells which shared one HLA-Dw specificity. The PLT reagents, as well as T-cell clones derived from these reagents, helped to characterize the T-cell recognized epitopes associated with the Class II molecules.[73–79]

Eventually there was increasing evidence from PLT data for the existence of additional D-region specificities.[79–84] Shaw *et al.*[85–88] initially characterized a new allelic series they designated SB (secondary B-cell alloantigen), now termed HLA-DP (defined in the 9th International Workshop 1984). Utilizing unrelated priming cells phenotypically identical for HLA-A, -B, -C, -DR, -Dw and -DQ specificities, Shaw generated reagents which defined five antigens of a single segregant series mapped between DR and glyoxylase.[88]

DNA Methodologies to Define HLA Alleles

DNA methods were initially used to identify Class II alleles, and more recently have also been used to test for Class I alleles. The cellular methodologies showed that serological methods were often inaccurate for typing unrelated and even many related donors for bone marrow transplantation. First, cellular and then DNA methodologies showed extensive polymorphism within specificities termed "serologically identical". Introduction of DNA techniques and demonstration of their general applicability were accomplished through the 10th International Histocompatibility Workshop.[89–92] Prior to that workshop, Wake *et al.*[93] described a restriction fragment length polymorphism (RFLP) technique using probes that recognized the HLA-DRB gene. Other investigators made probes for DRA, DQB and DQA genes.[94–96] This technique involved the use of various enzymes to produce restriction fragments that were detected by full length HLA Class II cDNA probes. The technique was more accurate than serology[97, 98] but was not without disadvantages, not the least of which was the fact that this cumbersome method did not identify the polymorphic coding sequences directly.

Sequencing of the Class I and II HLA genes[99–102] was initially accomplished by cDNA sequencing techniques. Once genomic sequencing was applied, the number of different sequences that were obtained increased significantly.[103] Sequence analysis revealed that cells considered to be homozygous by serological methods were in fact heterozygous. A summary of Class II sequences generated from reference cell lines was presented at the 1987 International Histocompatibility Workshop.[104, 105]

In the mid-1980s, introduction of the polymerase chain reaction (PCR),[106, 107] together with the additional new allele sequences available, facilitated the molecular typing of the HLA alleles. The PCR based techniques, including sequence specific primers (SSP) and probes (SSO) as well as sequence based typing (SBT), became accepted techniques in the clinical as well as research laboratories. Although the SSO method was used initially to detect HLA Class II alleles,[108–113] in the early 1990s, the SSP method was further developed and found to be faster for typing individual samples.[114] Thus, the SSP method could be used to type cadaveric donors for transplantation.[115] The SBT method,[116–118] although very accurate, requires expensive equipment for automation. Recently, these methods have also been applied to identify Class I alleles.[119–124] The SSO, SSP and SBT techniques are now being used routinely in many histocompatibility testing laboratories. Other methods also in use include: complementary strand analysis,[125] ligation-based typing,[126] PCR-RFLP,[127–129] heteroduplex analysis,[130] and single strand conformation polymorphism.[131, 132] The World Health Organization Committee for HLA nomenclature is responsible for naming alleles and periodically publishing an update of new alleles. Regardless of the method used for HLA typing, the challenge continues to be the interpretation of results in light of the available sequence information.

Monoclonal Antibodies Detecting HLA-D Region Molecules

The development of monoclonal antibodies (mAb) recognizing human Class II molecules facilitated the biochemical characterization of these molecules. By protein gel electrophoresis of the molecules precipitated with these mAbs, three distinct subtypes of Class II molecules were identified.[133–135] Each of the Class II products (HLA-DR, -DQ and -DP) consists of an α-β heterodimer with the α-chains having molecular weights ranging between 31,000 and 34,000 Da and the β-chains having molecular weights ranging between 25,000 and 29,000 Da; the Class I genes (HLA-A, -B and -C) encode a 45,000 Da chain found at the cell surface associated with β_2 microglobulin. By using sequential immunodepletion techniques and amino acid sequencing, Hurley *et al.*[136] identified subsets of the DR molecules homologous to murine I-E alloantigens, demonstrated the presence of multiple DR β-chains, and also demonstrated that the α- and β-chains of DP are similar to but distinct from those of DR molecules.[137] HLA-K

Utilizing two-dimensional gel electrophoresis and selected monoclonal antibodies, Charron *et al.*[138] demonstrated hybrid (F1) DQ molecules as well as parental forms in heterozygous cells. Karr *et al.*[139] similarly found two DQ molecules expressed by a homozygous typing cell; Giles *et al.*[140] demonstrated similar findings. The generation of these types of molecules may be important in generating diversity of the Class II antigens.

Genomic Organization of the HLA Complex

Understanding of the HLA complex was expanded with studies of the protein products and genomic analyses. The HLA complex on chromosome 6 has been shown to contain Class I, II and III regions. The Class I region contains genes encoding the classical transplantation antigens, HLA-A, -B and -C. Each of the genes encodes an α-chain gene found in association with β_2 microglobulin at the cell surface of nucleated cells (reviewed in Wells and Parham, 1996).[141] Other functional Class I loci, HLA-E, -F and -G,[142] are termed non-classical since their function is not known. These loci are less polymorphic than HLA-A, -B and -C, and have a restricted tissue distribution. Also contained in this region are a number of other Class I genes or gene fragments that are most likely pseudogenes, HLA-H, -J, -K.[142–145]

The Class II region encodes the α- and β-chain genes for HLA-DR, -DQ, -DP glycoproteins (DRA, DRB, DQA, DQB, DPA, DPB) expressed predominantly on antigen presenting cells. The number of DRB genes and pseudogenes differ, depending on the haplotype.[146–149] The DQ and DP regions each include a pseudogene pair (DQA2, DQB2, DPA2, DPB2).[150–152] Two other Class II genes, HLA-DNA and -DOA are also probably a pair, but no protein has been identified.[153, 154] The products of HLA-DMA and -DMB, two Class II-related genes, are found mostly in intracellular vesicles and at the site of loading of peptides onto Class II molecules.[155] Also inserted in this region is a tight cluster of genes involved in processing antigens for presentation by Class I molecules. Two transporter associated with antigen processing genes, TAP1 and TAP2, are members of the ATP-binding cassette (ABC) family.[156] The products of these genes form a complex in the endoplasmic reticulum (ER) and translocate peptides from the cytoplasm to the lumen of the ER.[157, 158] The LMP genes, LMP2 and LMP7 encode components of a proteasome complex[159, 160] and cleave peptides which enter this barrel structure. The Class III region encodes a number of genes with various functions including some components of the complement system and will not be discussed further.

The function of these MHC molecules is to present self or foreign peptide to the T-cell receptor. In 1987, Borkman *et al.*[161, 162] published the crystal structure of the HLA-A2 molecule which provided the first insight into the structure and function of the MHC proteins. The structure of the peptide binding site provided evidence as to how the MHC molecules bind and present different peptides. The Class II crystalline structure[163] showed a three-dimensional structure similar to that of the Class I molecule although there is little sequence homology between these molecules. These early studies provided the basis for extensive and ongoing studies into the nature of the peptides bound to MHC molecules, MHC restricted recognition, and the molecular basis of allorecognition.

International Histocompatibility Workshops

Data presented at the National Academy of Sciences conference on human kidney transplantation held in 1963 demonstrated improved allograft survival when donor and recipient were genetically related. In response to the obvious need for methods to accurately assess histocompatibility among related individuals, the 1st International Histocompatibility Testing Workshop was held at Duke University, June 9–12, 1964.[164] Organized by D. Bernard Amos, the wet-bench workshop provided an opportunity for investigators to demonstrate and compare different methodologies than being used to identify and characterize histocompatibility antigens. These included leukocyte agglutination, indirect antiglobulin consumption, complement-fixation, normal leukocyte transfer, mixed leukocyte cultures, mixed agglutination, and skin grafting. In contrast to later international histocompatibility workshops, this one was preceded by a one day conference in Washington, D.C. sponsored by the National Academy of Sciences, National Research Council. The conference on Histocompatibility Testing included among its invited participants three scientists who subsequently received Nobel Prizes — Gertrude B. Elion, Jean Dausset and Joseph E. Murray — as well as many other pioneers in histocompatibility testing and transplantation immunology.

During the following three years, investigations in a number of the international laboratories demonstrated that alloantibodies detected a complex antigenic system on leukocytes. One or more linked genes appeared to genetically control multiple antigens detected by alloantisera. At the Histocompatibility Testing Conference and Workshop held in Leiden in 1965,[165] several investigators presented data on the relationship between this antigenic system, non-serologic methods such as mixed leukocyte cultures and the normal lymphocyte transfer test, and transplant rejection. Graft survival appeared to be associated with genetic relationship of donor and recipient and with ABO compatibility.

Other species appeared to have a similar antigenic system. Since the problem of nomenclature was already apparent, a nomenclature committee was formed but no action taken as no two sera gave identical reaction patterns and there was insufficient knowledge regarding the genetic basis of the antigenic system.

By the next international workshop and conference held in Turino in 1967,[166] however, it seemed likely that the human histocompatibility antigens might belong to a single genetic system analogous to the H-2 system in the mouse. Therefore, the collaborative wet workshop, organized by Ruggero Ceppellini, focused on testing 11 selected families. Analysis of family data "fully confirmed the view that most of the leukocyte antigens, at present known, are controlled by one chromosome region, now identified as the HL-A system".[166] The workshop also showed that all 44 haplotypes gave different antigenic patterns, leading to the conclusion that "This seems to represent an extraordinary case of genetic polymorphism of a degree almost unknown in biology". This 3rd international workshop and conference also included initial attempts to biochemically characterize histocompatibility antigens.

Genetic analysis of family data was a major focus of the 4th International Histocompatibility Workshop organized by Paul Terasaki in 1967.[167] Although a number of investigators had evidence for at least two segregating series of alleles prior to the workshop, this hypothesis was not universally held. Therefore, mendelian segregation of individual antigenic specificities was tested in families. No significant deviations from two segregating series were observed. Further, results confirmed that three families, tested in the workshop, included individuals with recombinant haplotypes, providing additional evidence for the two locus hypothesis. At the conference held in Los Angeles, data were presented that leukocytes from patients with Hodgkin's disease reacted more frequently with six anti-HL-A 4c sera than did those from healthy donors. Interestingly, this reported association, although later shown in a larger collaborative study to be insignificant, sparked great interest in a wide variety of HLA and disease association studies.

Anthropological studies became the focus of the 1972 histocompatibility workshop and conference, held in Evian.[168] Organized by Jean

Dausset, this workshop demonstrated that the highly polymorphic histocompatibility antigens were very useful markers for the study of human evolution. The population studies further showed that some variants of previously defined antigenic specificities primarily occur in particular ethnic populations.

A major emphasis of all international histocompatibility workshops has been standardization of antigen definition and nomenclature. New alleles, defined by reagents from several different laboratories, are described in the workshop reports. At the 6th International Histocompatibility Workshop and Conference held in 1975 in Aarhus,[169] the nomenclature committee dropped the hyphen from HL-A, and the human histocompatibility genes henceforth have been officially called HLA. The studies included in this workshop emphasized the newly recognized third serologically defined locus (HLA-C); a cellularly defined locus detected by mixed lymphocyte reactivity using homozygous typing cells (HLA-D); and detection of other cell surface determinants using cell mediated lympholysis and primed lymphocyte tests. Of particular interest was the possible role that these gene products and assays might play in allotransplantation and disease. Genetic studies demonstrated linkage disequilibrium between the serologically and cellularly defined gene products and also with various complement proteins.

The next workshop, held at Oxford in 1977,[170] focused on the serologic detection and characterization of antigens present on B-cells but not T-cells, and on the genetic relationship of these genes to the mixed lymphocyte defined products of HLA-D genes. Although a strong association between the B-cell serologic and HLA-D specificities was found, they were not identical. Therefore, the B-cell specificities became officially known as D-related, or HLA-DR, specificities. This workshop also addressed the relationship between several diseases and the various HLA genes, in particular the HLA-D and -DR genes.

A major contribution of the 8th International Workshop and Conference held in 1980 in Los Angeles was detailed population analyses which provided HLA antigen and haplotype frequencies for ten distinct ethnic populations.[171] This workshop also confirmed preliminary evidence for B-cell specificities in addition to those encoded by HLA-DR. Hindsight reveals that these specificities were products of Class II genes we now call HLA-DRB3, HLA-DRB4 and HLA-DQ. It became a primary goal of the next, 9th international workshop and conference, held four years later in Munich,[172] to analyze biochemically the molecules precipitated by monoclonal antibodies that recognized these new B-cell specificities.

During the three years following the 9th International Workshop, significant scientific advances were made in the molecular characterization of the HLA complex. Therefore, the principle topic of the 10th International Histocompatibility Workshop and Conference held in New York in 1987 was molecular genetics of the HLA system.[173] It developed standardized assays for characterizing HLA genes and their products and also characterized a reference cell panel of B-lymphoblastoid cell lines that could be shared among investigators. Further, the workshop accomplished a worldwide transfer of molecular technology and other modern techniques from experienced laboratories to their less experienced colleagues. Further technological transfer and standardization occurred during the 11th International Histocompatibility Workshop and Conference[174] held four years later in Yokohama, particularly with respect to the development of reagents and methods for DNA typing. The 11th workshop also utilized molecular methods to examine HLA genes in a variety of anthropological, clinical and basic science studies.

The most recent international histocompatibility workshop and conference, apart from the one currently in progress to be held in 2002 in Seattle, took place in Paris in 1996[175] and had as its primary focus molecular polymorphism of HLA genes with emphasis on anthropologic studies. The workshop report provides a reference tabulation of HLA allele and haplotype frequencies for diverse populations worldwide. A further focus of the workshop and conference was the role of HLA in the immune system and its clinical implications, particularly in the areas of transplantation, autoimmunity, infectious diseases and cancer.

The goal of relating new knowledge of the HLA complex to human immune responses, and to relevant areas of medicine, has been a common theme among all of the international histocompatibility workshops. Although it is clear that most of the scientific developments in histocompatibility and immunogenetics have occurred outside of these workshops, it is very apparent that these large worldwide scientific collaborations have contributed very significantly to the international advancement of both scientific areas. There has been an unparalleled, generous sharing of knowledge, technology and reagents. The workshops have served as a fertile training ground for less experienced colleagues, and have developed standardized methods and nomenclature, thus ensuring better basic and clinical scientific communication.

Organizations

Particularly beneficial to collaboration among early serologists was the establishment of a serum bank by the National Institute of Allergy and Infectious Diseases.[176] The NIH Serum Bank, housed at the NIH in Bethesda and headed by Don Kayhoe, served as a repository for sera contributed by recipients of NIH contracts specifically designed to procure typing reagents, as well as sera contributed by many other generous investigators. This repository provided start-up reagents to new typing laboratories and reference reagents for national and international workshops. It also supported training workshops for technologists and laboratory directors who were just starting new tissue typing laboratories.

The NIH Serum Bank also developed and distributed serum trays and held the first of many Tray Users meetings at the National Naval Medical Center in 1972. These tray users meetings were intended to evaluate the results obtained with the trays and to identify and characterize new antigenic specificities. Although the serum bank fulfilled these goals, it also provided a focus for communication among American tissue typers, thus contributing to the founding of the American Association for Clinical Histocompatibility Testing (AACHT), later named the American Society for Histocompatibility and Immunogenetics (ASHI). ASHI, which celebrated its 25th anniversary in 1999, and AACHT before it, has been a strong advocate for advancing the "science and application of histocompatibility and immunogenetics; …the exchange of information; and …the highest standards of laboratory testing in the interest of optimal patient care" (ASHI mission statement). In addition to sponsoring annual national scientific and regional educational meetings, it has co-developed with the College of American Pathologists a proficiency-testing program. It has developed laboratory standards for histocompatibility testing and has established a laboratory accreditation program that current accredits laboratories not only in the United States but in other parts of the world, as well. ASHI also has sponsored the American Board of Histocompatibility and Immunogenetics (ABHI), the national organization that certifies the competency of histocompatibility technologists, specialists and laboratory directors through examination. Together, ASHI and ABHI have established, and through educational endeavors improved, the standards for histocompatibility testing personnel and laboratories.

In the late 1960s, Bernard Amos and David Hume co-founded the Southeastern Regional Organ Procurement Program, now SEOPF, which shared cadaver organs regionally on the basis of tissue matching. At its beginning, SEOPF member laboratories used the same serum panel for typing donors and recipients and all participated in wet workshops designed to minimize technical variation. Recipient selection was accomplished through a computer program.[177] SEOPF became a model for the U.S. and international organ sharing organizations, such as the United Network for Organ Sharing, EuroTransplant and the National Marrow Donor Program.

Influence of HLA in Transplantation

The role of the leukocyte antigens in transplantation in other species[178, 179] and man[180, 181] as well as the function of these antigens[182] has been the subject of intense investigation. Antibodies to the HLA antigens of the donor were shown to cause hyperacute rejection in kidney transplantation and Terasaki *et al.*[183] proposed a cytotoxicity crossmatch to prevent such rejection. Kissmeyer-Nielsen *et al.*[184] also reported hyperacute rejection in two patients who had a positive leukoagglutinin test. Subsequently, HLA antibodies were shown also to cause early transplant failure.[185] Recipients' sera with HLA antibodies were shown to react with a panel of cells from random individuals; the test became known as the percent reactive antibody (PRA) test. Patients with a high PRA had lower graft survival rates than patients with a negative PRA.[185]

In 1966, Terasaki *et al.*[186] proposed the first trial of HLA matching for cadaver donor selection and van Rood *et al.*[187] proposed the first international organ exchange organization. The parallel development of the field of immunosuppression has allowed for the successful transplantation of recipient/donor combinations with all levels of HLA mismatching.[188] However, recipients of HLA-identical grafts had remarkably high graft survival rates.[189] As the methodology to define HLA antigens improved and as new immunosuppression regimens were identified, the role of matching in successful solid organ transplantation has sparked controversy. Several reports from scientific registries containing results from thousands of kidney recipients[190, 191] as well as reports from single centers[192] showed improved graft survival with higher grades of HLA matching, while reports from some centers showed minimal improvement of graft survival with improved HLA matching.[193, 194] All data do show the highest graft survival rate for recipients of HLA-A, -B, -DR matched (six antigens matched) and phenotypically matched kidney grafts. Thus, mandatory policies for sharing these well-matched grafts have been implemented currently by the United Organ Sharing Network in the U.S. as well as sharing policies based on HLA antigen matching by the Eurotransplant Network in Europe.

Genetic HLA matching is essential for successful hematopoietic cell transplantation (HCT). Early studies in experimental animals established the importance of matching for the MHC genes and provided rationale for donor selection in clinical HCT.[195–198] The first successful marrow transplants were carried out with identical twins; however, serological typing for HLA antigens and the one-way MLC enabled the selection of MHC identical allogeneic donor/recipient pairs for HCT.[199–203] As HLA typing techniques and clinical care advanced over the past 30 years, comparative studies showed that phenotyping by serological methods and MLC failed to detect Class I and Class II disparities.[204–209] With the advent of molecular typing for HLA, definitive analysis of HLA genes and alleles is possible and needed to facilitate donor selection. Since the probability of having an HLA identical sibling in the average family is less than 30%, suitable partially matched related donors were considered as an option. These studies of partially matched related donors provided valuable information regarding the limits of HLA mismatching.[210–214] Matching based on serological HLA typing and MLC typing showed the risk of acute graft versus host disease (GVHD) and graft rejection correlated with the degree of HLA disparity.[210, 215, 216] Although transplants from HLA-matched unrelated donors were attempted in the early 1970s, identifying suitably matched donors was not feasible until large registries of donors were established by the Anthony Nolan Appeal in London[217, 218] and the National Marrow Donor Program.[219–221] Over 7000 unrelated donor transplants have been performed in over 200 transplant centers worldwide. Currently, HCT donors are also utilized from umbilical cord blood and growth factor mobilized peripheral blood stem cells.[222]

Predicting Long-Term Graft Outcome

Cellular assays once used to type for HLA Class II compatibility are currently being used to monitor changes, quantitative and qualitative, in the T-cell recognition process during the post-transplant period, as a means to predict long-term graft outcome. Identification of immune parameters that predict long-term graft outcome would aid in identifying those recipients at low versus high risk for immune complications and chronic rejection. Those recipients at low risk presumably could have their immunosuppression successfully lowered or even weaned, thereby avoiding the many side-effects of immunosuppressive therapy. Recipients who are at high risk for chronic rejection may benefit from changing or even increasing the immunosuppressive therapy.

Throughout the course of clinical transplantation, several investigators have reported that some recipients develop a tolerance-like state, as evidenced by the recipient becoming less reactive to the disparate foreign donor antigen, i.e. donor antigen-specific hyporeactivity.[223–226] The earliest account that kidney recipient-donor pairs with different HLA antigens (then termed HL-A) showed unidirectional non-stimulation in MLC was reported by Amos and Bach.[223] In the early 1970s, Bach *et al.*[224] postulated that the observed non-reactivity was caused by the deletion of clones that reacted to the foreign antigens on the donor graft. They speculated that these non-reactive recipients may be candidates for withdrawal of immunosuppression. Other investigators later showed that some long-term renal transplant recipients were hyporesponsive or non-responsive toward disparate HLA antigens expressed by the donor kidney as detected by MLC.[225–227]

This early literature dealing with donor antigen-specific reactivity as assayed by MLC was not consistent. Some investigators reported hyporeactive[227–232] and others normal[233–235] MLC reactivity to donor cells. These early studies were hampered by the lack of definition of the HLA antigens and by the relative insensitivity of the newly developing MLC technique. The inconsistent MLC results may be attributed, in part, to the use of patients' serum versus pooled normal sera as a protein source in the cultures. The serum of a tolerant patient serum was reported to contain blocking factor activity shown to be an immunoglobulin G antibody with anti-idiotypic activity.[236, 237] When cultures containing pooled human sera were tested,[229] low donor antigen-specific response was observed in recipients with good graft tolerance and an increased response was observed during periods of acute rejection.

The cell mediated lympholysis (CML) assay seemed to give relatively consistent results in several different studies[238–245] and, thus, many investigators began to rely on the CML technique more heavily than the MLC technique. In general, a higher percentage of recipients were found to be non-responsive to donor antigens in the CML assay than in the MLC assay.[246] Several investigators concluded that suppressor cells could be involved in reduced CML activity in some recipients[229, 240, 241, 247–249] while others proposed clonal deletion[223, 236–239, 250] as a possible mechanism. Importantly, these results showed that different percentages of recipients could be classified as hyporesponsive, depending on the assay used.

Goulmy *et al.*[233] studied large numbers of recipients, some of whom were tested longitudinally in CML and MLC assays and concluded that the correlation of CML responder status and graft outcome was of limited prognostic value.[250] They subsequently reported that successful complete steroid withdrawal seemed merely to coincide with CML donor antigen-specific hyporeactivity, because partial and complete steroid withdrawal was also achieved in CML-responsive recipients.[251]

More recently, donor antigen-specific hyporeactivity has also been observed in recipients who have received donor-specific transfusions (DST),[252–255] donor-specific marrow,[256, 257] and total lymphoid irradiation (TLI).[258–261] Alexander[255] reported that recipients of peritransplant DST developed donor antigen-specific hyporeactivity more often than those in the control group; however, there was no difference between the DST versus control groups in the ability to withdraw steroids. Barber *et al.*[256] reported similar findings in recipients who received combined cadaver donor kidney and bone marrow transplants. These recipients showed a trend toward less reactivity in MLC to donor antigens when compared to the control recipient group who received contralateral kidneys. Some of these marrow/solid organ recipients could be tapered off immunosuppression; however, the MLC results were not used as the basis for patient selection. Zeevi *et al.*[257] also reported recipients who received bone marrow augmented solid organ transplantation were more likely to develop donor antigen-specific hyporeactivity than the control solid organ recipients who did not receive bone marrow. Strober[260, 261] reported that 82% cadaver transplant recipients who had undergone TLI acquired donor antigen-specific unresponsiveness, as assayed by MLC, and maintained adequate graft function after completely stopping immunosuppression.

Other centres have reported donor antigen-specific hyporeactivity correlating with improved graft outcome in their recipients;[262–268] however, only one transplant center has used donor antigen-specific hyporeactivity to select recipients for altering of immunosuppression.[269] A subsequent study of a larger number of these recipients confirmed that the MLC donor antigen-specific hyporesponsive group had excellent tolerance to steroid withdrawal[270] and a better graft survival rate than recipients who were responsive to donor antigen.

Donor antigen-specific hyporeactivity has been observed in other solid organ recipient groups including lung, liver and heart[257, 271–278] In general, similar percentages of kidney and heart recipients develop hyporeactivity, while a higher percentage of liver and lung recipients develop donor antigen-specific hyporeactivity.[271, 272, 276] Reinsmoen *et al.*[272] have reported that the predictive value of donor antigen-specific hyporeactivity defining a chronic rejection-free state is high for kidney, heart and lung recipients. Although the sensitivity of the hyporeactivity assay is high, the specificity is low; i.e. a responsive result does not predict a recipient will develop chronic rejection. Investigators have focused on more quantitative measures and testing different pathways of antigen recognition to determine if these approaches can better identify recipients at high risk for developing chronic rejection.

Frequency analysis by limiting dilution analysis (LDA) allows the quantitation of the alloantigen-reactive cytotoxic and helper cells.[279–281] Other studies of long-term kidney allograft survivors showed functional reduction of alloreactive cells can occur at the level of the cytotoxic cells[253, 282–284] and/or helper T-cells.[285–287] De Haan *et al.*[273] reported, the development of CTLp hyporeactivity is independent of the development of HTLp hyporeactivity which occurs less frequently. Some observed results appear to be organ-specific in nature. Mathew *et al.*[288] postulated that CML unresponsiveness in renal allograft tolerance and in liver allograft tolerance is different, based on different effects of IL-2 to overcome unresponsiveness.

In more recent studies, the LDA has been proposed as a useful tool to predict graft survival and selection of recipients for immunosuppression withdrawal. Beik *et al.*[289] concluded that the donor antigen-specific hyporeactivity assay could be useful in the prediction of graft outcome and the selection of kidney recipients for reduced immunosuppression. Mason *et al.*[290] reported that primed HTLp frequencies against donor DR antigens were undetectable in half of recipients with chronic loss of function. Hornick *et al.*[291] also observed hyporesponsiveness of T-helper and T-cytotoxic in a substantial number of heart recipients with progressive chronic rejection. These authors concluded that an antigen-recognition pathway other than the direct pathway (i.e. indirect pathway) was involved in the development of chronic rejection. Recently, cytotoxic precursor frequencies have been shown to be predictive of which recipients can successfully decrease immunosuppression.[292]

These discordant observations may be the result of the relatively small number of recipients tested in each study, the differing number of HLA disparities between the recipient and donor, different solid organ groups tested, different immunosuppressive regimens used, different recipient/donor ethnic populations, and varying LDA methodologies. None of these studies has analyzed the predictive values of these assays.

Thus donor antigen-specific hyporeactivity, which predominantly measures direct allorecognition, appears to be a good predictor of recipients most likely to remain chronic rejection-free, but those recipients who remain responsive to donor antigen will not necessarily develop chronic rejection. In an effort to identify which recipients are at high risk for developing chronic rejection, several investigators have focused on studies of the indirect pathway of recognition. The direct pathway implies the recognition of intact donor MHC antigens expressed on the membranes of the graft while the indirect pathway implies recognition of donor-derived peptide presented by the recipient's own MHC molecules. Although direct recognition has been postulated to be more involved with acute rejection, and indirect recognition implicated with chronic rejection, these pathways may not be mutually exclusive. Shearer and colleagues[293, 294] first reported that transplant recipients who exhibit depressed CD4 T-helper/self APC function (i.e. indirect presentation pathway), as determined by response to nominal antigens (tetanus, flu) or nylon wool-purified allogeneic cells, were at low risk for acute and chronic rejection;[294, 295] however, donor cells were not used in these studies. Suciu-Foca and her colleagues have presented numerous studies of the indirect recognition of donor peptides and the possible identification of recipients at risk for developing chronic rejection.[296–299] Sayegh and his colleagues[300, 301] have also presented studies of recipients with chronic rejection who exhibit specific proliferation to the mismatched allopeptides presented by autologous APC's. They concluded that this assay may specifically identify recipients who are at risk for developing chronic graft dysfunction. These studies provide provocative evidence for the indirect pathway of allorecognition in the development of chronic rejection; however, hyporeactivity of this pathway may likewise predict a chronic rejection-free state.

Cellular responses of graft infiltrating cells and recognition of donor peptides via the indirect pathway of antigen presentation have also been shown to correlate with acute and chronic rejection. Woan *et al.*[302] and Tugulea *et al.*[303] have reported prediction of acute and chronic rejection in heart recipients by the correlation of cardiac biopsy growth and histological rejection. Numerous other studies have shown an association between the detection of anti-donor directed cellular reactivity of cells propagated from biopsies with the diagnosis of acute or chronic rejection.[304] These studies are of particular importance because the responding cells recovered from the graft are likely involved in the pathological processes occurring within the graft.

Thus, no one assay appears able to predict which recipients are at risk for developing chronic rejection and which are not. The best approach may be to assess the comprehensive immune state of the recipients by using assays that test several different mechanisms in a sequential analysis of a quantitative, donor antigen-specific response.

Conclusion

During the past half century, knowledge of the structure, function and regulation of the human histocompatibility complex has increased at a rapid rate. Yet, its precise role in allotransplantation remains elusive and controversial. The genetic complexity of the MHC; the incompleteness of our knowledge of it; the great variety of antigens which it presents to the immune system; the interaction of the MHC with a myriad of other immune effector molecules, some of which are encoded by genes closely linked to the MHC Class I and Class II genes — all of these factors, and more, greatly complicate our ability to fully understand the relative importance of MHC antigen-matching and avoidance of donor-specific anti-HLA antibodies for improving the likelihood of successful allotransplantation. Hopefully, this challenge will be met by clinical and basic scientists during the next half century.

References

1 Landsteiner K (1901). Agglutinationserocheineigen normalen menschlichen blutes. *Wein Klin Wochenschr* **14**, 1132.

2 Landsteiner K, Levine P (1927). A new agglutinable factor differentiating individual human bloods. *Proc Soc Exp Biol NY* **24**, 600.

3. Landsteiner K, Wiener AS (1940). An agglutinable factor in human blood recognized by immune sera for rhesus blood. *Proc Soc Exp Biol NY* **43**, 233.

4. Tyzzer EE (1916). Tumor immunity. *J Cancer Res* **1**, 125.

5. Little CC (1924). The genetics of tissue transplantation in mammals. *Cancer Res* **8**, 75.
6. Snell GD (1948). Methods for the study of histocompatibility genes. *J Genet* **49**, 87.
7. Gorer PA (1937). Further studies on antigenic differences in mouse erythrocytes. *Br J Exp Path* **18**, 31.
8. Gorer PA, Lyman S, Snell GD (1948). Studies on the genetic and anti-genic basis of tumor transplantation: linkage between a histocompatibility gene and "fused" in mice. *Proc Roy Soc London SB* **135**, 499.
9. Gibson T, Medawar PB (1943). The fate of skin homografts in man. *J Anat* **77**, 299.
10. Owen R (1945). Immunogenetic consequences of vascular anastomoses between bovine twins. *Science* **102**, 400.
11. Anderson D, Billingham RE, Lampkin GH, Medawar PB (1951). The use of skin grafting to distinguish between monozygotic and dizygotic twins in cattle. *Heredity* **5**, 379.
12. Bauer KH (1927). Homoiotransplantation von epidermis bei eineiigen zwillingen. *Bruns' Beitrage zur Klinishen Chirurgie* **141**, 442.
13. Carrel A (1910). Remote results of the replantation of the kidney and the spleen. *J Exp Med* **12**, 146.
14. Murray JE, Merrill JP, Harrison JH (1958). Kidney transplantation between seven pairs of identical twins. *Ann Surg* **148**, 343.
15. Dausset J (1952). Leuco-agglutinins. IV. Leuco-agglutinins and blood transfusion. *Vox Sang* **4**, 190.
16. Dausset JA, Nenna A (1952). Presence d'une leuco-agglutinine dans le serum d'un cas d'agranulocytose chronique. *Compt Rend Soc Bio* **146**, 1539.
17. Miescher P, Fauconnet M (1954). Mise en evidence de different groupes leucocytoures chez l'homme. *Schweiz Med Wochenschr* **84**, 597.
18. Payne R, Rolfs MR (1958). Fetomaternal leukocyte incompatibility. *J Clin Invest* **37**, 1756.

19. Payne R, Tripp M, Weigle J, Bodmer W, Bodmer J (1964). A new leukocyte isoantigenic system in man. In *Cold Spring Harbor Sym Quant Biol "Human Genetics"* **29**, 285.
20. van Rood JJ, van Leeuwen A (1963). Leucocyte grouping: a method and its application. *J Clin Invest* **42**, 1382.
21. Dausset JA, Brecy H (1957). Identical nature of the leukocyte antigens detectable in monozygotic twins by means of immune iso-leuco-agglutinins. *Nature* **180**, 1430.
22. Lalezari P, Spaet TH (1959). Studies on the genetics of leukocyte antigens. *Blood* **14**, 748.
23. Payne R, Hackel E (1961). Inheritance of human leukocyte antigens. *Am J Hum Genet* **13**, 306.
24. van Rood JJ, van Leeuwen A (1965). Defined leukocyte antigenic groups in man. In Russell PS, Winn HJ (eds.) *Histocompatibility Testing*. Washington, D.C.: National Academy of Science, Publication 1229, p. 21.
25. Dausset J, Ivanyi P, Ivanyi D (1965). Tissue alloantigens in humans: identification of a complex system (HU-1). In Balner H, Cleton FJ, Eernisse JG (eds.) *Histocompatibility Testing*. Copenhagen: Munksgaard, p. 51.
26. Bach FH, Amos DB (1967). HU-1: major histocompatibility locus in man. *Science* **156**, 1506.
27. Amos DB, Bach FH (1968). Phenotypic expressions of the major histocompatibility locus in man (HL-A): leukocyte antigens and mixed leukocyte culture reactivity. *J Exp Med* **128**, 623.
28. Kissmeyer-Nielsen F, Svejgaard A, Ahrons S, Staub NL (1969). Crossing-over within the HLA system. *Nature* **224**, 75.
29. Ward FE, Southworth JG, Amos DB (1969). Recombination and other chromosomal aberrations within the HL-A locus. *Transpl Proc* **1**, 352.
30. Sandberg L, Thorsby E, Kissmeyer-Nielsen F, Lindholm A (1970). Evidence of a third sublocus within the HL-A chromosomal region. In Terasaki PI (ed.) *Histocompatibility Testing*. Copenhagen: Munksgaard, p. 165.

31. Svejgaard A, Staub-Nielsen L, Ryder L, Kissmeyer-Neilsen F, Sandberg L, Lindholm A, Thorsby E (1973). Subdivision of HL-A antigens. Evidence of a new segregant series. In Dausset J, Colombani J (eds.) *Histocompatibility Testing*. Copenhagen: Munksgaard, p. 465.
32. Low B, Messeter L, Manssom S, Lindrolm T (1974). Crossing-over between the SD2 (FOUR) and SD3 (AJ) loci of the human major histocompatibility chromosomal region. *Tissue Antigens* **4**, 405.
33. Bijnen AB, Schreuder I, Meera Kahn P, Allen FH, Giles CM, Los WRT, Volkers WS, van Rood JJ (1976). Linkage relationships of the loci of the major histocompatibility complex in families with a recombination in the HLA region. *J Immunogenet* **3**, 171.
34. Belvedere MC, Curtoni ES, Dausset J, Lamm LU, Mayr W, van Rood JJ, Svejgaard A, Piazza A (1975). On the heterogeneity of linkage estimations between LA and FOUR of the HL-A system. *Tissue Antigens* **5**, 99.
35. Gorer PA, O'Gorman P (1956). The cytotoxic activity of isoantibodies in mice. *Transpl Bull* **3**, 142.
36. Terasaki PI, McClelland JD (1964). Microdroplet assay of human serum cytotoxins. *Nature (London)* **204**, 998.
37. Walford R (1964). Serologic typing of human lymphocytes with immune serum obtained after homografting. *Science* **144**, 868.
38. Yunis EJ, Amos DB, Eguro SY, Dorf ME (1972). Cross-reactions of HL-A antibodies. I. Characterizations by absorption and elution. *Transplantation* **14**, 474.
39. Amos DB (1969). Genetic and antigenic aspects of human histocompatibility systems. *Adv Immunol* **10**, 251.
40. Johnson AH, Rossen RD, Butler WT (1972). Detection of allo-antibodies using a sensitive antiglobulin microcytotoxicity test: identification of low levels of preformed antibodies in accelerated allograft rejection. *Tissue Antigens* **2**, 215.
41. Bain B, Vas MR, Lowenstein L (1964). The development of large immature cells in mixed leukocyte cultures. *Blood* **23**, 108.

42. Bach FH, Hirschhorn K (1964). Lymphocyte interaction: a potential histocompatibility test *in vitro*. *Science NY* **143**, 813.
43. Yunis EJ, Amos DB (1971). Three closely linked genetic systems relevant to transplantation. *Proc Natl Acad Sci* **68**, 3031.
44. Bach FH, Albertini RJ, Amos DB, Ceppellini K, Mattiuz PL, Miggiano VC (1969). Mixed leukocyte culture studies in families with known HLA-genotypes. *Transpl Proc* **I**, 339.
45. Eijsvoogel VP, DuBois M, Melief C, DeGroot-Kooy M, Koning C, van Rood JJ, van Leeuwen A, DuToit E, Schellekens P (1973). Position of a locus determining mixed lymphocyte reaction (MLR), distinct from the known HL-A loci, and its relation to cell-mediated lympholysis (CML). In J. Dausset J, Colombani J (eds.) *Histocompatibility Testing*. Copenhagen: Munksgaard, p. 501.
46. Bach FH, van Rood JJ (1976). The major histocompatibility complex-genetics and biology. *N Engl J Med* **295**, 806; 872; 927 (Oct 7, 14, 21, 1976).
47. Dupont B, Jersild C, Hansen GS, Nielsen S, Thomsen M, Svejgaard A (1973). Typing for MLC determinants by means of LD homozygous and LD-heterozygous test cells. *Transpl Proc* **5**, 1543.
48. Jorgensen F, Lamm L, Kissmeyer-Nielsen F (1973). Mixed lymphocyte cultures with inbred individuals: an approach to MLC typing. *Tissue Antigens* **4**, 323.
49. van den Tweel JG, Blusse van Oud Alblas A, Keuning JJ, Goulmy E, Termijtelen A, Bach ML, van Rood JJ (1973). Typing for MLC (LD). I. Lymphocytes from cousin marriage offspring as typing cells. *Transpl Proc* **5**, 1535.
50. Dausset J, Sasportes M, Lebrum A (1973). Mixed lymphocyte culture (MLC) between HL-A serologically identical parent-child and between HLA homo and heterozygous individuals. *Transpl Proc* **5**, 1511.
51. Mempel W, Grosse-Wilde H, Baumann P, Netzel B, Albert ED (1973). Population genetics of the MLC response: typing for

MLC determinants using homozygous and heterozygous reference cells. *Transpl Proc* **5**, 1529.

52. McDevitt HO, Benacerraf B (1969). Genetic control of specific immune responses. Immunoregulation. *Adv Immunol* **11**, 31.
53. Ceppellini R, Bonnard GD, Coppo F, Miggiano VC, Pospisil M, Curtoni ES, Pellegrino M (1971). Mixed leukocyte cultures and HL-A antigens. II. Inhibition by anti HL-A sera. *Transpl Proc* **3**, 63.
54. Greenberg L, Reinsmoen N, Yunis EJ (1973). Dissociation of stimulation (MLR-S) and response (MLR-R) in mixed leukocyte culture by serum blocking factors. *Transplantation* **168**, 520.
55. Sengar D, Opelz G, Terasaki PI (1973). Suppression of mixed leukocyte response by plasma from hemodialysis patients. *Tissue Antigens* **3**, 22.
56. Walford RL, Gossett T, Smith GS, Zeller E, Wilkinson J (1975). A new alloantigenic system on human lymphocytes. *Tissue Antigens* **5**, 196.
57. van Leeuwen A, Winchester R, van Rood JJ (1975). Serotyping for MLC. II. Technical aspects. *Ann NY Acad Sci* **254**, 289.
58. Duquesnoy RJ, Marrari M, Annen K (1979). Identification of a HLA-DR associated system of B-cell alloantigens. *Transpl Proc* **11**, 1757.
59. Tosi R, Tanigaki N, Centis D, Ferrara GB, Pressman D (1978). Immunological dissection of human Ia molecules. *J Exp Med* **148**, 1592.
60. Goyert SM, Shively JE, Silver J (1982). Biochemical characterization of a second family of human Ia molecules, HLA-DS, equivalent to murine I-A subregion molecules. *J Exp Med* **156**, 550.
61. Termijtelen A, van Leeuwen A, van Rood JJ (1982). HLA-linked lymphocyte activating determinants. *Immunol Rev* **66**, 79.
62. Park MS, Terasaki PI, Nakata S, Aoki D (1980). Supertypic DR groups: MT1, MT2 and MT3. In Terasaki PI (ed.) *Histocompatibility Testing*. Los Angeles: UCLA Press, p. 854.

63. Markert ML, Cresswell P (1982). Human B-cell alloantigens: alpha subunit variability. *J Immunol* **128**(5), 999.
64. Shackelford DA, Mann DL, van Rood JJ, Ferrara GB, Strominger JL (1981). Human B-cell alloantigens DC1, MT1 and LB12 are identical to each other but distinct from the HLA-DR antigen. *Proc Natl Acad Sci* **78**, 4566.
65. Tanegaki N, Tosi R, Duquesnoy RJ, Ferrara GB (1983). The distribution of DR5, MT2 and MB3 specificities on human Ia subsets. *Immunogenet* **17**, 37.
66. Sheehy MJ, Sondel PM, Bach ML, Wank R, Bach FH (1975). HL-A LD (lymphocyte defined) typing: a rapid assay with primed lymphocytes. *Science* **188**, 1308.
67. Sheehy MJ, Bach FH (1976). Primed LD typing (PLT) — technical considerations. *Tissue Antigens* **8**, 157.
68. Mawas CE, Charmot D, Sasportes M (1975). Secondary responses of *in vitro* primed human lymphocytes to allogenic cells. I. Role of HL-A antigens and mixed lymphocytes reaction stimulating determinants in secondary *in vitro* proliferative response. *Immunogenet* **2**, 449.
69. Bach FH, Sondel PM, Sheehy MJ, Wank R, Alter BJ, Bach ML (1975). The complexity of the HL-A LD system: a PLT analysis. In Kissmeyer-Nielsen F (ed.) *Histocompatibility Testing*. Copenhagen: Munksgaard, p. 576
70. Hartzman RJ, Scher I, Ahmed A, Sell KW (1977). The primed LD typing test: analysis of individuals immunized to GLT. *Transpl Proc* **9**(Supp. 1), 195.
71. Suciu-Foca N (1977). Complete typing of the HLA region in families. IV. The genetics of HLA-D as seen by HTC and PLT methods. *Transpl Proc* **9**, 1751.
72. Thompson JS, Easter CJ, Blaschke JW (1977). Use of primed lymphocyte (PLT) in unrelated individuals to identify 11 antigenic clusters. *Transpl Proc* **9**(4), 1759.
73. Fradelizi D, Nunex-Roldan A, Sasportes M (1978). Human Ia-like DRw lymphocyte antigens stimulating activity in primary mixed lymphocyte reaction. *Eur J Immunol* **8**, 88.

74. DeWolf WC, Carroll PG, Mehta CK, Martin SL, Yunis EJ (1979). The genetics of PLT response. II. HLA-DRw is a major PLT-stimulating determinant. *J Immunol* **123**, 37.
75. Morling N, Jakobsen BK, Platz P, Ryder LP, Svejgaard A, Thomsen M (1980). A "new" primed lymphocyte typing (PLT) defined DP-antigen associated with a private HLA-DR antigen. *Tissue Antigens* **16**, 95.
76. Suciu-Foca N, Susinno E, McKiernan P, Rohowsky C, Weiner J, Rubinstein P (1978). DRw determinants on human T-cells primed against allogeneic lymphocytes. *Transpl Proc* **10**, 845.
77. Hartzmann RJ, Pappas F, Romano PJ, Johnson AH, Ward FE, Amos DB (1978). Dissociation of HLA-D and HLA-DR using primed LD typing. *Transpl Proc* **10**, 809.
78. Reinsmoen NL, Noreen HJ, Sasazuki T, Segall M, Bach FH (1979). Roles of HLA-DR and HLA-D antigens in haplotype-primed LD typing reagents. Proceedings of the 13th International Leukocyte Culture Conference. In *The Molecular Basis of Immune Cell Function.* Amsterdam, North Holland: Elsevier, p. 529.
79. Sasportes M, Nunez-Roldan A, Fradelizi D (1978). Analysis of products involved in primary and secondary allogeneic proliferation in man. III. Further evidence for products different for Ia-like DRw antigens, activating secondary allogeneic proliferation in man. *Immunogenet* **6**, 55.
80. Nunez-Roldan A, Sasportes M, Fradelizi D (1978). Analysis of products involved in primary and secondary allogeneic proliferation in man. I. Detection secondary allogeneic proliferation in man. II. Detection of products different from Ia-like, DRw antigens, activating secondary allogeneic proliferation in man. *Immunogenet* **6**, 43.
81. Termijtelen A, Bradley BA, van Rood JJ (1980). A new determinant, defined by PLT, coded for in the HLA region and apparently independent of the HLA-D and DR loci. *Tissue Antigens* **15**, 267.

82. Wank R, Schendel DJ, Hansen JA, Dupont B (1978). The lymphocyte restimulation system: evaluation by intra-HLA-D group priming. *Immunogenet* **6**, 107.
83. Segall M, Reinsmoen NL, Noreen HJ, Bach FH (1980). Complexity of the HLA-D region studied by primed lymphocyte test. *J Exp Med* **152**, 156s.
84. Mawas C, Charmot D, Mercier P (1980). Split of HLA-D into two regions α and β by a recombination between HLA-D and GLO. I. Study in a family and primed lymphocyte typing for determinants coded by the β region. *Tissue Antigens* **15**, 458.
85. Shaw S, Johnson AH, Shearer GM (1980). Evidence for a new segregant series of B-cell antigens that are encoded in the HLA-D region and that stimulate secondary allogeneic proliferative and cytotoxic responses. *J Exp Med* **152**, 565.
86. Shaw S, Pollack MS, Payne SM, Johnson AH (1980). HLA linked B-cell alloantigens of a new segregant series: population and family studies of the SB antigens. *Hum Immunol* **1**, 177.
87. Shaw S, Duquesnoy R, Smith P (1981). Population studies of the HLA-linked SB antigens. *Immunogenet* **14**, 153.
88. Shaw S, Kavathas P, Pollack MS, Charmot D, Mawas C (1981). Family studies define a new histocompatibility locus, SB, between HLA-DR and GLO. *Nature* **293**, 745.
89. Bignon JD, Semana G, Tiercy M, Simons M, Lalouel JM, Cohen D (1989). DNA-RFLP analysis: HLA-DR beta workshop report. In Dupont B (ed.) *Immunobiology of HLA*. New York: Springer-Verlag, p. 851.
90. Trucco M, *et al.* (1989). RFLP analysis of DQ-beta chain gene: workshop report. In Dupont B (ed.) *Immunobiology of HLA*. New York: Springer-Verlag, p. 860.
91. Simons MJ, Erlich HA (1989). RFLP: sequence interrelations at the DPA and DPB loci. In Dupont B (ed.) *Immunobiology of HLA*. New York: Springer-Verlag, p. 952.
92. Kalil J, *et al.* (1989). Final workshop report: HLA class I RFLP. In Dupont B (ed.) *Immunobiology of HLA*. New York: Springer-Verlag, p. 877.

93. Wake CT, Long EO, Mach B (1982). Allelic polymorphism and complexity of the genes for HLA-DR β-chains — direct analysis by DNA-DNA hybridization. *Nature* **300**, 372.
94. Andersson M, *et al.* (1984). Genomic hybridization with class II transplantation antigen cDNA probes as a complementary technique in tissue typing. *Hum Immunol* **11**, 57.
95. Larhammar D, *et al.* (1982). Complete amino acid sequence of an HLA-DR antigen-like β-chain as predicted by the nucleotide sequence: similarities with immunoglobulins. *Proc Natl Acad Sci* **79**, 3687.
96. Auffray C, Lillie JW, Arnot D, Grossberger D, Kappes D, Strominger JL (1984). Isotypic and allotypic variation on human class II histocompatibility antigen α-chains. *Nature* **308**, 327.
97. Bidwell JL, Bidwell EA, Savage DA, Middleton D, Klouda PT, Bradley BA (1988). A DNA-RFLP typing system that positively identifies serologically well-defined and ill-defined HLA-DR and DQ alleles, including DRw10. *Transplantation* **45**, 640.
98. Carlsson B, Wallin J, Bohme J, Moller E (1987). HLA-DR-DQ haplotypes defined by restriction fragment analysis. Correlation to serology. *Hum Immunol* **20**, 95.
99. Ploegh HL, Orr HT, Strominger JL (1980). Molecular cloning of a human histocompatibility antigen cDNA fragment. *Proc Natl Acad Sci* **77**, 6081.
100. Lee JS, Trowsdale J, Bodmer WF (1982). cDNA clones coding for the heavy chain of human HLA-DR antigen. *Proc Natl Acad Sci* **79**, 545.
101. Long EO, Wake CT, Gorski J, Mach B (1983). Complete sequence of an HLA-DR β-chain deduced from a cDNA clone and identification of multiple non-allelic DR β-chain genes. *EMBO J* **2**, 389.
102. Wiman K, *et al.* (1982). Isolation and identification of a cDNA clone corresponding to an HLA-DR antigen β-chain. *Proc Natl Acad Sci* **79**, 1703.
103. Parham P, Lawlor DA, Salter RD, Lomen CE, Bjorkman PJ, Ennis PD (1989). HLA-A, -B and -C: patterns of polymorphism

in peptide-binding proteins. In Dupont B (ed.) *Immunobiology of HLA*. New York: Springer-Verlag, p. 10.

104. Larhammar D, *et al.* (1983). Molecular analysis of human class II transplantation antigens and their genes. *Hum Immunol* **8**, 95.
105. Gregersen PK, *et al.* (1989). First domain sequence delivery of DR and DQ subregion alleles. In *Immunobiology of HLA*. Dupont B (ed.) New York: Springer-Verlag, p. 1027.
106. Saiki R, *et al.* (1985). Enzymatic amplification of β-globin genomic sequences and restriction site analysis for diagnosis of sickle cell anemia. *Science* **230**, 1350.
107. Mullis KB, Faloona F (1987). Specific synthesis of DNA *in vitro* via a polymerase catalyzed chain reaction. *Methods Enzymol* **155**, 335.
108. Saiki R, Chang CA, Levenson CH, Boehm CD, Kazazian HH, Erlich HA (1987). Rapid genetic analysis of enzymatically-amplified DNA with non-radioactive allele-specific oligonucleotide probes. *Am J Hum Genet* **41**, A237.
109. Bugawan TL, *et al.* (1988). Analysis of HLA-DP alleleic sequence polymorphism using the *in vitro* enzymatic DNA amplification of DP-α and DP-β loci. *J Immunol* **141**, 4024.
110. Horn GT, Bugawan TL, Long CM, Erlich HA (1988). HLA class II antigens and susceptibility to insulin-dependent diabetes mellitus. *Proc Natl Acad Sci* **85**, 6012.
111. Morel C, Zwahlen F, Jeannet M, Mach B, Tiercy JM (1990). Complete analysis of HLA-DQB1 polymorphism and DR-DQ linkage disequilibrium by oligonucleotide typing. *Hum Immunol* **29**, 64.
112. Erlich HA, Scharf SJ, Long CM, Horn GT (1989). Analysis of isotypic and allotypic sequence variation in the HLA-DRb region using the *in vitro* enzymatic amplification of specific DNA segments. In Dupont B (ed.) *Immunobiology of HLA*. New York: Springer-Verlag, p. 181.
113. Baxter-Lowe LA, Hunter JB, Casper JT, Gorski J (1989). HLA gene amplication and hybridization analysis of polymorphism. *J Clin Invest* **84**, 613.

114. Park MS, Tonai R (1992). Phenotype frequencies of the class II (DR, DQ) DNA alleles by the patterns of sequence-specific primer mixtures (SSPM) in four different populations and the probable haplotypes between DRB1 allele and DQB1 allele. In Terasaki PI, Cecka M (eds.) *Clinical Transplants*. Los Angeles, CA: UCLA Tissue Typing Laboratory, p. 475.
115. Savel-Koul PHM, de Bruyn-Geraets DP, van den Berg-Loonen EM (1995). High resolution HLA-DRB1 SSP typing for cadaveric donor transplantation. *Tissue Antigens* **45**, 41.
116. Santamaria P, Boyce-Jacino MT, Lindstrom AL, Barbosa JJ, Faras AJ, Rich SS (1992). HLA class II "typing": direct sequencing of DRB, DQB and DQA genes. *Hum Immunol* **33**, 69.
117. Spurkland A, Knutsen I, Markussen G, Vartdal F, Egeland T, Thorsby E (1993). HLA matching of unrelated bone marrow transplant pairs: direct sequencing of *in vitro* amplified HLA-DRB1 and -DQB1 using magnetic beads as solid support. *Tissue Antigens* **41**, 155.
118. Knipper AJ, Hinney A, Schuch B, Enczmann J, Uhrberg M, Wernet P (1994). Selection of unrelated bone marrow donors by PCR-SSP typing and subsequent non-radioactive sequence-based typing for HLA DRB1/3/4/5/, DQB1 and DPB1 alleles. *Tissue Antigens* **44**, 275.
119. Oh SH, Fleischhauer K, Yang SY (1993). Isoelectric focusing subtypes of HLA-A can be defined by oligonucleotide typing. *Tissue Antigens* **41**, 135.
120. Yoshida AJ, Kimura A, Numano F, Sasazuki T (1992). Polymerase chain reaction based analysis of polymorphism in the HLA-B gene. *Hum Immunol* **34**, 257.
121. Petersdorf EW, Stanley JF, Martin PJ, Hansen JA (1994). Molecular diversity of the HLA-C locus in unrelated marrow transplantation. *Tissue Antigens* **44**, 93.
122. Browning MJ, Krausa P, Rowan A, Bicknell DC, Bodmer JG, Bodmer WF (1993). Tissue typing the HLA-A locus from genomic DNA by sequence-specific PCR: comparison of HLA

genotypes and surface expression on colorectal tumor cell lines. *Proc Natl Acad Sci* **90**, 2842.

123. Krausa P, Browning MJ (1996). A comprehensive PCR-SSP typing system for identification of HLA-A locus alleles. *Tissue Antigens* **47**, 237.
124. Sadler AM, *et al.* (1994). Low resolution DNA typing for HLA-B using sequence-specific primers in allele- or group-specific ARMS/PCR. *Tissue Antigens* **44**, 148.
125. Arguello R, Avakian H, Goldman JM, Madrigal A (1996). A novel method for simultaneous high resolution identification of HLA-A, HLA-B, and HLA-Cw alleles. *Proc Natl Acad Sci* **93**, 10961.
126. Fischer GF, Fae I, Petrasek M, Moser S (1995). A combination of two distinct *in vitro* amplification procedures for DNA typing of HLA-DRB and -DQB1 alleles. *Vox Sang* **69**, 328.
127. Uryu N, Maeda M, Ota M, Tsuji K, Inoko H (1990). A simple and rapid method for HLA-DRB and -DQB typing by digestion of PCR-amplified DNA with allele specific restriction endonucleases. *Tissue Antigens* **35**, 20.
128. Maeda M, *et al.* (1989). A simple and rapid method for HLA-DQA1 genotyping by digestion of PCR-amplified DNA with allele specific restriction endonucleases. *Tissue Antigens* **34**, 290.
129. Maeda M, *et al.* (1990). A simple and rapid method for HLA-DP genotyping by digestion of PCR-amplified DNA with allele specific restriction endonucleases. *Hum Immunol* **27**, 111.
130. Bidwell JL, Hui KM (1990). Human HLA-DR/Dw allotype matching by analysis of HLA-DRB gene PCR product polymorphism (PCR "fingerprints"). *Technique* **2**, 93.
131. Carrington M, *et al.* (1992). Typing of HLA-DQA1 and DQB1 using DNA single-strand conformation polymorphism. *Hum Immunol* **33**, 208.
132. Summers C, Fergusson W, Gokhale D, Taylor M (1992). Donor-recipient bone marrow matching by single strand conformation polymorphism analysis. *Lancet* **339**, 621.

133. Hurley CK, Giles RC, Capra JD (1983). The human MHC: evidence for multiple HLA-D region genes. *Immunol Today* **4**(8), 219.
134. Lampson L, Levy R (1980). Two populations of Ia-like molecules on a human B-cell line. *J Immunol* **125**(1), 293.
135. Dick H (1982). HLA-DR: more than three loci? *Immunol Today* **3**, 8.
136. Hurley CK, Nunez G, Winchester R, Finn OJ, Levy R, Capra JD (1982). The human HLA-DR antigens are encoded by multiple β-chain loci. *J Immunol* **129**(5), 2103.
137. Hurley CK, Shaw S, Nadler L, Schlossman S, Capra JD (1982). The α- and β-chains of SB and DR antigens are structurally distinct. *J Exp Med* **156**, 1557.
138. Charron DJ, Lotteau V, Turmel P (1984). Hybrid HLA-DC antigens provide molecular evidence for gene transcomplementation. *Nature* **312**, 157.
139. Karr RW, Kannpell CC, Zangara RJ, Goyert SM, Silver J Jr, Schwartz BD (1983). An HLA-DR5 homozygous cell line expressed two DS (I-A-like) molecules. *J Exp Med* **158**, 1374.
140. Giles RC, DeMars R, Chang CC, Capra JD (1985). Allelic polymorphism and transcomplementation of molecules encoded by the HLA-DQ subregion. *Proc Natl Acad Sci* (in press).
141. Wells RS, Parham P (1996). HLA class I genes: structure and diversity. In Browning MJ, McMichael AJ (eds.) *HLA and MHC Genes, Molecules and Function*. Oxford, United Kingdom: BIOS Scientific Publishers, Ltd, p. 77
142. Geraghty DE (1993). Structure of the HLA class I region and expression of its resident genes. *Curr Opin Immunol* **5**, 3.
143. Geraghty DE, Pei J, Lipsky B, Handsen JA, Taillon-Miller P, Bronson SK, Chaplin DD (1992a). Cloning and physical mapping of the HLA class I region spanning the HLA-E to HLA-F interval by using yeast artificial chromosomes. *Proc Natl Acad Sci* **89**, 2669.

144. Geraghty DE, Koller BH, Hansen JA, Orr HT (1992b). The HLA class I gene family includes at least six genes and twelve pseudogenes and gene fragments. *J Immunol* **149**, 1934.
145. Le Bouteiller P (1994). HLA class I chromosomal region, genes and products: facts and questions. *Crit Rev Immunol* **14**, 89.
146. Andersson G, Larhammar D, Widmark E, Servenius B, Peterson PA, Rask L (1987). Class II genes of the human major histocompatibility complex. Organization and evolutionary relationship of the DRbeta genes. *J Biol Chem* **262**, 8748.
147. Gorski J, Rollini P, Mach B (1987). Structural comparisons of the genes of two HLA-DR supertypic groups: the loci encoding DRw52 and DRw53 are not truly allelic. *Immunogenet* **25**, 397.
148. Kawai J, Ando A, Sato T, Nakatsuji T, Inoko H (1989). Analysis of gene structure and antigen determinants of DR2 antigens using DR gene transfer into mouse L cells. *J Immunol* **142**, 312.
149. Rollini P, Mach B, Gorski J (1985). Linkage map of three HLA-DR beta chain genes: evidence for a recent duplication event. *Proc Natl Acad Sci* **82**, 7197.
150. Gustafsson K, Widmark E, Jonsson A-K, Servenius B, Sachs DH, Larhammar D, Rask L, Peterson PA (1987). Class II genes of the human major histocompatibility complex. Evolution of the DP region as deducted from nucleotide sequences of the four genes. *J Biol Chem* **262**, 8778.
151. Jonsson A-K, Hyldig-Nielsen J-J, Servenius B, Larhammar D, Andersson G, Jorgensen F, Peterson PA, Rask L (1987). Class II genes of the human major histocompatibility complex. Comparisons of the DQ and Dx alpha and beta genes. *J Biol Chem* **262**, 8767.
152. Trowsdale J (1993). Genomic structure and function in the MHC. *Trends Genet* **9**, 117.
153. Trowsdale J, Kelly A, Lee J, Carson S, Austin P, Travers P (1984). Linkage map of two HLA-SBbeta-related genes: an intron in one of the SBbeta-related genes contains a processed pseudogene. *Cell* **38**, 241.

154. Tonnelle C, DeMars R, Long EO (1985). DO beta: a new beta chain gene in HLA-D with a distinct regulation of expression. *EMBO J* **4**, 2839.

155. Sanderson F, Kleijmeer MJ, Kelly AP, Verwoerd D, Tulp A, Neefjes JJ, Geuze HJ, Trowsdale J (1994). Accumulation of HLA-DM, a regulator of antigen presentation, in MHC class II compartments. *Science* **266**, 1566.

156. Townsend A, Trowsdale J (1993). The transporters associated with antigen processing. *Sem Cell Biol* **4**, 53.

157. Abderrahim H, Sambucy JL, Iris F, Ougen P, Billault A, Chumakov IM, Dausset J, Cohen D, Le Paslier D (1994). Cloning the human major histocompatibility complex in YACs. *Genomics* **23**, 520.

158. Cresswell P (1994). Assembly, transport and function of MHC class II molecules. *Ann Rev Immunol* **12**, 259.

159. Glynne R, Powis SH, Beck S, Kelly A, Kerr L-A, Trowsdale J (1991). A proteasome-related gene between the two ABC transporter loci in the class II region of the human MHC. *Nature* **353**, 357.

160. Kelly A, Powis SH, Kerr LA, Mockridge I, Elliott T, Bastin J, Uchanska-Ziegler B, Ziegler A, Trowsdale J, Townsend A (1992). Assembly and function of the two ABC transporter proteins encoded in the human major histocompatibility complex. *Nature* **355**, 641.

161. Bjorkman PJ, Saper MA, Samraoui B, Bennett WS, Strominger JL, Wiley DC (1987a). Structure of the human class I histocompatibility antigen, HLA-A2. *Nature* **329**, 506.

162. Bjorkman PJ, Saper MA, Samraoui B, Bennett WS, Strominger JL, Wiley DC (1987b). The foreign antigen binding site and T-cell recognition regions of class I histocompatibility antigens. *Nature* **329**, 512.

163. Brown JH, Jardetzky TS, Gorga JC, Stern LJ, Urban RG, Strominger JL, Wiley DC (1993). Three-dimensional structure of the human class II histocompatibility antigen HLA-DR1. *Nature* **364**, 33.

164. Russell PS, Winn HJ, Amos DB (eds.) (1965). In *Histocompatibility Testing*. Washington, D.C.: National Academy of Sciences-National Research Council, Publication 1229.
165. Amos DB, van Rood JJ (eds.) (1966). In *Histocompatibility Testing*. Baltimore: The Williams and Wilkins Company, p. 435.
166. Curtoni ES, Mattiuz PL, Tosi RM (eds.) (1967). In *Histocompatibility Testing*. Baltimore: The Williams and Wilkins Company, p. 435.
167. Terasaki PI (ed.) (1970). In *Histocompatibility Testing*. Copenhagen: Munksgaard.
168. Dausset J, Colombani J (eds.) (1973). In *Histocompatibility Testing*. Copenhagen: Munksgaard.
169. Kissmeyer-Nielsen F (ed.) (1975). In *Histocompatibility Testing*. Copenhagen: Munksgaard.
170. Bodmer WF, Batchelor JR, Bodmer JG, Festenstein H, Morris PJ (eds.) (1978). In *Histocompatibility Testing*. Copenhagen: Munksgaard.
171. Terasaki PI (ed.) (1980). In *Histocompatibility Testing*. Los Angeles, CA: UCLA Tissue Typing Laboratory.
172. Albert ED, Baur MP, Mayr WR (eds.) (1984). In *Histocompatibility Testing*. Berlin: Springer-Verlag.
173. Dupont B (ed.) (1987). In *Immunobiology of HLA Histocompatibility Testing*. New York: Springer Verlag, Vols. 1 and 2.
174. Tsuji K, Aizawa M, Sasazuki T (eds.) (1992). In *HLA*. Oxford: University Press, Vols. 1 and 2.
175. Charron D (ed.) (1997). In *HLA: Genetic Diversity of HLA-Functional and Medical Implications*. E.D.K. Publishers, Vols. 1 and 2.
176. Amos DB (1990). HLA-A Mouser's Recollections. In Terasaki PI, Dausset J, Payne R, van Rood JJ, van Leeuwen A, Amos DB, Bodmer W, Bodmer JJ, Walford RL, Kissmeyer-Nielsen F, Svejgaard A, Batchelor JR, Bach FH (eds.) *History of HLA: Ten Recollections*. Los Angeles, CA.: UCLA Tissue Typing Laboratory, p. 61.

177. Stulting RD, Ward FE (1975). A computerized system for the selection of organ transplant recipients. *Transplantation* **19**(1), 27.
178. Medawar PB (1946). Immunity to homologous grafted skin. II. The relationship between the antigens of blood and skin. *Br J Exp Path* **27**, 15.
179. Amos DB (1953). The agglutination of mouse leukocytes by iso-immune sera. *Br J Exp Path* **34**, 464.
180. Friedman EA, Retan JW, Marshall DC, Henry L, Merrill JP (1962). Accelerated skin graft rejection in humans preimmunized with homologous peripheral leukocytes. *J Clin Invest* **40**, 2162.
181. Rapaport FG, Lawrence HS, Converse JM, Mulholland JH (1964). Homograft antigen studies in man. *Ann NY Acad Sci* **120**, 280.
182. Amos DB, Kostyu DD (1980). HLA — a central immunological agency of man. *Adv Hum Genet* **10**, 137.
183. Terasaki PI, Marchioro TL, Starzl TE (1965). Sero-typing of human lymphocyte antigens: preliminary trials on long-term kidney homograft survivors. In *Histocompatibility Testing*. Washington, D.C.: National Academy of Science-National Research Council, p. 83.
184. Kissmeyer-Nielsen F, Olsen S, Petersen VP, Fjelborg O (1966). Hyperacute rejection of kidney allografts associated with pre-existing humoral antibodies against donor cells. *Lancet* **2**, 662.
185. Terasaki PI, Mickey MR, Kreisler M (1971). Presensitization and kidney transplant failures. *Postgrad Med J* **47,** 89.
186. Terasaki PI, Vredevoe DL, Mickey MR, Porter KA, Marchioro TL, Faris TD, Starzl TE (1966). Serotyping for homotransplantation. VI. Selection of kidney donors for thirty-two recipients. *Ann NY Acad Sci* **129**, 500.
187. van Rood JJ (1967). A proposal for international cooperation in organ transplantation: eurotransplant. In Curtoni ES, Mattiuz PL, Tosi RM (eds.) *Histocompatibility Testing*. Copenhagen: Munksgaard, p. 451.

188. Starzl TE, Porter KA, Andres G, Halgrimson CG, Hurwitz R, Giles G, Terasaki PI, Penn I, Schroter GT, Lilly J, Starkie SJ, Putnam CW (1970). Long-term survival after renal transplantation in humans: with special reference to histocompatibility matching, thymectomy, homograft glomerulonephritis, heterologous ALG, and recipient malignancy. *Ann Surg* **172**, 437.
189. Mickey MR, Kreisler M, Albers ED, Tanaka N, Terasaki PI (1971). Analysis of HLA incompatibility in human renal transplants. *Tissue Antigens* **2**, 57.
190. Terasaki P, Takemoto S (1999). Impact of HLA system on organ transplantation: a critical review before the end of the millennium. *Adv Nephrol* **29**, 163.
191. Takemoto S, Terasaki PI, Cecka JM, *et al.* (1992). Survival of nationally shared, HLA-matched kidney transplants from cadaveric donors: the UNOS Scientific Renal Transplant Registry (see comments). *N Engl J Med* **327**, 834.
192. Reisaeter AV, Leivestad T, Vartdal F, Spurkland A, Fauchald P, Brekke IB, Thorsby E (1998). A strong impact of matching for a limited number of HLA-DR antigens on graft survival and rejection episodes. *Transplantation* **66**, 523.
193. Matas AJ, Najarian JS (1993). The role of HLA typing in clinical kidney transplants: 30 years later. In Terasaki PI (ed.) *Clinical Transplants*. Los Angeles, CA: UCLA Tissue Typing Laboratory, p. 428.
194. Stratta RJ, Armbrust MJ, Lorentzen DF, Hoffman RM, D'Alesssandro AM, Sollinger HW, Pirsch JD, Kalayoglu M, Belzer FO (1987). Cadaveric renal transplantation in the cyclosporine and OKT3 eras: the University of Wisconsin-Madison experience. In Terasaki PI (ed.) *Clinical Transplants*. Los Angeles, CA: UCLA Tissue Typing Laboratory, p. 183.
195. Uphoff DE, Law LW (1958). Genetic factors influencing irradiation protection by bone marrow. II. The histocompatibility-2 (H-2) locus. *J Natl Cancer Inst* **20**, 617.

196. Snell GD (1992). The Nobel Lectures in Immunology. Lecture for the Nobel Prize for Physiology or Medicine, 1980: Studies in histocompatibility. *Scand J Immunol* **36**, 513.
197. Epstein RB, Storb R, Ragde H, Thomas ED (1968). Cytotoxic typing antisera for marrow grafting in littermate dogs. *Transplantation* **6**, 45.
198. Thomas ED, Storb R (1999). The development of the scientific foundation of hematopoietic cell transplantation based on animal and human studies. In Forman S, Blume K, Thomas ED (eds.) *Hematopoietic Cell Transplantation.* Malden, MA: Blackwell Science Inc, Vol. 2, p. 1.
199. Gatti RA, Meuwissen HJ, Allen HD, Hong R, Good RA (1968). Immunological reconstitution of sex-linked lymphopenic immunological deficiency. *Lancet* **2**, 1366.
200. Bach FJ, Albertini RJ, Joo P, Anderson JL, Bortin MM (1968). Bone marrow transplantation in a patient with the Wiskott-Aldrich syndrome. *Lancet* **2**, 1364.
201. De Koning J, van Bekkum DW, Dicke KA, Dooren LJ, van Rood JJ (1969). Transplantation of bone marrow cells and fetal thymus in an infant with lymphopenic immunological deficiency. *Lancet* **2**, 1223.
202. Thomas ED, *et al.* (1971). Allogeneic marrow grafting using HL-A matched donor-recipient sibling pairs. *Trans Assoc Am Physicians* **84**, 248.
203. Santos GW, *et al.* (1971). Marrow transplantation in man following cyclophosphamide. *Transpl Proc* **3**, 400.
204. Tiercy JM, *et al.* (1991). Selection of unrelated donors for bone marrow transplantation is improved by HLA class II genotyping with oligonucleotide hybridization. *Proc Natl Acad Sci* **88**, 7121.
205. Petersdorf EW, Stanley JR, Martin PJ, Hansen JA (1994). Molecular diversity of the HLA-C locus in unrelated marrow transplantation. *Tissue Antigens* **44**, 93.
206. Santamaria P, *et al.* (1994). Frequent HLA class I and DP sequence mismatches in serologically HLA-A, HLA-B, HLA-DR)

and molecularly (HLA-DRB1, HLA-DQA1, HLA-DQB1) HLA-identical unrelated bone marrow transplant pairs. *Blood* **83**, 280.

207. Tiercy JM, Djavad N, Rufer N, Speiser DE, Jeannet M, Roosnek E (1994). Oligotyping of HLA-A2, -A3, and B44 subtypes: detection of subtype incompatibilities between patients and their serological matched unrelated bone marrow donors. *Hum Immunol* **41**, 207.
208. Nagler A, Brautbar C, Slavin S, Bishara A (1996). Bone marrow transplantation using unrelated and family related donors: the impact of HLA-C disparity. *Bone Marrow Transplant* **18**, 891.
209. Reinsmoen NL, Bach FH (1982). Five HLA-D clusters associated with HLA-DR4. *Hum Immunol* **4**, 249.
210. Anasetti C, *et al.* (1989). Effect of HLA compatibility on engraftment of bone marrow transplants in patients with leukemia or lymphoma. *N Engl J Med* **320**, 197.
211. Beatty PG, *et al.* (1985). Marrow transplantation from related donors other than HLA-identical siblings. *N Engl J Med* **313**, 765.
212. Powles RL, *et al.* (1983). Mismatched family donors for bone-marrow transplantation as treatment for acute leukaemia. *Lancet* **1**, 612.
213. Anasetti C, *et al.* (1990). Effect of HLA incompatibility on graft-versus-host disease, relapse, and survival after marrow transplantation for patients with leukemia or lymphoma. *Hum Immunol* **29**, 79.
214. Szydlo R, *et al.* Results of allogeneic bone marrow transplantation for leukemia using donors other than HLA-identical siblings. *J Clin Oncol* **15**, 1767.
215. Anasetti C, Hansen JA (1994). Effect of HLA incompatibility in marrow transplantation from unrelated and HLA-mismatched related donors. *Transfus Sci* **15**, 221.
216. Anasetti C (1999). Hematopoietic cell transplantation from HLA partially matched related donors. In Forman S, Blume K, Thomas E. (eds.) *Hematopoietic Cell Transplantation*. Malden, MA.: Blackwell Science Inc, 2nd rev., p. 904.

217. James DCO (1984). Organization of a hospital bone marrow panel. In Smit Sibinga CTH, Das PC, Opelz G (eds.) *Transplantation and Blood Transfusion*. Boston: Martinus Nijoff, p. 131.
218. Cleaver SA (1993). The Anthony Nolan Research Centre. *Bone Marrow Transpl* **11**(S1), 38.
219. McCullough J, Hansen J, Perkins H, Stroncek D, Bartsch G (1989). The National Marrow Donor Program: how it works, accomplishments to date. *Oncology* **3**, 63.
220. Howe CE, Radde-Stepaniak T (1999). Hematopoietic cell donor registries. In Thomas ED, Blume KG, Forman SJ (eds.) *Hematopoietic Cell Transplantation*. Malden, MA: Blackwell Science Inc, p. 503.
221. National Marrow Donor Program (NMDP). http:/www.marrow.org.
222. International Bone Marrow Transplant Registry (IBMTR). http://www.IBMTR.org.
223. Amos DB, Bach FH (1968). Phenotypic expressions of the major histocompatibility locus in man (HL-A): leukocyte antigens and mixed leukocyte culture reactivity. *J Exp Med* **128**, 623.
224. Bach ML, Engstrom MA, Bach FH, Etheredge EE, Najarian JS (1972). Specific tolerance in human kidney allograft recipients. *Cell Immunol* **3**, 161.
225. Carpenter C (1974). Transplantation: immunogenetics and effector mechanisms. In Gottlieb A (ed.) *Developments in Lymphoid Cell Biology*. Cleveland, OH.: CRC Press, p. 133.
226. Etheredge E, Shons A, Najarian J (1973). Mixed lymphocyte culture changes following allotransplantation. In Daguillard F (ed.) *7th Leucocyte Culture Conference*. New York, NY: Academic Press, p. 399.
227. Revillard JP, Robert M, DuPont E, Betuel H, Rifle G, Traeger J (1973). Inhibition of mixed lymphocyte culture by alloantibodies in renal transplantation and in pregnancy. *Transpl Proc* **5**, 331.
228. Garovoy MR, Phillips SM, Carpenter CB, Merrill JP (1973). Antibody modulation of cellular reactivity post renal transplantation. *Transpl Proc* **5**, 129.

229. Charpentier BM, Lang P, Martin B, Fries D (1982). Specific recipient-donor unresponsiveness mediated by a suppressor cell system in human kidney allograft tolerance. *Transplantation* **33**, 470.
230. Shons AR, Etheredge EE, Schmidtke JR, Najarian JS (1973). The mixed leukocyte response in human renal transplantation. *Transpl Proc* **5**, 337.
231. Johnson AH, Rossen RD, Hersh EM, Farrow SS, Butler WT, Suki WN (1974). Blastogenic suppression and alloantibody activity of sera from renal allograft recipients. *J Clin Invest* **53**, 270.
232. Rashid A, Sengar DP, Harris JE (1979). Relationship of donor-specific mixed lymphocyte culture reactivity to graft function in recipients of cadaveric renal allografts. *Clin Exp Immunol* **136**, 521.
233. Goulmy E, Persijn G, Blokland E, D'Amaro J, van Rood JJ (1981). Cell-mediated lympholysis studies in renal allograft recipients. *Transplantation* **31**, 210.
234. Stiller CR, Sinclair NR, Abrahams S, McGirr D, Singh H, Howson WT, Ulan RA (1976). Anti-donor immune responses in prediction of transplant rejection. *N Engl J Med* **294**, 978.
325. Ellis TM, Lee HM, Mohanakumar, T (1981). Alterations in human regulatory T lymphocyte subpopulations after renal allografting. *J Immunol* **127**, 2199.
236. Hattler BG, Miller J (1973). Blocking antibody in the sera and on the kidney of successful human transplants. *Transpl Proc* **5**, 655.
237. Miyajima T, Higuchi R, Kashiwabara H, Yokoyama T, Fujimoto S (1980). Anti-idiotypic antibodies in a patient with a functioning renal graft. *Nature* **283**, 306.
238. Pfeffer PF, Thorsby E, Hirschberg H (1983). Cell-mediated cytotoxicity toward the donor in patients with well-functioning kidney grafts. *Transplantation* **35**, 546.

239. Pfeffer PF, Hirschberg H, Thorsby E (1981). Donor-specific decreased primary and secondary cell-mediated immune responses in patients with well functioning grafts. *Transpl Proc* **13**, 1604.
240. Fujii Y, Dohi K, Takai M, Kanauchi M, Hanatani M, Hamaguchi T, Hayashi M, Ishikawa H (1983). Cell mediated immunity in primary renal diseases: concanavalin A induced suppressor T-cell function and autologous mixed lymphocyte reaction. *Japan J Nephrol* **25**(11), 1325.
241. Seki Y, Sakagami K, Takeuchi H, Tanaka K, Namba A, Fuchimoto S, Orita K (1983). Evidence for donor-specific cell-mediated lympholysis unresponsiveness and suppressor cells in well functioning kidney transplants. *Transpl Proc* **15**, 2131.
242. Wramner L, Olausson M, Soderstrom T, Lindholm L, Rydberg L, Brynger H (1987). Evidence of donor-specific cellular suppressor activity in donor-specific cell-mediated lympholysis responsiveness in renal transplant patients. *Transplantation* **44**, 390.
243. Herzog WR, Zanker B, Irschick E, Huber C, Franz HE, Wagner H, Kabelitz D (1987). Selective reduction of donor-specific cytotoxic T-lymphocyte precursors in patients with a well-functioning kidney allograft. *Transplantation* **43**, 384.
244. Cohen DJ, Lee JM, Mohanakmar T (1985). Mechanisms of CML hyporesponsiveness in long-term renal allograft recipients. *Hum Immunol* **14**, 279.
245. Wonigeit K, Pichmayr R (1978). Posttransplant monitoring of donor-specific T-cell reactivity at the precursor cell level. *Transpl Proc* **10**, 563.
246. Thomas J, Thomas F, Mendez-Picon G, Lee H (1977). Immunological monitoring of long-surviving renal transplant recipients. *Surgery* **81**, 125.
247. Thomas J, Johns C, Hoffmann S, Thomas F, Lee HM (1978). Suppressor cells inhibiting the generation of cytotoxic T lymphocytes in long-term human transplant recipients. *Proc Dial Transpl Forum* **8**, 87.

248. Liburd EM, Pazderka V, Kovithavongs T, Dossetor JB (1978). Evidence for suppressor cells and reduced CML induction by the donor in transplant patients. *Transpl Proc* **10**, 557.
249. Dossetor JB, Liburd EM, Kovithavongs T, Pazderka V (1979). Suppressor cells and loss of *in vitro* killer cell inductiveness to donor cells in the peripheral blood lymphocytes of allograft recipients. *Transpl Proc* **1**, 1260.
250. Goulmy E, Stijnen T, Groenewoud AF, Persijn GG, Blokland E, Pool J, Paul LC, van Rood JJ (1989). Renal transplant patient monitored by the cell-mediated lympholysis assay: evaluation of the clinical value. *Transplantation* **48**, 559.
251. Goulmy E, Bittner K, Blokland E, Pool J, Persijn G, van Rood JJ, Lange H (1991). Renal transplant patients with steroid withdrawal evaluated longitudinally for their donor-specific cytotoxic T-cell reactivity. *Transplantation* **52**, 1083.
252. Bean MA, Mickelson E, Yanagida J, Ishioka S, Brannen GE, Hansen JA (1990). Suppressed antidonor MLC responses in renal transplant candidates conditioned with donor-specific transfusions that carry the recipients noninherited maternal HLA haplotype. *Transplantation* **49**, 382.
253. Grailer AP, Sollinger HW, Kawamura T, Burlingham WJ (1991). Donor-specific cytotoxic T lymphocyte hyporesponsiveness following renal transplantation in patients pretreated with donor-specific transfusions. *Transplantation* **51**, 320.
254. Mickelson EM, Bean MA, Ishioka S (1985). Monitoring alloimmune responses *in vivo* following donor-specific transfusion: generation of suppressor T-cells and alterations in alloreactive precursor frequencies. *Transpl Proc* **17**, 2389.
255. Cooperative Clinical Trials in Transplantation Research Group, Alexander JW, Light JA, Donaldson LA, Delmonico, FL, Diethelm AG, Wilkinson A, Rosenthal JT, Thistlewaite JR, Hunsicker LG, Matas AJ, First MR, Reinsmoen NL, Rose SM (1991). Evaluation of pre- and post-transplant donor-specific transfusion/cyclosporine A in non-HLA identical living donor kidney transplant recipients. *Transplantation* **68**(8), 1117.

256. Barber WH, Mankin JA, Laskow DA, Deierhoi MH, Julian BA, Curtis JJ, Diethelm AG (1991). Long-term results of a controlled prospective study with transfusion of donor-specific bone marrow in 57 cadaveric renal allograft recipients. *Transplantation* **51**, 70.
257. Zeevi A, Pavlick M, Lombardozzi S, Banas R, Pappo O, Rao AS, Fontes P, Demetris J, Shapiro R, Dodson F (1995). Immune status of recipients following bone marrow-augmented solid organ transplantation. *Transplantation* **59**(4), 616.
258. Saper V, Chow D, Engleman ED, Hoppe RT, Levin B, Collins G, Strober S (1998). Clinical and immunological studies of cadaveric renal transplant recipients given total lymphoid irradiation and maintained on low-dose prednisone. *Transplantation* **45**, 540.
259. Chow D, Saper V, Strober S (1987). Renal transplant patients treated with total lymphoid irradiation show specific unresponsiveness to donor antigens in the mixed leukocyte reaction (MLR). *J Immunol* **138**, 3746.
260. Strober S, Dhillon M, Schubert M, Holm B, Engleman E, Benike C, Hoppe R, Sibley R, Myburgh JA, Collings G, Levin B (1989). Acquired immune tolerance to cadaveric renal allografts: a study of three patients treated with total lymphoid irradiation. *N Engl J Med* **321**, 28.
261. Strober S, Slavin S, Gottlieb M, Zan-Bar I, King DP, Hoppe RT, Fuks Z, Grumet FC, Kaplan HS (1979). Allograft tolerance after total lymphoid irradiation (TLI). *Immunol Rev* **46**, 87.
262. Bas J, Mestre M, Grino JM, Massip E, Castelao AM, Romeu A, Gonzalez L, Valls A, Buendia E (1993). *In vitro* donor-specific hyporesponsiveness and T-cell subsets in renal allograft recipients. *Allergol Et Immunopathol* **21**(4), 136.
263. Bas J, Masip E, Mestre M, Grino JM, Gonzalez C, Gonzalez L, Buendia E (1992). Donor-specific hyporesponsiveness in renal transplantation. *Transpl Proc* **24**(1), 76.
264. Kerman RH, Katz SM, Schoenberg L, Baraket O, van Buren CT, Kahan BD (1997). Ten-year follow-up of mixed lymphocyte reaction hyporesponsive living related cyclosporine monotherapy-treated renal allograft recipients. *Transpl Proc* **29**(1–2), 198.

265. Kerman RH, Susskind B, Katz SM, van Buren CT, Kahan BD (1997). Postrenal transplant MLR hypo-responders have fewer rejections and better graft survival than MLR hyper-responders. *Transpl Proc* **29**(1–2), 1410.
266. Ghobrial II, Morris AG, Booth LJ (1994). Clinical significance of *in vitro* donor-specific hyporesponsiveness in renal allograft recipients as demonstrated by the MLR. *Transpl Int* **7**, 420.
267. Reinsmoen NL, Kaufman D, Matas A, Sutherland DER, Najarian JS, Bach FH (1990). A new *in vitro* approach to determine acquired tolerance in long-term kidney allograft recipients. *Transplantation* **50**, 783.
268. Reinsmoen NL, Matas AJ (1993). Evidence that improved late renal transplant outcome correlates with the development of *in vitro* donor antigen-specific hyporeactivity. *Transplantation* **55**, 1017.
269. Flechner SM, Kerman RH, van Buren CT, Epps L, Kahn BD (1984). The use of cyclosporine in living-related renal transplantation: donor-specific hyporesponsiveness and steroid withdrawal. *Transplantation* **38**, 60.
270. Kahan BD, Kerman RH, van Buren CT, Flechner SM, Golden DL, Lewis RM (1989). Clinical outcome in 36 patients after at least one and up to 5 years of steroid withdrawal based upon specific mixed lymphocyte reaction hyporesponsiveness toward the living related donor. *Transpl Proc* **2**, 1579.
271. Reinsmoen NL, Bolman RM, Savik K, Butters K, Matas AJ, Hertz MI (1994). Improved long-term graft outcome in lung transplant recipients who have donor antigen-specific hyporeactivity. *J Heart Lung Transpl* **13**, 30.
272. Reinsmoen NL, Gillingham KJ, Jackson AM, Butters K, McSherry C, Diko M, Hertz MI (1998). Can cellular assays predict chronic rejection? *Graft* **1**(2), 60.
273. De Haan A, van Den Berg AP, Hepkema BG, van Dijk E, Haagsma EB, Sloof MJH, Lems SPM, de Leij LFMH, Prop J (1998). Donor-specific hyporeactivity after liver transplantation. *Transplantation* **66**(4), 516.

274. Mathew JM, Marsh JW, Susskind B, Mohanakumar T (1993). Analysis of T-cell responses in liver allograft recipients. Evidence for deletion of donor-specific cytotoxic T-cells in the peripheral circulation. *J Clin Invest* **91**, 900.

275. O'Connell PJ, Mba-Jonas A, Leverson GE, Heisey DM, Meyer KC, Love RB, Burlingham WJ (1998). Stable lung allograft outcome correlates with the presence of intragraft donor-derived leukocytes. *Transplantation* **66**(9), 1167.

276. Reinsmoen NL, Jackson AM, McSherry C, Ninova D, Weisner RH, Kondo M, Krom RAF, Hertz MI, Bolman RM, Matas AJ (1995). Organ-specific patterns of donor antigen-specific hyporeactivity and peripheral blood allogeneic microchimerism in lung, kidney, and liver transplant recipients. *Transplantation* **60**, 1546.

277. van Twuyver E, de Hoop J, ten Berge RJ, Wilmink JM, Lems SP, van de Berg AP, Slooff MJ, de Waal LP (1996). Comparison of T-cell responses in patients with a long-term surviving renal allograft versus a long-term surviving liver allograft. It's a different world. *Transplantation* **61**(9), 1392.

278. van Den Berg AP, de Haan A, Hepkema BG, van Dijk E, van Twuyver E, Klompmaker IJ, de Leij LH, de Waal LP, Slooff MJ (1998). Donor-specific immunological non-responsiveness after liver transplantation. *Transpl Int* **11**(Suppl. 1), S239.

279. Breur-Vriesendrop BS, Vingerhoed J, van Twuyver E, de Waal LP, Ivanyi P (1991). Frequency analysis of HLA-specific cytotoxic T lymphocyte precursors in humans. *Transplantation* **51**, 1096.

280. Orosz CG, Adams PW, Ferguson RM (1987). Frequency of human alloantigen-reactive T lymphocytes. *Transplantation* **43**, 718.

281. Bishop DK, Orosz CG (1989). Limiting dilution analysis for alloreactive TCGF-secretory T-cells. Two related LDA methods that discriminate between unstimulated precursor T-cells and *in vivo*-alloactivated T-cells. *Transplantation* **47**, 671.

282. Hadley GA, Anderson CG, Mohanakumar T (1992). Selective loss of functional antidonor cytolytic T-cell precursors following

donor-specific blood transfusions in long-term renal allograft recipients. *Transplantation* **54**, 333.

283. Mohanakumar T, Naziruddin B (1993). Posttransplant loss of response to donor antigens. *Transpl Sci* **3**, 202.
284. Eberspacher MLL, Otto G, Herfarth C, Kabelitz D (1994). Frequency analysis of donor-reactive cytotoxic T lymphocyte precursors in liver allograft recipients. *Transplantation* **57**, 1746.
285. Zanker B, Jooss-Rudiger J, Franz HE, Wagner H, Kabelitz D (1993). Evidence that functional deletion of donor-reactive T lymphocytes in kidney allograft recipients can occur at the level of cytotoxic T-cells IL-2 producing T-cells or both. *Transplantation* **56**, 628.
286. Herzog W, Zanker B, Irschick E, Huber C, Franz HE, Wagner H, Kabelitz D (1987). Selective reduction of donor-specific cytotoxic T lymphocyte precursors in patients with a well-functioning kidney allograft. *Transplantation* **43**, 384.
287. Irschick E, Hladik F, Berger M, Niederwieser D, Herzog W, Zanker B, Kabelitz D, Margreiter K, Aubock J, Miller K (1990). Clonal reduction of CTL-P and acquired allograft tolerance in various human transplantation models. *Transpl Proc* **22**, 1869.
288. Mathew JM, Marsh JW, Susskind B, Mohanakumar T (1993). Analysis of T-cell responses in liver allograft recipients. Evidence for deletion of donor-specific cytotoxic T-cells in the peripheral circulation. *J Clin Invest* **91**, 900.
289. Beik AI, Higgins RM, Lam FT, Morris AG (1997). Clinical significance of selective decline of donor reactive IL-2 producing T lymphocytes after renal transplantation. *Transpl Immunol* **5**, 89.
290. Mason PD, Robinson CM, Lechler RI (1996). Detection of donor-specific hyporesponsiveness following late failure of human renal allografts. *Kidney Int* **50**, 1019.
291. Hornick PI, Mason PD, Yacoub MH, Rose ML, Batchelor R, Lechler RI (1998). Assessment of the contribution that direct allorecognition makes to the progression of chronic cardiac transplant rejection in humans. *Circulation* **97**(13), 1257.

292. van Besouw NM, van Der Mast BJ, De Kuiper P, Smak Gregoor PJH, Vaessen LMB, Ijzermans JNM, van Gelder T, Weimar W (2000). Donor-specific T-cell reactivity identifies kidney transplant patients in whom immunosuppressive therapy can be safely reduced. *Transplantation* **70**, 136.
293. Clerici M, Shearer GM (1990). Differential sensitivity of human T helper cell pathways by *in vitro* exposure to cyclosporine A. *J Immunol* **144**, 2480.
294. Schulick RD, Muluk SC, Clerici M, Bermas BL, Via CS, Weir MR, Shearer GM (1994). Value of *in vitro* CD4+ T helper cell function test for predicting long-term loss of human renal allografts. *Transplantation* **57**, 480.
295. Muluk SC, Clerici M, Via CS, Weir MR, Kimmel PL, Shearer GM (1991). Correlation of *in vitro* CD4+ T helper cell function with clinical graft status in immunosuppressed kidney transplant recipients. *Transplantation* **52**(2), 284.
296. Ciubotariu R, Liu Z, Colovai AI, Ho E, Itescu S, Ravalli S, Hardy MA, Cortesini R, Rose EA, Suciu-Foca N (1998). Persistent allopeptide reactivity and epitope spreading in chronic rejection of organ allografts. *J Clin Invest* **101**(2), 398.
297. Tugulea S, Ciubotariu R, Colovai AI, Liu Z, Itescu S, Schulman LL, Fisher PE, Hardy MA, Rose EA, Michler RE, Cortesini R, Suciu-Foca N (1997). New strategies for early diagnosis of heart allograft rejection. *Transplantation* **64**, 842.
298. Liu Z, Tugulea T, Cortesini R, Suciu-Foca N (1998). Specific suppression of T helper alloreactivity by allo-MHC class I-restricted CD8 + CD28– T-cells. *Int Immunol* **10**(6), 775.
299. Molajoni ER, Cinti P, Orlandini A, Molajoni J, Tugulea S, Ho E, Liu Z, Suciu-Foca N, Cortesini R (1997). Mechanism of liver allograft rejection: the indirect recognition pathway. *Hum Immunol* **53**(1), 57.
300. Vella JP, Spadafora-Ferreira M, Murphy B, Alexander SI, Harmon W, Carpenter CB, Sayegh MH (1997). Indirect allorecognition of major histocompatibility complex allopeptides

in human renal transplant recipients with chronic graft dysfunction. *Transplantation* **64**(6), 795.

301. Knoflach A, Chandraker A, Sayegh MH (1997). Chronic rejection from bedside to bench: role of T-cells. *Ann Transpl* **2**(2), 53.
302. Woan MC, Moliterno RM, Zeevi A, Mallick C, Zerbe TR, Fricker J, Griffith BP, Kormos RL, Armitage J, Duquesnoy RJ (1993). Characterization of the biopsy growth assay in monitoring cardiac transplant patients for rejection. *Transpl Proc* **25**(1), 946.
303. Tugulea S, Ciubotariu F, Colovai AI, Liu Z, Itescu S, Schulman LL, Fisher PE, Hardy MA, Rose EA, Michler RE, Cortesini R, Suciu-Foca N (1997). New strategies for early diagnosis of heart allograft rejection. *Transplantation* **64**(6), 842.
304. Duquesnoy RJ, Trager JD, Zeevi A (1991). Propagation and characterization of lymphocytes from transplant biopsies. *Crit Rev Immunol* **10**(6), 455.

II

THE HISTORY OF ORGAN DONATION AND SHARING

Gauke Kootstra

Introduction

Organ donation and transplantation between humans started some 60 years ago. One is allowed to speak of history, simply because in the early days every transplant was an experiment, while in the day of today all of it has become routine; from exciting frontier work to routine and actually so much of routine that today a transplant has to compete with any procedure for facilities in the operating room and intensive care! The early days were unique, the pioneers paved the path for today's successes, and it is history that gives credit to this pioneer's work.

I remember the first human kidney transplant in 1966 at Groningen University Hospital. I was a second year resident and I had the honorable job to sit at the telephone to coordinate the donor and the recipient procedure. When the severely brain injured young man was brought in the evening to the theatre, the ventilator was switched off, cardiac arrest was awaited for and then the laparotomy was performed. Only the left kidney was removed and transplanted. Nobody thought of the right kidney: donation yes, sharing no. The recipient was selected based on fitness, strength, not on basis of best match. Tempora mutantur.

The Early Kidney Programs

Transplantation of the kidney paved the way for the transplantation of other organs. The first human-to-human kidney transplant was from a postmortem donor to a sublimate-intoxicated female. Y.Y. Voronoy performed the operation in 1933 in the Ukraine. In the United States, David Hume started in 1947 followed by three teams in Paris in 1950. Most of the transplants were cadaver and for a short time successful but not more than a few weeks. The French were the first to share: the kidneys of a decapitated (guillotine) criminal were procured right after the execution and divided over two teams in two separate hospitals.

In the early days, it were the transplant teams that procured the kidney themselves, mostly from a dead person in the center itself. Then, later on the surrounding hospitals joined and a regional procurement organization became reality. The facilities to establish such a postmortem program required intensive care facilities and well-equipped emergency rooms. This was initially only available in the U.S., Canada, Europe and Australia. In the other parts of the world where intensive care facilities were less available and maybe legal and social circumstances less ideal, the accent was on donation by family members.

The Exchange Programs

In 1967, the first international exchange organization was founded in Western Europe: Eurotransplant.[1] This exchange program involved West Germany, Luxembourg, Belgium and the Netherlands. There was no governmental involvement. The basis of the collaboration was the finding with HLA-typing that a good match gave a better result. An exchange of kidneys was realized with tissue typing of the donor and performance of the crossmatch in the procurement center, and then allocation through the Eurotransplant Office to the best-matched recipient of the waiting list. So the exchange was set up to enlarge the pool size for a greater chance for a good match.

France followed in 1969, Scandiatransplant, the United Kingdom Transplant and Ireland in 1972. An Interstate kidney-sharing program was established in Australia in the early 1980s.[2]

In the U.S., regional exchange program's f. i. the New England Organ Bank were organized like Eurotransplant for several years. At the beginning of the nineties, the United Network of Organ Sharing was established in the U.S. This meant government involvement. The same happened in France: in 1996 the government took the organization of Francetransplant over.

One of the side-effects of these exchange programs was that standardization and quality control were introduced because one had to accept an organ procured by another team.

In order to be able to transport kidneys over longer distances in these exchange programs, the technique of machine preservation (MP) was developed. Belzer made machine preservation popular.[3] In the U.S., the development of MP had an additional reason as well in the efforts to reduce delayed graft function. One has to realize that in the early days the concept of brain death did not exist, and therefore most kidneys were procured after ventilator switch off and subsequent cardiac arrest. Parameters for immediate or delayed graft function were developed. Ligandin,[4] an enzyme in the perfusate and currently called glutathion-s-transferase, was investigated to predict immediate or delayed graft function. In the San Francisco-based kidney transplant program of Belzer, a delayed graft function of only 5% was claimed with MP. At the European side of the Atlantic, MP was developed by Claes in Sweden, specifically in the lab of Gelin in Gothenburg.[5] Again, the intention was to improve early graft function. Of importance is to realize that the ventilator switch off procedure had to be followed in Sweden until 1988 because of non-acceptance of the concept brain death.

The Donors

The first kidney transplants were from postmortem donors, from hydrocephalic children who had one kidney taken out for drainage of cerebral fluid via the ureter (Matson procedure) or from relatives, although rare. Christmas 1952 saw a mother-to-son kidney transplant in Paris wherein the son had lost a kidney at an accident. It was a solitary kidney! The transplant of the mother functioned well for three weeks, then anuria. The boy died a few days later.[6]

The first human kidney transplant with a long-lasting success was performed in 1954 at Boston, USA.[7] The crucial point in the Brigham case was the fact that donor and recipients were identical twins, proven by skin transplants. The recipient was prepared by hemodialysis. The technique was as developed by the Paris' teams: end-to-side artery and vein to the external iliac vessels and an ureteroneocystostomy. Basically, this technique has survived up till now. The kidney lasted for eight years.

The first case of sharing and preservation was located in Paris, in 1951, as mentioned. On January 12, a person sentenced to death, was decapitated by the guillotine. Both kidneys were procured on the spot and one was flushed with plasma, the other with "serum de Ringer". One kidney was transplanted by Dubost and Oeconomos in Hospital Necker, the other in either the same hospital or somewhere else by Servelle and Rougeulle. The ischemia times were for current standards very short: 1 hour and 50 minutes and 3 hours and 20 minutes, respectively. Both kidneys developed function, but the patients died at day 17 and 20. It is the first documented case of sharing and it is an early example of Non-Heart beating donation as well.

At the first informal meeting of transplanters and immunologists in 1963 in Washington, one had to conclude that the results of kidney transplantation were up to then rather disappointing. Some questioned whether one could continue to use living donors as source of course with exception of twin donors.

By 1963, 244 kidney transplants of which 28 of identical twins had been performed in the U.S., U.K. and France, while in Belgium and Canada programs had just started. In Denver, Tom Starzl started a voluntary living donor programs, mainly prisoners and related family members. The results were above standards by those days.

The following years saw a revival of animal donors, like in the beginning of the century. Reemtsma in New Orleans used chimpanzee kidneys *en bloc*, with a longest survival of nine months; Starzl used baboons with poor results and Hume in Richmond reported a similar case that failed. In 1964, Starzl transplanted a baboon liver, Hardy in Jackson a baboon heart, both failed.

Brain Death

The concept of "coma depasse" comes from France and it opened the possibility to procure organs for transplantation while the heart still beats. Alexandre from Leuven, Belgium was in 1963 the first to start the harvest with beating heart, followed by the team coordinated by Hamburger in Paris in 1964.[8] Still kidneys were not taken out with beating heart. All preparations were done and the kidneys were dissected free and then the ventilator was stopped. One waited for the organ extirpation until the heart had stopped. The kidneys were flushed with a solution to which lidocaine and heparine were added and then the kidneys were cold stored.

Heartbeating donation was officially allowed in France in 1968; brain death was an accepted concept in the U.S. since in the same year the committee *ad hoc* of Harvard Medical School published their report on the criteria for brain death.

The United Kingdom is late when in 1976 the Royal Code on brain death is published. On the continent, probably Sweden is the latest with accepting brain death in 1988.

The Non-Renal Organs

Successful liver transplantation became possible only when heartbeating donation became a reality and when cyclosporin stronger immunosuppression became available since 1980 onwards. When a liver was too large for a small recipient, reduced sized livers were used. By reducing the liver in such a way that the removed segment could be used for another patient, split liver transplantation was thus born. This was a challenge for exchange programs because one had to rely on perfect information relating to the anatomy in making the decision to accept the offer for a given patient or refuse it.

Shortage implicated new inventions as the living related donor in which a segment of the liver is taken away and transplanted.

Sharing of the liver is performed but on a rather small scale. The problem is that preservation time for the liver is very restricted and the need to share for a better match is not necessary: there is hardly an influence of matching in liver transplantation.

The heart takes a special position in organ donation and sharing; only after acceptance of the concept of brain death heart transplantation became possible. Short preservation time restricts sharing and again there is not much to win with matching. The fact that the heart is a single organ might have contributed as well to the fact that mostly the transplant center procures and transplants without involvement of another group.

The pancreas has gone through a very interesting and complex history of surgical techniques. One of the first human transplants was with duodenum and one end was anastomosed to the skin as a duodenostoma, to enable drainage of exocrine fluid. In the current technique, the duodenum is used for connection with bladder or intestine. In between is a whole development with segmental transplantation, with or without duct occlusion. Max Dubernard in Lyon[9] elaborated the technique of duct occlusion, and although it has in some aspect definite advantages, it is now fully replaced by full size pancreas transplantation. Several models of reducing the initial high incidence of graft thrombosis have been developed and abandoned in the following years. A major contribution was the development of the UW preservation solution by Southard and Belzer[10] specifically made for the pancreas. It did not perse extended the preservation time; it improved the quality of the graft and reduced the incidence of thrombosis.

In the early eighties, liver and pancreas transplant surgeons[11] disputed the length of the portal vein on either organ. This dispute has quietened since short portal veins can either be used directly or extended with donor veins.

Cluster transplantation, i.e. liver with either intestine or pancreas or both, has a rather short history. It is developed in Pittsburgh by Starzl and successors and in Europe in Innsbruck by Margreiter.

The Transplant Coordinator

The transplant coordinator is an American invention. It was recognized early in the U.S. that paramedical personnel could be very supportive in the procurement of kidneys and successively in machine preservation and transplantation. In Europe, these procedures were kept in the hands of the doctors until in the mid-seventies. At the introduction in Europe initially doctors were recruited, later on paramedical personnel mainly from nursing staff was employed. Especially in the case of a multi organ donor (MOD), no team could currently do without a transplant coordinator.

The main task of a transplant coordinator is communication, between the procurement teams in the donor hospital and the recipient teams at "home". This has become a crucial item in the process of successful organ retrieval. Transplant coordinators are involved in the preservation and transplantation of the organs; they are the key figures in the sharing process.

While in Europe and Australia, the transplant coordinators are based at the transplant center, in the U.S. the whole process of donation has become more or less independent of a transplant center. There are regional procurement programs with coordinators, having to report to the United Network of Organ Sharing. Every organ has a fixed price and the Organ Procurement Organization (OPO) charges that amount of money to the transplant center. The more organs are procured by an OPO, the larger is the budget of that OPO.

Conclusion

The history of organ donation is a fascinating one, especially for the kidney, which has functioned as the pilot for the other organs. Because the kidney is a paired organ there is the option for live donation and this has been the case for a whole scala of donors from genetically related family members to prisoners, partners, friends and spouses. The technique for the transplantation of a kidney, the heart, lung and liver has been standardized quite early. This was not the case for the pancreas.

Sharing is basically only done for the kidney, because preservation allows it and matching is of advantage especially in the sensitized recipient. Nevertheless, the continuing shortage of donor organs might reduce the willingness to share any organ. Only when organ donation has reached a situation wherein demand and supply are in balance, sharing will be unrestricted.

References

1. Rood JJ, van Freudenberg J, Leeuwen A, van Schippers HM, Zweerus R, Terpstra JL (1971). Eurotransplant, an organisation to arrive at the best possible results in kidney transplantation. *Transpl Proc* **3**, 933–941.
2. McBride M, Chapman JR (1995). An overview of transplantation in Australia. *Anaesth Intens Care* **23**, 60–64.
3. Belzer FO, Ashby BS, Gulyassy PF, Powell M (1968). Successful seventeen hour preservation and transplantation of human cadaver kidney. *N Engl J Med* **278**, 608–610.
4. Cho SI, Zalneraitis B, Ohmi N, Arias JM (1981). Prediction of cadaver kidney function by ligandin analysis. *J Surg Res* **30**, 361–364.
5. Claes G, Blohme I, Stenberg K (1976). Kidney preservation by continuous hypothermic albumin perfusion without membrane oxygenation. *Proc Eur Dial Transpl Assoc* **12**, 483–491.

6. Michon L, Hamburger J, Oeconomos N, Delinotte P, Richet G, Vaysse J, Antoine B (1953). Une tentative de transplantation renale chez l'homme: aspects medicaux et biologiques. *Presse Med* **61**, 1419–1423.
7. Murray JE, Merrill JP, Harrison JH (1955). Renal homotransplantations in identical twins. *Surg Forum* **6**, 432.
8. Kuss R, Bourget P (1992). Une histoire illustree de la greffe dórganes. Laboratoire Sandoz, Rueil-Malmaison, France, p. 65.
9. Dubernard JM, Traeger J, Neyra P, Touraine JL, Tranchant D, Blanc-Brunat N (1978). A new method of preparation of segmental pancreatic graft for transplantation. Trials in dog and man. *Surgery* **84**, 633–639.
10. Belzer FO, Kalayoglu M, D'Alessandro AM, Pirsch JD, Sollinger HW, Hoffmann R, Boudjema K, Southard JH (1990). Organ preservation: experience with University of Wisconsin solution and plans for the future. *Clin Transpl* **4**, 73–77.
11. Bunzendahl H, Ringe B, Meyer HJ, Gubernatis G, Pichlmayer R (1988). Combination harvesting procedure for liver and whole pancreas. *Transpl Int* **1**, 99–102.

III

THE HISTORY OF THE CONCEPT OF BRAIN DEATH AND ORGAN PRESERVATION

Vassilios E Papalois
Arthur J Matas

Introduction

The acceptance of the concept of brain death and the development of organ preservation techniques were two of the major cornerstones on which organ transplantation was built. The achievement of a uniform definition of brain death, the broadened application of brain death laws, and the establishment of methods for prompt brain death diagnosis (before the onset of hemodynamic or respiratory collapse) is a very interesting chapter in the history of the great medical revolution of organ transplantation. Equally exciting is the chapter on the development of organ preservation: as a result, organs that were procured thousands of miles away from the recipient can be used for transplantation allowing time to identify the right recipient based on immunologic tissue matching, size of organ, and medical urgency for the transplant.

Brain Death

The concept

For many centuries, death was determined based on cessation of cardiorespiratory function but advances in medical technology have allowed for cardiac and respiratory functions to be maintained artificially, even in the presence of irreversible loss of brain function. As early as 1959, published reports described clinical conditions similar to what we currently view as brain death. Mollaret and Goulon[1] coined the term "coma depasse" (state beyond coma) to describe a neurologic state associated with deficits beyond what had been previously described as coma. Such patients were unresponsive, had no brainstem reflexes, and showed progressive hemodynamic collapse. Mollaret and Goulon reflected on the prognosis of such a condition, but did not equate it with death and did not recommend withdrawal of ventilatory support. During the same year, Jouvet[2] described "death of the central nervous system". Such patients also lacked brainstem reflexes, were apneic in spite of hypercapnea, and had an electrically silent electroencephalogram (EEG) with scalp and deep thalamic electrodes. Jouvet concluded that, under such conditions, recovery was not possible and that respiratory support was to be discontinued.

These two early reports were followed by several studies that attempted to define clinical criteria for brain death and methods for diagnosis. As investigators moved toward more objective evidence and supportive studies, the concept of brain death became somewhat blurred. The trend was toward a more clinical approach in determining brain death and in efforts to accept brain death as patient death.

In 1968, the Ad Hoc Committee of the Harvard Medical School,[3] in charge of examining the definition of brain death, published its report, "A Definition of Irreversible Coma", defining irreversible coma as a new criterion for death. The definition included the criteria of absence of spontaneous respirations and movements, unresponsiveness, dilated pupils, loss of cephalic and deep tendon reflexes, and absence of postural responses. A flat or isoelectric EEG was considered of great confirmatory value. Testing was to be repeated at least 24 hours later with no change. The final diagnosis of brain death depended on ruling out hypothermia (temperature below 32.2°C) or central nervous system depressants (e.g. barbiturates). The committee did not describe what would happen if mechanical ventilation was discontinued.

In 1971, Mouhandas and Chou[4] formulated criteria for brain death based on clinical grounds. They did not feel an EEG was mandatory and thought that repeat examinations could be performed at 12 hours. However, they proposed that irreparable brain lesions should be identified "with presently available means". Their criteria emphasized the clinical diagnosis of brain death and set the foundation for more modern criteria by defining a clear irreparable cause of brain death, emphasizing brainstem death with loss of brainstem reflexes, and questioning the value of an EEG.

In 1976, the Conference of Royal Colleges and Faculties of the United Kingdom endorsed a document outlining diagnostic criteria for brain death.[5] Their criteria were also based purely on clinical grounds, with strong emphasis on brainstem death. They wrote a detailed outline of pre-conditions and cautionary notes while considering brain death. This outline became known as the "U.K. code" and was discussed in much greater detail by Pallis[6] who emphasized the conceptual evolution from classical death to brain death (as an initial step) and subsequently from total brain death to brainstem death.

In the U.S. in 1981, the President's Commission for the Study of Ethical Problems in Medicine and Biomedical and Behavioral Research published its report, "Guidelines for the Determination of Death"[7] (Table 1). The Commission stated that "an individual with irreversible cessation of the entire brain, including the brainstem, is dead". The report, similar in many ways to the U.K. code, outlined complicating conditions that may affect the accuracy and reliability of brain death diagnosis. The emphasis was on cessation of cerebral and brainstem functions, establishment of irreversibility, and evaluation for complicating conditions. When criteria were not met as outlined, confirmatory studies (e.g. EEG and cerebral blood flow) were suggested. The Commission's criteria subsequently became the text for brain death legislation in all 50 states.

Table 1 1981 President's Commission criteria for brain death

1. *Absence of*:
   (a) Cerebral functions: deep coma
   (b) Brainstem function: pupillary, corneal, oculovestibular, gag, cough, and respiratory (apnea) reflexes
2. *Irreversibility*:
   (a) Cause of coma is established
   (b) Possibility of recovery of any brain function is excluded
   (c) Findings confirmed on a second exam after an appropriate period of observation
3. *Complicating conditions*:
   (a) Drug and metabolic intoxication
   (b) Hypothermia: below 32.2°C core temperature
   (c) Children: younger than five years require longer periods of observation and supportive studies
   (d) Shock
4. *Confirmatory studies*:
   Not required. EEG or blood-flow study may be used as indicated by medical circumstances

Legislation

Acceptance of brain death as clinical death depended on integrating both medical and legal criteria. Again, "death" had long been accepted as meaning "cessation of respiratory and cardiac functions". Such functions can be replaced by machines and, to some extent, by pharmacologic agents, but functions of the brain cannot. Therefore, the evolution toward the concept of brain death as an acceptable diagnosis of death was a natural one. It was not until 1971 in Finland (even though reports as early as 1959 had described it clinically) that brain death was first legally accepted as death.

In the U.S., legislative activity in support of organ donation and brain death diagnosis started at the state level. The Uniform Anatomical Gift Act (UAGA) of 1968 established the legality of anatomic gifts after death, and of organ donor cards; the act also prohibited the sale of organs. By 1972, the UAGA had been adopted in all 50 states. However, without brain death laws, removal of organs was possible only after a prolonged period of circulatory ischemia. In 1970, Kansas became the first state to adopt a brain death statute. By 1977, 12 states had enacted laws to accept brain death as sufficient for declaration of a patient's death. In 1978, the National Conference of Commissioners on Uniform State Laws drafted and established the Uniform Brain Death Act which expanded the traditional definition of death (cardiopulmonary) to include brain death. That same year, the Uniform Determination of Death Act was introduced and finally passed by congress in 1980. Today, all 50 states have adopted some form of brain death legislation. Similarly, brain death legislation is established in western European countries. In some countries like Japan and India although brain death laws were recently established, the rate of social acceptance of the concept of brain death is very low.

Organ Preservation

The beginnings of organ preservation can be traced to the physiologic studies of isolated organ perfusion during the early part of this century. Best known is the work of Carrel and Lindbergh, who designed a rather complex perfusion circuit and were able to maintain a variety of organs and endocrine tissues for several days in normothermic conditions.[8]

It was apparent from the experiments of Medawar in the 1940s that tissues and organ allografts would be vigorously attacked by the recipient's immune system and destroyed in a short time. Early attempts at kidney transplantation failed in animals due to rejection. At that time, there was little concern about preserving organs because of the immunologic hurdles that needed to be overcome. In the 1950s and 1960s, studies aimed at understanding the nature of allograft rejection and the discovery of agents (e.g. beta-mercaptopurine) to block rejection helped to make organ preservation a clinical reality. Clinicians and basic scientists recognized that anti-rejection drugs made it possible to use immunologically mismatched tissues from cadaver organ donors. Using cadaveric organs as the major source of organs for transplantation would be the basic way to make transplantation therapy widespread.

In the 1960s, after Murray demonstrated successful kidney transplantation in a set of twins, several laboratories became active in pursuing organ preservation. Many laboratories made contributions, especially those of Pegg and Calne in Cambridge, England, Belzer in UCSF and later in Wisconsin, Collins in UCLA, Starzl in Denver and Pittsburgh, Marshall in Melbourne, and Halasz in San Diego. At that time, the primary interest was kidney preservation.

The concept of metabolic inhibition by hypothermia as a key to organ preservation was quickly accepted. Pegg and colleagues[9] attempted to preserve kidneys by simple cooling in cold saline or blood; transplants were successful with preservation times of about 12 hours. Belzer[10] reasoned that continuous hypothermic perfusion with a biologically based fluid (plasma) would be superior to simple cold storage because the kidney would be continuously supplied with metabolism-stimulating substrates and the end-products of metabolism would be removed. Hypothermic metabolism would be stimulated and supported and tissues that were injured during the agonal period and during procurement would be repaired. Dog kidneys were successfully transplanted after preservation for three days using low perfusion pressure (50 to 60 mm Hg), low flow (about 0.6 to 1.0 mL/min/g), and oxygenated cryoprecipitated plasma (CPP). CPP was prepared by freezing and thawing autologous dog plasma to remove cold-sensitive lipoproteins (by millipore filtration). These lipoproteins, which were found to aggregate in the kidney glomerular vessels,[11] cause an increase in renal vascular resistance, endothelial cell injury, and possibly occlusion and ischemia in portions of the kidney.

Belzer's method of machine perfusion using CPP as the perfusate became standard for human kidney preservation in the early 1970s,[12] dramatically increasing the number of kidney transplants performed. At that time, there was no clear definition or acceptance of brain death as a criterion for organ removal. Therefore, most organs were obtained from donors who were allowed to undergo death as judged by cessation of heart function; consequently, their kidneys were not well-perfused for varying lengths of time. Machine perfusion allowed successful preservation of these "less than ideal" kidneys damaged by hypotension or ischemia. It has since been shown[13, 14] that machine perfusion for "less than ideal" kidneys is superior to other methods such as simple cold storage.

In the late 1970s, Collins *et al.*[15] showed that dog kidneys could be safely preserved for 30 hours by simple cold storage. This method relied on refrigerated storage after flushout with a solution composed primarily of glucose, potassium (K^+) and phosphate. The Collins solution contained a high concentration of K^+, used to suppress the loss of intracellular K^+ that occurred in metabolically depressed cells in exchange for extracellular cations, principally sodium (Na^+).[16] This so-called "extracellular-type solution" became the standard for human kidney preservation over the next eight to ten years, primarily because of its simplicity. By that time, brain death had become a criterion for organ donation; thus, organs could be retrieved from heart-beating cadavers. No machine was required, the solution was simple to prepare and shelf-stable, and the time of safe preservation (about 24 hours) was sufficient for use of most cadaver kidneys.

Since those two landmark developments in organ preservation first by Belzer and then by Collins, many modified preservation solutions have been tested, but only a few have been found to be significantly better. The various preservation solutions that have been used clinically, their basic metabolic components and the reference of their original description are listed in Table 2.

Two solutions developed in the late 1980s at the University of Wisconsin (UW) have had much success in advancing organ preservation. The composition of these solutions is given in Tables 3 and 4. The UW cold storage solution (ViaSpan, DuPont) extended preservation times for the liver[17] and pancreas[18] from about four hours (with Collins solution) to between 30 and 72 hours. It also appears to facilitate better, longer-term kidney preservation[19] than the Collins type solutions, and is safe for six to 12 hours of heart and lung preservation.[20, 21] In addition, the UW gluconate solution (Belzer machine perfusion solution) has found widespread use in the few centres that continue to

Table 2 Some preservation solutions used clinically

Solution Name	Active Components	Reference for Original Description
Collins	PO_4 buffered glucose solution	*Lancet* (1969)
Silica-gel plasma	Perfusion solution Plasma defatted with silica gel	*Surg Gyn Obstet* (1974)
Hypertonic citrate	Citrate, mannitol	*Transplantation* (1976)
Euro-Collins	Same as Collins (no Mg^{++})	*Euro Surg Res* (1980)
PBS	PO_4 buffered sucrose	*Transplantation* (1989)

Table 3 UW cold storage solution

Substance	Amount
K-lactobionate	100 mM
KH_2PO	25 mM
$MgSO_4$	5 mM
Raffinose	30 mM
Adenosine	5 mM
Glutathione	3 mM
Insulin	100 U/L
Bactrim	0.5 ml/L
Dexamethasone	8 mg/L
Allopurinol	1 mM
Hydroxyethyl starch	50 g/L

This solution is brought to pH 7.4 at room temperature with NaOH. The final concentrations are as follows: $Na^+ = 30 \pm 5$ mM; $K^+ = 120 \pm 5$ mM; and mOsm/L = 320 ± 5.

Table 4 Composition of the UW gluconate solution

Substance	Concentration
K-gluconate	100 mmol/L
Raffinose	35 mmol/L
Ribose	1 mmol/L
NaH_2PO_4	25 mmol/L
Mg-gluconate	5 mmol/L
Adenine	1 mmol/L
Adenosine	5 mmol/L
Hydroxyethyl starch	50 g/L
Glutathione	3 mmol/L
Glucose	5 mmol/L
HEPES	10 mmol/L
Insulin	100 U/L
Dexamethasone	16 mg/L
Cotrimoxazole	2 m/L
$CaCl_2$	1.5 mmol/L

The pH is adjusted by the addition of NaOH/KOH (1 : 3) to pH = 7.7 (at room temperature).

use machine perfusion of kidneys.[22] This solution is similar to the UW cold storage solution, except that lactobionate has been replaced with gluconate and the Na^+/K^+ ration is higher. The UW gluconate solution has been found to successfully preserve dog kidneys for five[23] and seven days.[24]

Epilogue

The availability of organs was, and still is, one of the major problems in organ transplantation. The acceptance of brain death and the advancements in organ preservation were major steps forward but still not the final solution to the problem. It took a lot of effort and courage by scientists dedicated to transplantation and medical ethics and by inspired legislators to establish the concept of brain death and the laws to support it. As a result, highly motivated and brave families around the world donated the organs of their loved ones, changing the lives of others forever. Likewise, the history of the development of organ preservation is one of the most characteristic examples of human effort to solve complicated medical problems with well-designed, tedious and honest experimentation. It will take similar, perhaps even higher scientific and moral standards and values to deal with today's more complicated issues, such as xenotransplantation and the cloning of humans or human organs, which are proposed as possible answers to the problem of organ availability.

References

1. Mollaret P, Goulon F (1959). Le Coma Depasse (Memoire Preliminaire). *Rev Neurol* **101**, 3–15.
2. Jouvet M (1959). Diagnostic electro-sous-corticographique de la mort du systeme nerveux central au cours de certain comas. *Electroenceph Clin Neurophysiol* **11**, 805–808.
3. Ad Hoc Committee of the Harvard Medical School (1968). A definition of irreversile coma. *JAMA* **205**, 85–88.
4. Mouhandas A, Chou SN (1971). Brain death: a clinical and pathological study. *J Neurosurg* **35**, 211–218.
5. Conference of the Royal College and Faculties of the United Kingdom (1976). Diagnosis of brain death. *Lancet*, **2**, 1069–1070.
6. Pallis C (1982). ABC of brainstem death. *Br Med J* **285**, 1409–1490.
7. President's Commission for the Study of Ethical Problems in Medicine and Biomedical and Behavioral Research (1981). Guidelines for the determination of death. *JAMA* **246**, 2184–2186.
8. Carrel A, Lindbergh CA (1935). The culture of whole organs. *Science* **81**, 621.
9. Pegg DE, Calne RY, Pryse-Davies J, Leigh-Brown F (1964). Canine renal preservation using surface and perfusion cooling techniques. *Ann NY Acad Sci* **120**, 506–523.

10. Belzer FO, Ashby BS, Dmphy JE (1967). Twenty-four and 72-hour preservation of canine kidneys. *Lancet* **2**, 536–539.
11. Belzer FO, Ashby BS, Huang JS, Dumpy JE (1968). Etiology of rising perfusion pressure in isolated organ perfusion. *Ann Surg* **168**, 382.
12. Belzer FO, Ashby BS, Gulyassy PF, Powerll M (1968). Successful 17-hour preservation and transplantation of human-cadaver kidney. *N Engl J Med* **278**, 608–610.
13. Booster MH, van der Vusse GJ, Wijnen RMH, Stubenitsky BM, Kootstra G, Yin M (1994). University of Wisconsin solution is superior to histidine tryptophan ketoglutarate for preservation of ischemically damaged kidneys. *Transplantation* **58**, 979–983.
14. Light JA, Gage F, Kowalski AE, Sasaki TM, Callender CO (1996). Immediate function and cost comparison between static and pulsatile preservation in kidney recipients. *Clin Transpl* **10**, 233–236.
15. Collins GM, Bravo-Shugarman MB, Terasaki PI (1969). Kidney preservation for transplantation: initial perfusion and 30-hour ice storage. *Lancet* **21**, 1219–1225.
16. Macknight ADC, Pilgrim JP, Robinson BA (1974). The regulation of cellular volume in liver slices. *J Physiol* **238**, 279–294.
17. Jamieson NV, Sundberg R, Lindell S, Southard JH, Belzer FO (1988). Preservation of the canine liver for 24–48 hours using simple cold storage with UW solution. *Transplantation* **46**, 517–525.
18. Wahlberg JA, Love R, Lanegaard L, Southard JH, Belzer FO (1987). Seventy-two hour preservation of the canine pancreas. *Transplantation* **43**, 5–8.
19. Ploeg RJ, Goossens D, NcAnulty JF, Southard JH, Belzer FO (1988). Successful 72-hour cold storage of dog kidneys with UW solution. *Transplantation* **46**, 191–196.
20. Jeevanandam V, Barr ML, Auteri JS *et al.* (1991). University of Wisconsin solution for human donor heart preservation: initial clinical experience. *Ann Thorac Surg* **52**, 1213–1216.

21. Hardesty RL, Aeba R, Armitage JM, Kormes RL, Griffith BP (1993). A clinical trial of University of Wisconsin solution for pulmonary preservation. *J Thorac Cardiovasc Surg* **105**, 660-666.
22. Barber WH, Deierhoi MH, Phillips MG, Diethelm AG (1988). Preservation by pulsatile perfusion improves early renal allograft function. *Transpl Proc* **20**, 865–868.
23. McAnulty JF, Ploeg RJ, Southard JH, Belzer FO (1989). Successful five-day perfusion preservation of the canine kidney. *Transplantation* **47**, 37–41.
24. Schilling M, Saunder A, Southard JH, Belzer FO (1993). Long-term renal preservation and prevention of acute tubular necrosis by inhibition of arachidonate metabolism. *Transpl Proc* **25**, 2534-2535.

IV

THE HISTORY OF KIDNEY TRANSPLANTATION

Vassilios E Papalois
Nadey S Hakim
John S Najarian

Introduction

Kidney transplantation is now well-established as the treatment of choice for end-stage renal failure. The long-term experimental and clinical efforts that led to the amazing results we enjoy today represent

Table 1 Historic first kidney transplants

Year	Name of Surgeon and Place	Type of Surgery
1906	Jaboulay, France	First documented kidney transplant from an animal to a human.
1908	Carrel, USA	First kidney autotransplant in the canine model with long-term graft survival.
1936	Voronoy, Russia	First (unsuccessful) kidney transplant between humans.
1954	Murray, USA	First successful kidney transplant in humans between identical twins.
1958	Murray, USA	First kidney transplant in humans using immunosuppression.
1959	Murray, USA	First successful kidney transplant between dizygotic twins.
1960	Kuss, France	First successful kidney transplant between non-twin siblings
1961	Kuss, France	First successful kidney transplant between non-siblings.
1962	Murray, USA	First successful cadaver kidney transplant using immunosuppression.

one of the most fascinating stories in the history of medicine in the 20th century. We herein present the highlights of this great scientific journey focusing on the first heroic strides of the pioneers in the field (Table 1).

1902–1912: Experimental Efforts of Ullmann and Carrel

At the Vienna Medical Society meeting in January 1902, Hungarian surgeon Emerich Ullmann reported the first case of kidney transplantation, in which a dog's kidney was implanted into another dog's neck.[1] The ureter was brought out to the skin, and some urine was excreted over a period of five days. In the same year, in front of an assembly of physicians, Ullmann presented a goat that had had a dog's kidney transplanted into its neck and all witnessed the excretion of urine. However, excretion stopped on the next day because of thrombosis of the vessels. Ullmann continued to conduct transplant experiments between various animal species. He later claimed that he had attempted to transplant a pig's kidney into the elbow of a young uraemic human female, but was unable to overcome the technical difficulties. Ullmann eventually abandoned his transplant research and directed his work to other surgical areas.

Also in 1902, French surgeon Alexis Carrel, apparently unaware of Ullmann's work, published a paper in the journal *Lyon Medecine* entitled "The operating technique for vascular anastomosis and organ transplantation".[2] He reported the case of a kidney removed from a dog and transplanted into its neck. After the vascular clamps were removed, circulation was immediately restored in the graft; several hours later, urine was excreted at the end of the ureter. However, this transplant soon failed because of infectious complications.

In 1904, Carrel left Lyon for the United States. From 1905 to 1906, he worked in Chicago in close collaboration with Charles Guthrie. The two investigators made significant contributions in the field of vascular grafts and organ transplantation. They developed the clinically applicable technique of *en bloc* transplantation of two kidneys with an excretory apparatus sutured to the skin. They also perfused the kidneys with Locke's solution before reimplantation; the same principle is still used today. For the first time, in 1906, they compared the function of the transplanted kidney with that of the contralateral kidney. They reported that, three days after transplanting a dog's kidney into its own neck, they opened the wound in the abdomen to compare the autotransplanted kidney with the remaining native kidney. The transplanted kidney had normal size and consistency, with normal arterial pulsations and a slightly increased circulation. The urine excretion by the transplanted kidney was five times as fast and the composition of the urine was only slightly different, compared with the remaining native kidney. The transplanted kidney had lower urea and higher chloride concentrations. This was clearly an early remarkable description of a form of acute tubular necrosis, an entity still unknown at that time. Carrel and Guthrie published a series of papers and had a great impact on the international scientific community.[3] Their achievements also met with considerable enthusiasm and admiration among the public and the media. However, at the same time, some of the media described the anger and hostility of anti-vivisection groups, which considered Carrel and Guthrie to be criminals who should be condemned to the electric chair.

After the initial years of their collaboration, Guthrie left for St. Louis, Missouri, and Carrel was asked to direct the Department of Experimental Surgery at the Rockefeller Institute in New York, where he continued his transplant research. In February 1908, Carrel performed a successful autotransplant of a kidney in a dog with, for the first time, long-term graft survival.[4] Two weeks after the transplant, the remaining native kidney was removed. The transplanted kidney had normal function. In March 1909, the dog gave birth to 11 puppies and in December 1909, to another three puppies. In July 1910, the dog, in excellent health until then, died suddenly from small bowel obstruction. At autopsy, the transplanted kidney had normal vessels, a normal excretory tract, and a normal histopathologic structure.

Carrel's results with kidney allotransplants in dogs were not as good. About one week post-transplant, the kidney allografts were

Fig. 1 Alexis Carrel pioneered the field of transplantation. He perfected the vascular anastomosis techniques, performed the first experimental kidney autotransplants with long-term graft survival and conceived the idea of rejection by attributing the failure of experimental kidney allotransplants to what he termed "biological incompatibility". He was awarded the Nobel Prize in 1912.

oedematous; two to three weeks later, haematuria and proteinuria were clearly evident. Eventually, the allografts became diffusely sclerotic, with complete loss of renal function. In 1914, Carrel published his landmark conclusion: the success of the autografts proved that, from the technical point of view, transplantation was possible; the loss of the allografts was due to what he termed "biological incompatibility".[5] Obviously, Carrel had conceived the idea of rejection. For his great efforts and contributions to the field of organ transplantation, Carrel (Fig. 1) was awarded the Nobel Prize in 1912.

1906–1913: Initial Kidney Transplants from Animals to Humans

French surgeon Mathieu Jaboulay was the first to clearly document kidney transplants from animals to humans.[6] In 1906, he transplanted the left kidney of a pig into the left elbow of a woman suffering from nephrotic syndrome. However, this graft failed because of early vascular thrombosis. In a second case performed the same year, he transplanted the kidney of a goat into the cubital fossa of a 50-year-old cachexic woman who had undergone nephrectomy because of renal suppuration. This graft also failed because of vascular thrombosis.

Some years later, others attempted to transplant kidneys from monkeys into humans. In 1909, Unger transplanted *en bloc* two kidneys from a Borneo macaque monkey into a young girl with chronic nephritis in a pre-terminal condition.[7] He observed a small production of urine, but the girl died 32 hours post-transplant. In 1913, Schonstadt transplanted one kidney from a Japanese monkey into a young girl with mercury poison nephritis.[8] As with the transplant performed by Unger, the kidney produced a small amount of urine, but the girl died 60 hours post-transplant.

Those first unsuccessful attempts at kidney transplantation in humans were followed by the years of World War I, as well as by a certain loss of interest on the part of the early pioneers. It took more than two decades before new efforts arose in the adventurous field of kidney transplantation.

1936: First Unsuccessful Kidney Transplant Between Humans

In 1936, in Kherson, Russia, Voronoy reported the first kidney transplant between humans using a cadaver kidney.[9] The recipient was a 26-year-old woman who was admitted in a uraemic coma after swallowing mercury in a suicide attempt. Voronoy retrieved a kidney from a 60-year-old man who had died from a fracture of the base of the skull. Using local anaesthesia, he transplanted it on the recipient's right femoral vessels. The kidney was placed in a pouch in the groin and the ureter was brought out in the skin. After the vascular clamps were released, the kidney's blood supply was restored and several drops of urine appeared. But eventually, diuresis stopped and the woman died a few days post-transplant. The autopsy revealed degenerative lesions of the tubules and the glomeruli, but no vascular thrombosis. Within the next few years, Voronoy performed four other transplants of cadaver kidneys in humans, but all failed.

1943–1944: Medawar's Explanation of Graft Destruction Due to "Biological Incompatibility" as Described by Carrel

During World War II, the British Medical Research Council focused on the problem of skin grafting for the treatment of burns. They asked Peter Medawar, a British-Lebanese Professor of Zoology, to work with Thomas Gibson, a plastic surgeon, to attempt to perfect skin auto- and, if possible, allotransplantation in humans. In the laboratory, Medawar

Fig. 2 Sir Peter Medawar defined the immunological origin of rejection and unravel many aspects of tissue immunology. His historic achievements established the field of modern transplant immunology. He was awarded the Nobel Prize in 1960. This is a picture from the day he was knighted next to his wife Lady Jean Medawar.

observed a persistent finding in experiments with mice or rabbits: if an initial skin graft was placed from animal A to animal B, it survived about seven days. Then, if a second skin graft was placed from animal A to animal B in exactly the same fashion, it was destroyed in about half that period of time. Medawar characterized this finding as the "second set response" and defined its immunologic origin.[10–12] Carrel's description of graft destruction due to "biological incompatibility" was now explained as rejection due to an immunologic response. Medawar in collaboration with Brent and Billingham proceeded to unravel many aspects of tissue immunology; his historic achievements established the field of modern transplant immunology.[13–15] Medawar (Fig. 2) was awarded the Nobel Prize in 1960.

1947–1953: Initial Unsuccessful Allotransplants After World War II

World War II significantly delayed clinical efforts in the field of transplantation. After the war, the ambition, talent, and courage of young clinicians again turned to the drama of patients with end-stage renal failure.

In 1947, a young woman was admitted to the Peter Bent Brigham Hospital in Boston. She had been in a coma with anuria for ten days after an abortion, complicated by acute tubulopathy due to septic shock and an incompatible blood transfusion. The decision to have her undergo a kidney transplant was made by a team led by David Hume.[16] The hospital administration objected to the use of an operating theatre for an experimental procedure. So the operation was performed in the middle of the night under difficult conditions, in a small room using two lamps. A kidney was removed from a cadaver donor immediately after death and transplanted on the humeral vessels in the recipient's cubital fossa. The kidney produced urine for only two days post-transplant. Fortunately, soon after the transplant, the patient regained consciousness, natural diuresis was restored, and the transplanted kidney was removed.

In 1950, at the Presbyterian Hospital in Chicago, Lawler replaced the left kidney of a woman with a compromised prognosis because of complicated polycystic kidney disease. The kidney graft was retrieved from a woman who died from ruptured oesophageal varices.[17] Heparinized saline was perfused into the vessels of the graft before the transplant. The vessels of the graft were anastomosed to the vessels of the recipient's resected kidney, and the ureter was anastomosed to the recipient's ureter. Graft function was impossible to evaluate, because the right kidney was left in place. On day 63 post-transplant, during an operation to evacuate a perirenal abscess, the transplanted kidney was described as being of normal size and consistency, although no other details about its function were given. By six months post-transplant, the graft was completely destroyed, but the recipient lived for some time with her right polycystic kidney.

In January 1951, in Paris, two kidney transplants were performed the same day by two different teams. Charles Dubost and Nicolaos Oeconomos used a kidney that had been retrieved from a prisoner immediately after his execution by the guillotine.[18] The recipient was a 44-year-old woman with advanced renal failure due to chronic pyelonephritis and nephroangiosclerosis. The kidney was perfused with plasma, transferred to the operating theatre, and transplanted retroperitoneally into the iliac fossa, with anastomosis of the graft vessels to the recipient's iliac vessels. The kidney produced a small amount of urine, but the woman died on day 17 post-transplant from complications of terminal uraemia. The other kidney from the same executed prisoner was transplanted by Marceau Servelle into a 22-year-old woman who had end-stage hypertensive chronic nephritis in a solitary kidney.[19] The kidney graft was transplanted using the same technique as Dubost and Oeconomos used. The immediate postoperative course was extremely difficult. The kidney graft produced an increasing amount of urine between day 15 and 19 post-transplant, probably indicating that it was recovering from acute tubular necrosis, but the woman died suddenly.

Later that same month, also in Paris, Rene Kuss performed a kidney transplant in a 44-year-old woman with advanced renal failure.[20]

They used a kidney removed for therapeutic reasons after Matson's operation for hydrocephalus, in which the ureter acted as a drainage tract. The kidney was placed in the iliac fossa. Renal function was restored immediately after the vascular clamps were removed, but the quantity of urine remained minimal. On day 30 post-transplant, the woman developed severe haematuria and died several days later. On autopsy, the kidney graft showed signs of ischaemic necrosis with thrombosis of the renal artery at the hilum. This case was most likely the first kidney transplant from a living donor. During 1951, the same team performed three similar transplants using kidneys removed for therapeutic reasons, but none was successful. The operative conditions were favourable (use of living donors, flushing of the kidney, a properly equiped operating theatre), but rejection caused the grafts to fail.

Also in 1951, other attempts of allotransplants between humans were reported, particularly in the U.S. Scala, in Springfield, Illinois, transplanted a kidney that was removed for therapeutic reasons in a man with end-stage glomerulonephritis.[21] The vessels of the graft were anastomosed to the recipient's splenic vessels. Probably for the first time, the very recently developed artificial kidney was used to prepare this recipient pre-transplant and to cover the period of acute tubular necrosis post-transplant. Yet, this transplant was also a failure. The recipient died during the 5th week post-transplant. The autopsy revealed a rejected, infected kidney graft.

From 1951 to 1953, the greatest experience with kidney transplantation took place at the Peter Bent Brigham Hospital in Boston, where David Hume (Fig. 3) and John Merrill reported nine allotransplants using either cadaver kidneys or kidneys removed after Matson's operation in hydrocephalic children.[16] These kidneys were transplanted into the thigh on the femoral vessels, with the ureter brought out in the skin so that urine production could be monitored. Of those first nine allotransplants, eight failed; so did another three similar transplants performed by Murray (who replaced Hume, who left to join the Navy).

However, in one of Hume's cases, the graft survived for five months — a record at that time. In February 1952, a young doctor with

Fig. 3 In the early 1950s, David Hume performed at the Peter Bent Brigham Hospital in Boston some of the first difficult kidney transplants that open the road for the successful kidney transplants that followed a few years later at the same hospital.

end-stage glomerulonephritis and hypertension had finally received a kidney after waiting for many months. The graft became available following the death of a patient on the operating table during a heart operation. It was transplanted into the thigh, but failed to excrete urine after revascularization. The artificial kidney was used to compensate for the acute tubular necrosis period. Normal diuresis was restored on day 20 post-transplant. The recipient was eventually discharged home with normal renal function. However, five months post-transplant, hypertension recurred, and the recipient died of heart failure. Histologic examination of the kidney graft revealed signs of advanced arteriosclerosis, but very few signs of rejection. This initial success was attributed to chance "biological compatibility" between the donor and the recipient, and the delayed failure was attributed to hypertension. It was suggested that, in similar cases, the recipient's native kidneys should be removed pre-transplant. Doing so would, of course, require haemodialysis, which, at that time, was only exceptionally used by highly specialized centres.

Another kidney transplant that came close to a big success occurred at the Necker Hospital in Paris in December 1952. It was the first time that a healthy kidney from a healthy living donor was used.[22] A 16-year-old carpenter underwent nephrectomy of his ruptured right kidney after a fall. After the nephrectomy, the surgeons discovered that the boy had congenital absence of his left kidney. To satisfy his mother's insistent demand to offer her son his last chance of survival by giving him a kidney, the surgeons (Michon, Vaysse and Oeconomos) performed a transplant during the night of December 24. The graft was placed in the iliac region, according to the technique developed the previous year in Paris. The ischaemia time was limited to 55 minutes. The graft excreted urine immediately and laboratory parameters returned to normal within a few days. The boy's condition was excellent over the following days. The procedure appeared to be a great success. Then, suddenly, on day 21 post-transplant, the abrupt onset of anuria indicated rejection; the boy died several days later. This transplant — performed under the most favourable technical and intensive care conditions, using an organ from a living related donor in excellent health, with a very short ischaemia time — clearly demonstrated that allotransplants were destined to fail. The only hope for success was a transplant between monozygotic twins.

1954–1957: First Successful Kidney Transplants in Humans Between Identical Twins

On December 23, 1954, the first successful kidney transplant in humans was performed between identical twins. It took place at the Peter Bent Brigham Hospital in Boston, thanks to a team led by Joseph Murray.[23] The donor and the recipient were 23-year-old twins, one of whom was suffering from very advanced glomerulonephritis. The kidney was implanted according to the Paris technique in the iliac position, with implantation of the ureter into the bladder. After an ischaemia time of 90 minutes, the kidney functioned immediately. The blood urea level rapidly returned to normal. Because of persistent hypertension, the recipient's own kidneys were removed six months post-transplant. He returned to work, married, had a family, and led a normal life. The procedure, therefore, was a complete long-term success. This first successful transplant between monozygotic twins was followed by six similar transplants, which Murray reported in 1958, as well by many other such transplants around the world. However, success was possible only between identical twins (who represent only a small percentage of the population). Moreover, a new threat had been recognized: the risk of recurrence of glomerulonephritis in the transplanted kidney, which happened in about 30% of these cases.

1958: First Kidney Transplant in Humans Using Immunosuppression

In 1958, a young woman was admitted to the Peter Bent Brigham Hospital in Boston with anuria after her solitary kidney had been removed because of a life-threatening haemorrhage. It was impossible at that time to continue haemodialysis, which had kept her alive for several days. After serious deliberation, a team led by Murray decided to perform the first kidney transplant under immunosuppression, using total body irradiation.[24] Pre-transplant, the woman received a dose of 600 rads, considered to be sub-lethal. Maintained in a sterile environment for a month, she received bone marrow cells from 11 different donors, including several of her brothers, in hopes of obtaining a cross-acceptance to one of the donors. In the end, because of the uncertain result of a potential transplant, the idea of sacrificing the kidney of a healthy volunteer donor was judged to be unreasonable. However, a kidney that was removed after a Matson's operation became available and was transplanted into the woman's thigh. The ureter was brought out into the skin. Her post-operative course was complicated by marrow aplasia and leucopenia, causing massive haemorrhage. Many marrow and cell transfusions were performed. On day 15 post-transplant, the kidney graft produced a significant amount of urine and haemodialysis was stopped. Unfortunately, the woman eventually died because of massive haemorrhage due to platelet depletion. On histologic examination of the transplanted kidney, the crucial finding was only minor evidence of rejection. Similar transplants were performed at the same hospital using very high doses of irradiation for immunosuppression. In view of the very disappointing results, the decision was made to use significantly lower doses of irradiation in the future.

1959: First Successful Kidney Transplant Between Non-identical Twins

In January 1959, the first successful kidney transplant between dizygotic twins was performed, again at the Peter Bent Brigham Hospital in Boston, again by a team led by Murray.[25] The recipient was a 23-year-old man with end stage renal failure due to glomerulonephritis and pyelonephritis. He received two doses (250 and 200 rads) of irradiation one week apart, then received a kidney from his brother. The kidney was placed in the iliac region, with implantation of the ureter into the bladder. After several days of acute tubular necrosis, the kidney graft started to function, producing good volumes of urine. The recipient's native kidneys had to be removed on day 11 post-transplant, because of the development of bilateral pyonephritis. He recovered very well and maintained good renal function. However, a few months post-transplant, he had an episode of rejection, which was confirmed on biopsy and treated with steroids and low doses of total body irradiation. After overcoming this episode of rejection, he led a normal life with good renal function. He died 20 years later of heart failure. For his significant contributions in the field of transplantation, Murray (Fig. 4) was awarded the Nobel Prize in 1990.

Fig. 4 At the Peter Bent Brigham Hospital in Boston, Joseph Murray performed the first successful kidney transplants in humans between identical twins, the first successful kidney transplant between dizygotic twins as well as the first successful cadaveric kidney transplant using immunosuppression. He was awarded the Nobel Prize in 1990.

In June 1959, a similar success took place in Paris. The transplant was performed at the Necker Hospital by Jean Hamburger's team.[26] The recipient was a 37-year-old man with end-stage renal failure who received 460 rads of irradiation in two sessions pre-transplant. His non-identical twin brother's kidney was transplanted into the iliac fossa, with implantation of the ureter in the bladder. Normal renal function was soon restored, and the recipient led a normal life. He died 26 years later of bladder cancer.

Those two successful kidney transplants between dizygotic twins in 1959 represented the first major victory against the barrier of immunologic incompatibility and rejection.

1960: First Successful Kidney Transplant Between Non-twin Siblings

In January 1960, the first successful kidney transplant between non-twin siblings took place at the Foch Hospital in Surenes, France. It was performed by a team led by Rene Kuss.[27] This transplant was between a female donor and her brother, who was six years younger. The 36-year-old recipient initially underwent radical nephrectomy for renal cancer with no local or lymphatic invasion. Eventually, his remaining kidney developed a renal carcinoma, which was progressing very slowly without evidence of local invasion or distant metastasis. However, his clinical condition was dramatically deteriorating, with simultaneous development of end-stage renal failure. After serious deliberation and at the insistence of his family, surgeons decided to perform a nephrectomy and kidney transplant. After the nephrectomy (during which no evidence was seen of local invasion or distal metastasis), the man was put on haemodialysis for a few days. He was then irradiated and received one of his sister's kidneys, which was placed in the iliac region. The graft worked immediately. The recipient eventually returned home and maintained good renal function, but died 4½ months post-transplant because of disseminated liver metastasis. This first successful kidney transplant between non-twin siblings was also the first sobering lesson about the difficulty of transplantation and immunosuppression in recipients with malignant disease.

1960–1961: First Successful Kidney Transplant Between Non-siblings

In 1960 and 1961, at the Foch Hospital, Kuss (Fig. 5) performed two kidney transplants between non-siblings.[28] The first such recipient was a 26-year-old woman with end-stage chronic pyelonephritis and hypertension. Many members of her family were willing to give her a kidney, but finally her brother-in-law was selected as the most suitable donor. In June 1960, she was irradiated; the donor's left kidney was

Fig. 5 Rene Kuss was one of the French pioneers of kidney transplantation. He performed the first successful kidney transplant between non-twin siblings and between non-siblings.

transplanted, with immediate good renal function. An episode of rejection occurred at five weeks post-transplant. It was treated with low doses of total body irradiation, steroids, and based on results of the experimental studies conducted by Roy Calne, 6-mercaptopurine, a new agent with a highly immunosuppressive effect in animals.[29] This episode of rejection was successfully reversed. Good kidney function lasted for 16 months. The second transplant between non-siblings was performed in October 1961. The recipient was a 38-year-old woman who received a kidney that became available after Matson's operation. Satisfactory renal function was immediately restored. Induction immunosuppression consisted of irradiation, followed by maintenance on 6-mercaptopurine and steroids. At 18 months post-transplant, renal function was perfect, but the recipient died of septicaemia due to complete marrow aplasia.

1961–1962: First Kidney Transplants Using Azathioprine

Experimentation with 6-mercaptopurine began in 1959 in London: Roy Calne demonstrated prolonged kidney graft survival in dogs.[29] This drug was first used in humans in 1960, basically as a complement to initial total body irradiation. But it was rapidly abandoned, because the useful dose and the toxic, even lethal dose were too hard to distinguish. In 1960, Calne arrived at the Peter Bent Brigham Hospital in Boston. Together with Murray, he launched a long series of kidney transplants in dogs using azathioprine, the less toxic and more effective imidazole derivative of 6-mercaptopurine. In some dogs that received a kidney transplant using azathioprine for immunosuppression, graft survival was extended to several years.[30, 31] Soon after those experiments, azathioprine was used in clinical practice at the Peter Bent Brigham Hospital. After returning to London in 1961, Calne initiated a kidney transplant programme using azathioprine and steroids, which became the mainstream immunosuppressive regimen for almost 20 years and substantially improved the results of kidney transplantation.

1962: First Successful Cadaver Kidney Transplant Using Immunosuppression

In April 1962, at the Peter Bent Brigham Hospital, Murray performed the first successful human cadaver kidney transplant, using an immunosuppressive regimen of azathioprine and actinomycin C.[32] The graft functioned for more than one year — an absolute record for a cadaver kidney at that time. In October 1962, in Richmond, Virginia, Hume performed another cadaver kidney transplant, using a similar immunosuppressive regimen.[21] As with Murray's case, this kidney kept the recipient off dialysis for about one year. The success of those two transplants significantly encouraged cadaver kidney transplants.

Fig. 6 Jean Hamburger, one the French pioneers of kidney transplantation and organ procurement from heart-beating cadaver donors.

1963–1968: Progress in Cadaver Organ Procurement

The acceptance of the concept of irreversible coma, which was later referred to by the more appropriate term "brain death", allowed the removal of organs from heart-beating cadaver donors and dramatically increased the pool of available organs.[33] The first organ procurements from heart-beating cadaver donors were performed by Guy Alexander in 1963 and Jean Hamburger (Fig. 6) in 1964. By 1968, special laws were established in Europe and the U.S. that allowed surgeons to obtain organs from heart-beating cadaver donors.

Late 1960s–Early 1970s: Treatment of Choice for End-Stage Renal Failure

Immunosuppressive regimens were enriched by the introduction of anti-lymphocyte serum (ALS)[34] and anti-lymphocyte globulin (ALG),[35]

Fig. 7 Sir Roy Calne introduced azathioprine and cyclosporin in clinical kidney transplantation. Both medications improved dramatically the long-term graft survival. He is also recognized as one of the pioneers of liver transplantation. This is a picture from the day he was Knighted next to his wife Lady Patsy Calne.

by greater understanding of the importance of HLA matching,[36, 37] and by improvements in organ preservation, notably by Folkert Belzer[38] and Geoffrey Collins.[39] All these refinements helped the advancement of kidney transplantation. In the late 1960s and early 1970s, it was well-established as the treatment of choice for end-stage renal failure. The next critical step was the introduction of cyclosporin in clinical kidney transplantation by Calne (Fig. 7) in the late 1970s,[40, 41] which substantially improved long-term results. Since then, immunosuppression has been further refined by the use of new agents, such as monoclonal antibodies (OKT3),[42] tacrolimus,[43] mycophenolate mofetil,[44] the microemulsion formulation of cyclosporin,[45] and monoclonal antibodies against interleukin-2 receptors.[46, 47] All these agents have allowed more effective prevention and reversal of rejection, leading to excellent long-term graft survival and high quality of life.

The Future

The wonderful journey of kidney transplantation is anything but over. New challenges loom, such as the donor shortage, xenotransplantation, tolerance induction, and quality of life issues. But the achievements of the past allow optimism for the future. The driving force to face these new challenges will, once again, be the inspired and hard work of scientists combined with the faith and encouragement of patients and their families. New achievements in transplantation will prove again that, the greatness and the glory of science is that what yesterday was a dream and today is an ambitious research project, tomorrow will be established clinical practice that alleviates human pain.

Acknowledgements

The authors wish to thank Mary Knatterud for her editorial assistance.

References

1. Peer LA (1955). *Transplantation of Tissues*. Baltimore: Williams and Wilkins, pp. 25–29.
2. Carrel A (1902). La technique operatoire des anastomoses vasculaires et la transplantation des visceres. *Lyon Med* **98**, 859–864.
3. Carrel A, Guthrie CC (1905). The transplantation of veins and organs. *Am Med* **10**, 1101–1102.
4. Carrel A (1908). Results of the transplantation of blood vessels, organs and limbs. *JAMA* **51**, 1662–1667.
5. Carrel A (1914). The transplantation of organs. *NY Med J* 839.
6. Jaboulay M (1906). Greffe de reins au pli du coude par soudures arterielles et veineuses. *Lyon Med* **107**, 575–577.

7. Unger E (1910). Nierentransplantation. *Wien Klin Wochenschr* **47**, 573–578.
8. Reemtsma K, McCracken BH, Schlegel JU, *et al.* (1964). Renal heterotransplantation in man. *Ann Surg* **160**, 384–410.
9. Voronoy Y (1936). Sobre el bloqueo del aparato reticuloendotelial del hombre en algunas formas de intoxicacion por el sublimado y sobre la transplantacion del rinon cadaverico como metodo de tratamiento de la anuria consecutiva a aquella intoxicacion. *Siglo Medico* **97**, 296–298.
10. Gibson T, Medawar PB (1943). The behavior of skin homografts in man. *J Anat* **77**, 299–310.
11. Medawar PB (1944). The behavior and fate of skin homografts in rabbits. *J Anat* **78**, 176–199.
12. Medawar PB (1945). Second study of behavior of skin homografts in rabbits. *J Anat* **79**, 157.
13. Billingham RE, Krohn PI, Medawar PB (1951). Effect of cortisone on survival of skin homografts in rabbits. *Br Med J* **1**, 1157–1163.
14. Billingham RE, Brent L, Medawar PB (1953). Actively acquired tolerance of foreign cells. *Nature* **172**, 603–606.
15. Billingham RE, Brent L, Medawar P (1956). Quantitative studies on tissue transplantation immunity. III. Actively acquired tolerance. *Philos Trans Roy Soc London Biol Sci* **239**, 357–412.
16. Hume DM, Merril JP, Miller BF, Thorn GW (1955). Experiences with renal homotransplantation in the human: report of nine cases. *J Clin Invest* **34**, 327–382.
17. Lawler RH, West JW, McNulty PH, *et al.* (1950). Homotransplantation of the kidney in the human. *JAMA* **144**, 844–845.
18. Dubost C, Oeconomos N, Nenna A, Milliez P (1951). Resultats d'une tentative greffe renale. *Bull Soc Med Hop Paris* **67**, 1372–1382.
19. Servelle M, Soulie P, Rougeulle J (1951). Greffe d'une rein de supplicie a une malade avec rein unique congenital, atteinte de nephrite chronoque hypertensive azatemique. *Bull Soc Med Hop Paris* **67**, 99–104.

20. Kuss R, Teinturier J, Millieaz P (1951). Quelques essais de greffe rein chez l'homme. *Meme Acad Chir* **77**, 754–764.
21. Groth CG (1972). Landmarks in clinical renal transplantation. *Surg Gynecol Obstet* **134**, 323–328.
22. Michon L, Hamburger J, Oeconomos N, *et al.* (1953). Une tentative de transplantation renale chez l'homme: aspects medicaux et biologiques. *Presse Med* **61**, 1419–1423.
23. Murray JE, Merrill JP, Harrison JH (1958). Kidney transplantation between seven pairs of identical twins. *Ann Surg* **148**, 343–359.
24. Murray JE, Merrill JP, Dammin GJ, *et al.* (1960). Study of transplantation immunity after total body irradiation: clinical and experimental investigation. *Surgery* **48**, 272–284.
25. Merrill JP, Murray JE, Harrison JH, *et al.* (1960). Successful homotransplantation of the human kidney between nonidentical twins. *New Engl J Med* **262**, 1251–1260.
26. Hamburger J, Vaysse J, Crosnier J, Auvert J, Lalanne CM, Hopper J, Jr (1962). Renal homotransplantation in man after radiation of the recipient: experience with six patients since 1959. *Am J Med* **32**, 854–870.
27. Kuss R, Legrain M, Mathe G, *et al.* (1962). Homologous human kidney transplantation: experience with six patients. *Postgrad Med J* **38**, 528–531.
28. Kuss R (1991). Human renal transplantation memories, 1951–1981. In Terasaki PI (ed.) *History of Transplantation: Thirty-Five Recollections.* Los Angeles: UCLA Tissue Typing Laboratory, pp. 37–59.
29. Calne RY (1960). The rejection of renal homografts: inhibition in dogs by 6-mercaptopurine. *Lancet* **1**, 417–418.
30. Calne RY (1961). Inhibition of the rejection of renal homografts in dogs with purine analogues. *Transpl Bull* **28**, 445.
31. Calne RY, Murray JE (1961). Inhibition of the rejection of renal homografts in dogs by Burroughs Wellcome 57–222. *Surg Forum* **12**, 118.

32. Murray JE, Merrill JP, Harrison JH, Wilson RE, Dammin GJ (1963). Prolonged survival of human kidney homografts by immunosuppressive drug therapy. *New Engl J Med* **268**, 1315–1323.
33. *Ad Hoc* Committee of the Harvard Medical School (1968). A definition of irreversible coma. *JAMA* **205**, 85–88.
34. Starzl TE, Marchioro TL, Porter KA, Iwasaki Y, Cerilli GJ (1967). The use of heterologous antilymphoid agents in canine renal and liver homotransplantation and in human renal transplantation. *Surg Gynecol Obstet* **124**, 301–318.
35. Najarian JS, Merkel Moore GE (1969). Clinical use of antilympoblasts. *Transpl Proc* **1**, 460–462.
36. Dausset J, Ivanyi P, Ivanyi D (1965). Tissue alloantigens in humans: identification of a complex system (Hu-1). In Balner H (ed.) *Histocompatibility Testing*. Copenhagen: Ejnar Munksgaard, pp. 51–62.
37. Terasaki PI, Vredevoe DL, Porter KA, Mickey MR, Marchioro TL, Faris TD, Herrmann TJ, Starzl TE (1966). Serotyping for homotransplantation. V. Evaluation of a matching scheme. *Transplantation* **4**, 688–699.
38. Belzer FO, Ashby BS, Gulyassy PF, Powerll M (1968). Successful 17-hour preservation and transplantation of human cadaver kidney. *New Engl J Med* **278**, 608–610.
39. Collins GM, Bravo Shugarman MB, Terasaki PI (1969). Kidney preservation for transplantation: initial perfusion and 30-hour ice storage. *Lancet* **21**, 1219–1225.
40. Calne RY, White DJG, Thiru S, Evans DB, McMaster P, Dunn DC, Craddock GN, Pentlow BD, Rolles K (1978). Cyclosporin A in patients receiving renal allografts from cadaver donors. *Lancet* **2**, 1323–1327.
41. Calne RY, Rolles K, White DJG, Thiru S, Evans DB, McMaster P, Dunn DC, Craddock GN, Henderson RG, Aziz S, Lewis P (1979). Cyclosporin A initially as the only immunosuppressant in 34 recipients of cadaveric organs: 32 kidneys, 2 pancreases, and 2 livers. *Lancet* **2**, 1033–1036.

42. Ortho Multicenter Transplant Study Group (1985). A randomized clinical trial of OKT3 monoclonal antibody for acute rejection of cadaveric renal transplants. *New Engl J Med* **313**, 337–342.
43. Starzl TE, Fung J, Jordan M, Shapiro R, Tzakis A, McCauley J, Johnston J, Iwaki Y, Jain A, Alessiani M, Todo S (1990). Kidney transplantation under FK 506. *JAMA* **264**, 63–67.
44. European Mycophenolate Mofetil Cooperative Study Group (1995). Placebo-controlled study of mycophenolate mofetil combined with cyclosporine and corticosteroids for prevention of acute rejection. *Lancet* **345**, 1321–1325.
45. Wahlberg J, Wilczek HE, Fauchald P, Nordal KP, Heaf JG, Olgaard K, Hansen JM, Lokkegaard H, Mueller EA, Kovarik JM (1995). Consistent absorption of cyclosporine from a micro-emulsion formulation assessed in stable renal transplant recipients over a one-year study period. *Transplantation* **60**, 648–652.
46. Vincenti F, Kirkman R, Light S, Bumgardner G, Pescovitz M, Halloran P, Neylan J, Wilkinson A, Ekberg H, Gaston R, Backman L, Burdick J (1998). Interleukin-2 receptor blockade with daclizumab to prevent acute rejection in renal transplantation. Daclizumab Triple Therapy Study Group. *New Engl J Med* **338**, 161–165.
47. Kahan BD, Rajagopalan PR, Hall M (1999). Reduction of the occurrence of acute cellular rejection among renal allograft recipients treated with basiliximab, a chimeric anti-interleukin-2 receptor monoclonal antibody. U.S. Simulect Renal Study Group. *Transplantation* **67**, 276–284.

V

THE HISTORY OF LIVER TRANSPLANTATION

Roy Y Calne

Introduction

Transplantation of the liver was a formidable undertaking in the 1950s when surgeons started to experiment in this area. Compared with the kidney and the heart, the liver is a much more complicated organ from a surgical point of view. The dual blood supply of the liver and its close relationship to vital structures in the abdomen and chest raised many technical difficulties. In addition, the liver is sensitive to ischaemia and deteriorates rapidly when it is deprived of its extensive blood supply. The insertion of an extra liver was the initial approach by Welch in 1995.[1] At first glance, it would appear that inserting an extra liver, thereby saving the recipient the trauma and hazards of removal of the native liver, might be an advantage but in fact this has not proved to be true. The liver, being a large irregular organ, can easily become twisted so that the channels for blood inflow and outflow may be compromised. It was to be difficult to obtain a suitable inflow of portal blood to the auxiliary liver, and in the early cases there were no long-term survivors, but at that time there was no effective immunosuppression. In 1956, Cannon in Los Angeles[2] wondered whether the liver might be an immunologically privileged organ and reported attempts at liver replacement in the orthotopic position having removed the animal's own liver. However, there were no survivors from the operation.

In the late 1950s and early 1960s, serious attempts were made at orthotopic liver transplantation in the dog independently in two institutions, namely, Dr. Francis Moore at the Peter Bent Brigham Hospital in Boston and Dr. Thomas Starzl in the North Western University of Chicago.[3,4] As Starzl has pointed out, the Boston group were continuing their interest in organ transplantation following their unique and extensive experience of kidney transplantation, whereas in Chicago Dr. Starzl was particularly interested in the metabolism of the liver and especially a comparison of systemic venous blood and portal venous blood entering the liver. A major obstacle in performing the orthotopic operation was the fatal result of clamping the portal vein and inferior vena cava (IVC) in the dog during the recipient hepatectomy. Both groups developed strategies for overcoming this. In Boston, a passive veno/venous bypass was used between the inferior vena cava and portal vein to the superior vena cava. In Chicago, the portal venous flow was first diverted to the IVC via a porta-caval shunt and then the IVC was cannulated through an iliac vein to the superior vena cava. At the end of the procedure, the side-to-side porta-caval shunt was taken down.

A satisfactory, if difficult, technique was successfully achieved in both institutions and rejection of the liver was found to be similar in the dog to rejection of the kidney and immunosuppression with azathioprine and other agents used to treat recipients of kidney transplants were also effective in experimental liver grafting. One of Dr. Starzl's dogs, after initial treatment with immunosuppression, survived without maintenance immunosuppression for 11½ years — an "operational tolerance" that was seldom seen after experimental kidney grafting in dogs.[5]

I personally remember the liver transplant experiments performed by Dr. Moore at the Peter Bent Brigham Hospital in 1960–1961. They occurred once a week and were nicknamed the "sputnik" procedure as the Russians had recently launched their orbiting satellite, much to the anguish of the United States. The operation required a large team and many hours of work and then caring for the animal post-operatively through the night. It certainly required great perseverance as there were many failures for each success. Eventually, however, the experimental operation was sufficiently successful for Starzl to do the first human orthotopic liver transplantation in 1963.[6]

The early cases of liver transplantation in Denver, where Starzl had now moved, proved to be formidable surgical undertakings. The previous surgery that patients had been subjected to made the donor hepatectomy exceedingly difficult and dangerous. The liver failure had resulted in grossly impaired clotting and thrombocytopenia, the passive external shunts as used in dogs proved to be not only unnecessary, but also dangerous with clotting occurring and pulmonary emboli.

By October 1963, the Denver group had no long-term success and had decided to have a moratorium on orthotopic liver transplantation which lasted for more than three years. A few attempts at auxiliary liver transplantation in the U.S., the U.K. and Australia also failed.[1] In 1967,[7] the Denver group resumed orthotopic liver transplantation obtaining some long-term successes and we did the first technically successful liver transplant in Europe in 1968.[8]

Starzl's initial interest in liver transplantation had been physiological and Moore's was an extension of the clinical programme. My own interest was immunological because of the extraordinary behaviour of liver grafts in pigs that we had observed. My early reactions to liver transplantation had been negative, having witnessed the enormous effort and large personnel involved in experimental operations in Boston with an occasional success and many failures. However, when I moved to Cambridge in 1965 I started collaborating with Richard Binns, a distinguished immunologist who had written a thesis on Medawar-type tolerance in the pig following intra-uterine foetal injection of potential donor tissue. The tolerance was often only partial, when tested by skin grafts, but when we used kidneys we observed long-term survival. The species and organ variability of transplant response has always intrigued me and we learned of some long surviving liver grafts without immunosuppression performed in Paris and in Bristol. We repeated these experiments and were amazed to observe animals surviving years with orthotopic liver grafts from unrelated donors without the help of any immunosuppressive agents. In the past and still now, the liver has been a fascinating organ to transplantation immunologists.[9,10]

In 1969, an editorial in the *Lancet* entitled "Strange English Pigs"[11] referred to experiments by ourselves and others showing that the pig can survive indefinitely after receiving an orthotopic liver allograft from an unrelated pig without the need for immunosuppression. Moreover, other organs transplanted together with the liver, which would normally be rejected promptly, also enjoyed the privilege of prolonged survival. Similar observations were subsequently made in rats and it was then possible to study the phenomenon in grafts between inbred strains where the Major Histocompatibility Complex (MHC) of donor and recipient were known. Unfortunately, such unaided tolerance does not occur in all species. In man and the dog, immunosuppressive drugs are necessary to control rejection of liver allografts, and in pigs the outcome is variable. Some animals reject their livers promptly and some animals can survive indefinitely, our longest being 11 years after transplantation when the animal died of old age.

Without immunosuppressive treatment, all these experimental livers undergo a rejection reaction which has been well documented, histologically and biochemically and then the rejection, instead of progressing to destruction of the organs, recovered spontaneously. This is an extremely interesting phenomenon and cautions the excessive use of immunosuppression which might prevent this switch-off of the immune response.[12] In rats with transplants between MHC-incompatible donor and recipient there appeared to be an unstable period associated with the immunological engagement of donor and recipient. When the animal had passed through this unstable period successfully, then skin grafts from the donor strain would be accepted indefinitely. It was suggested that a reciprocal immunological engagement between donor passenger cells, especially Küpfer cells, but also other antigen presenting cells and recipient lymphocytes reacting against the donor resulted in an immunological truce in which potential combatants were either killed or became peaceful.

Tolerant recipients of liver allografts have a high blood concentration of Class 1 MHC molecules of donor-type. Injections of serum and lymph from such animals given daily successfully prolonged survival of skin or heart grafts in naïve animals of appropriate strains.

In the huge experience of clinical liver transplantation in Pittsburgh, a number of patients have demonstrated operational tolerance after varying periods of treatment with continuous immunosuppression after liver grafting. The longest survivor in the world has had no immunosuppressive drugs for 15 years.[13] She was non-compliant because she did not like the side-effects of the drugs. It is very unusual for recipients of other organs to be able to stop taking immunosuppression without paying the penalty of acute or chronic rejection. In our experience, combined liver and kidney allografts from the same donor had a beneficial effect on kidney survival, although the liver survival was unaffected by the addition of a kidney graft. Studies from Sydney show that in the rat, increasing amounts of transplantation antigen that resulted when multiple organ grafts were transplanted from the same donor to the same recipient could produce operational tolerance without any immunosuppressive treatment even when a liver was not one of the organs.[14]

The liver is a large MHC antigen source and it has remarkable reparative capability, even after severe damage. There are many bone marrow derived passenger cells in the liver which include lymphocytes, dendritic cells and Küpfer cells. The potential activity of passenger cells has been shown in fatal graft-versus-host disease and has occasionally followed liver grafting in man, especially when the recipients were already severely immunologically compromised. The passenger cells are shifting populations and after some weeks the Küpfer cells are replaced and become recipient in origin. A changeover occurs of passenger cells of donor origin to those of the recipient. Starzl and colleagues have shown that donor passenger cells migrate widely in the recipient and their survival is associated with long-term successful grafts of liver and kidney. The Pittsburgh workers believe that this microchimaerism is responsible for the long-term graft acceptance but other workers have disputed a causal relation, suggesting that microchimaerism could be an epi-phenomenon. Sritwatanawongsa and Davies[15] performed a series of experiments to determine the roles of microchimaerism and the cellular components of the liver in rats using strain combinations in which specific tolerance occurred. Thirty days after liver allografting, most of the passenger cells had been exchanged for those of the host. The livers were then re-transplanted into secondary recipients of either donor or recipient strain. It was possible then to assess the separate roles of passenger cells and parenchymal cells in the production of tolerance as tested by donor skin grafts. To obtain full tolerance, both parenchyma and passenger cells of donor type were required and it is possible that the export in large amounts of Class I antigens by the parenchymal cells was one factor important in maintaining tolerance. Davies and Pollard[16] had shown that in man, half the circulating Class I MHC antigen in the blood is produced by the liver and soluble forms of these molecules are synthesized by human liver grafts. This Class I MHC antigen, alone or complexed with antibody after transplantation could be an important factor in the immunological consequences of a liver graft but, as explained above, the liver has not been shown to induce tolerance on its own in man without immunosuppressive treatment.

In the course of experimental liver transplants in the pig, we had made every technical mistake possible but we seemed to appreciate the important aspects of the operation. Starzl's resumed programme was beginning to look potentially viable and I felt that our team was now in a position to look at orthotopic liver grafting having had a small but bitter experience of accessory liver transplantation.

In 1968, a lady with a primary malignant growth in her liver was referred to me. She was anxious to go through this new and experimental operation, even when we had explained the numerous dangers; she said she had nothing to lose. Most of my colleagues at Addenbrooke's were opposed to my doing a liver graft for a variety of reasons.

Then, on May 2, 1968 a child with a viral brain infection became comatose, with irreversible destruction of the brain stem. The ventilator was to be turned off, but the parents gave permission for the kidneys and liver to be used to help other patients in need.

I convened a meeting of colleagues to discuss the case. That morning, Francis Moore phoned me. By extraordinary good fortune, he happened to be in Cambridge visiting his son, a graduate student in molecular biology. I quickly asked Moore to join our meeting.

When the details of both the recipient and potential donor had been presented, I asked each of may colleagues for an opinion. They were unanimous in opposing the operation, each with a different argument — the illness of the recipient, the youth of the donor, the possibility of the donor's disease spreading to the recipient, the operation never having been done in our hospital (or even Europe) before, the prohibitive expense. After listening to this litany of pessimism, I introduced Dr. Moore, world-famous (even in Cambridge), and together with Starzl, one of the pioneers in liver grafting.

Moore's response was short and typical of him. "Roy, you have to do it." The opposition collapsed, and we made immediate plans for the surgery. With Moore as my first assistant, we began a long and, at that time, very difficult operation. Yet all went smoothly. I crossed the Rubicon, i.e. I removed the cancerous liver in what I call the "moment of truth", since the patient's life depends on immediate function of the new liver. As the donor was a child, the main vein draining the liver and the lower half of the body, the inferior vena cava, was very small, and I was worried that it had insufficient calibre for the adult recipient. I therefore left the IVC in the recipient intact and joined the end of the donor IVC to the side of that of the recipient, in what came to be known as the "piggy-back operation". This operation was re-invented years later by other teams, who had not read our 1968 report in the British Medical Journal.

Our patient woke up shortly after the end of the operation, and she and we were delighted. Sadly, two and a half months later, she developed a fatal pneumonia due to the immunosuppressive drugs given to prevent rejection. We had, however, demonstrated that the operation *could* be done in Addenbrooke's Hospital. Moore's simple, forthright advocacy had won the day.

After the first few liver transplants had been performed at Addenbrooke's, it became clear to me that, since nobody on the hospital's medical side was especially interested in the liver, I would need to seek help elsewhere. An old colleague from the Royal Free Hospital, Roger Williams, had established an extremely busy and successful liver unit at King's College Hospital in London. Williams had a surgeon's eminently practical and active mindset, and was used to making decisions and acting on them; so when I contacted him he was keen to collaborate and provide medical expertise. Because he understood prognoses and realized when conventional treatment would no longer help patients, he was well qualified to refer patients for transplantation.

We planned that I would direct the surgical procedure at Addenbrooke's while he oversaw the medical side at King's We would do the operations at both institutions, or wherever there was a donor, since the criteria of brain death had not yet been accepted and we had less time to do the donor operation.

Events would often take the following course. We would hear of a donor in another hospital, a patient on a ventilator. A doctor would have diagnosed brain stem destruction and decided to stop the ventilator with the relatives' consent. They would also have permission for organs to be removed for transplantation.

Williams and I meanwhile knew that the liver was extremely sensitive to a lack of blood supply, and would be damaged during the removal procedure. It required a very high degree of co-operation from doctors, nurses and the administration of the donor hospital to overcome this difficulty, and they almost always gave it. A team of surgeons would leave from Addenbrooke's, laden with special instruments and accompanied by an operating theatre nurse, a technician to cool the liver and an intensive care nurse. From King's, a medical registrar would come for a stint of 24 hours. He or she would then be replaced by another medical registrar, until the patient was well enough to be moved to King's or to Addenbrooke's. The liver patient, very ill, or even on a mechanical ventilator with liver failure, would be taken from King's or Addenbrooke's by special ambulance to the donor hospital.

With surgical preparations all ready, the donor patient would be moved into the operating theatre. The ventilator would be turned off, as had been planned whether or not there was to be a transplant operation. We would wait in sterile gowns, masks and gloves until the anaesthetist declared that the heart had stopped, at which point we would be galvanized into action, immediately starting the operation: the donor liver and kidneys were now devoid of circulation and delays would render them useless.

The removal of liver and kidneys without damaging them or their blood vessels is not a particularly easy operation, and it was necessary to develop the skills to perform it swiftly. Before they were removed, the liver and kidneys had to be cooled by infusing cold preservation fluid through the arteries, and they needed to cool further after removal. Then they had to be packed in sterile bags surrounded by ice. Under these circumstances, kidneys could be kept for 12 to 24 hours, but we had to transplant the liver within a mere five to eight hours.

Once we had the donor organs, the recipient would be prepared for surgery. This operation is a two-stage one. First the diseased liver must be removed, in what is often a very difficult operation, and then the new liver must be inserted in its normal anatomical space. One of the major complicating factors is that the liver is the main manufacturer of clotting agents in the blood: chronic liver disease causes a rise in pressure in the veins of the abdomen so that they may bleed even without being touched, just from the manoeuvres involved in preparing to take out the diseased liver. The remaining blood platelets, made by the bone marrow, are rapidly consumed by the body's attempts to clot small bleeding veins. So it is not surprising that serious blood loss, and difficulties in obtaining enough blood to replenish amounts lost during an operation, were major problems in the early days of liver transplantation. In fact, liver transplantation alone used to regularly wipe out the supplies at local blood banks — a state of affairs that almost led to a cessation of the liver transplant programme.

There was no room for error in the long surgical procedure; a mistake at any stage could have been fatal for the recipient. The operation was usually accompanied by the unpleasant whirring noise of the sucker removing the blood that was shed. As soon as the sucker could be turned off safely the surgeon knew that the patient was probably out of the wood. If possible, the operation was carried out in a systematic way, one part at a time, and if bleeding occurred in one area, then it was often best to leave that area temporarily and come back to it later.

Before inserting the new liver, a decision had to be made, in consultation with the anaesthetist, as to whether or not to bypass the blood from the big veins in the lower part of the body to the major draining vein in the upper part of the body.[17] This took extra time and had its own hazards. Bypass required the expertise of a special pump technician to make sure that the blood flow through the bypass system was adequate and safely administered. It was vital that no air got into the circuit, as it could be fatal.

At the end of the operation we knew the liver should look perfect, with good pulsation in the artery, a soft portal vein and soft vena cava above and below the liver, safe drainage of the bile duct to that of the recipient, and no bleeding. If everything had gone well, the abdomen was then closed and the patient was returned to the intensive care ward by the anaesthetist. Occasionally, if there was severe bleeding which could not be stopped, gauze packs were inserted and left in place for 48 hours and then removed. This lesson had been learned from the treatment of liver trauma. The gauze packs helped to stop the bleeding and have saved many lives over the years. The operation could take anything between just over two hours to more than 12. The average was about three to four hours in an uncomplicated case.

For patients who survived the operation, the experience could be surreal. They might wake up in a strange hospital, having never even been to that town before. The relatives had to come long distances to visit, and it could be ten days to two weeks before the patient was well enough to be transferred back to King's or Addenbrooke's. In truth, the logistics of this arrangement were impossibly complex: the disturbance of the local hospital and the reliance on the goodwill of the staff were too much to expect except in a few exceptional institutions. But there were very few donors in our own hospitals.

All this changed when it became possible to remove organs while the donor was still on the ventilator and the circulation of the vital organs was intact. We could then remove the livers in distant hospitals and bring them back to either Addenbrooke's or King's for grafting. But, while the practicalities became easier, there was still little general enthusiasm for liver transplantation in the 1960s, and the results remained extremely poor for some time. Nevertheless the dramatic successes we achieved, however few in number, convinced us that the procedure was possible and worthwhile. When a life-saving operation, despite an extremely high early mortality, is shown to be possible, it eventually becomes established, the errors are recognized and eliminated, and a new generation of surgeons wonders why the pioneers had such a hard time.

Mrs Winnie Smith was referred to me around this time. She had a large liver tumour involving both sides of the liver, so it could not be removed and leave her with enough liver to support life. She was a plump, cheerful lady of 48, living in the nearby town of Welwyn with her husband and two children, and worked as a hospital cleaner. We discussed the option of a liver transplant and she was keen to go ahead with it. The weather was deteriorating and heavy snow was falling when a road traffic accident victim suffering from brain damage was admitted to Newmarket hospital, 14 miles from Cambridge.

As I have explained, in those days we used to move the recipient to the donor hospital, so in this case our surgical team went to Newmarket and Mrs Smith was admitted there. The operation was long but went well and one of our surgeons stayed to take care of her. She made a good recovery and when the snow melted she was transferred to Addenbrooke's and later discharged. She went back to work and taking care of her family.

When the International Transplantation Society Meeting was held in San Francisco in 1972, I managed to raise funds so that Mrs Smith could go. She came, accompanied by her husband, and they found their first trip abroad a most exciting experience. Her case was presented and she took the platform and answered questions from the audience. At that time, it was nearly five years since her transplant and she was one of the longest survivors. However, shortly after returning home she developed jaundice, due to blockage of the bile duct, which led to a fatal infection in the bloodstream. It was a sad ending for a marvelous and inspiring patient who had shown us and the world that liver transplantation really could give dying patients a new lease of life.

Gordon Bridewell is our longest survivor with a liver graft. Here was a young man in a terrible situation, with a family history of liver disease. He had the dreaded hepatitis B virus which had caused cirrhosis of the liver, and on top of this he developed a huge tumour — also in his liver. His brother had died from the same disease a few months previously and, as a result, Gordon turned up at King's College in a desperate state. At that time, we had not defined our indications and contra-indications for liver transplantation, and we were prepared to take all comers. So we did a transplant on Gordon in King's, after finding that the huge tumour on his liver had apparently not spread elsewhere. We gave him γ-globulin injections against hepatitis B, which is often not effective, but we started the treatment when his diseased liver came out and for a short period after the new liver had been inserted.

Against all the odds, he did very well. He left hospital and was soon rehabilitated and back at work as a car engineer. Twenty-three years later, when he was knocked off his bicycle by a lorry and developed a pain over ribs that had been fractured, there was concern that he might have damaged his liver transplant. However, he recovered from this added injury and it transpired that we had a shared interest in painting — I did a portrait of Gordon and he presented me with a painting of the West Country where he lived. While he was sitting for the portrait he asked me a very pertinent question: "If I came to see you now with the condition I was suffering from 23 years ago, I don't think you'd accept me for a transplant, would you?" The truthful answer, which was the one I gave him, was that no, we would not.

He had two major contra-indications to a liver transplant, hepatitis B and the large tumour which we would have expected to recur. So he is the exception which proves the rule — our longest survivor with a liver transplant, still at work now, nearly 25 years after the operation. He taught us another important lesson — that we can make general assessments based on statistics but each patient is an individual, with his or her unique biological reaction to a transplant. For this reason, we can never say for certain what the outcome will be in a particular case. We can only give an overall statistical impression of what would be expected to happen, according to our experience of liver transplantation in our own unit and worldwide.

For reasons already mentioned, we used a vast amount of blood in nearly all the early liver transplant operations. A complicating factor was that the blood bank would not issue extra blood clotting factors unless there was a demonstrable defect in the specimens we had sent to the laboratory. Because there was a time lag between collecting the blood, sending it to the laboratory and getting the result, the patient could deteriorate to a critical state during this time, losing blood, consuming the few clotting factors that were left and producing the usually fatal condition we call "red ink syndrome", when the blood fails to clot successfully.

I asked for clotting factors urgently for a patient in this condition, and was told by the haematology lab that the clotting factors in the sample of the patient's blood were normal. The patient was meanwhile losing copious amounts of blood onto the floor. I spoke to the consultant haematologist and said that perhaps if she came to the operating theatre, she would be convinced that their assessment was erroneous. She reluctantly said she would come, but didn't know which theatre we would be in. I explained that this would be a very easy task for her: all she had to do was walk along the corridor, and enter the operating theatre with a pool of blood issuing from beneath the door. She came and was suitably impressed, realizing that the arrangements for blood clotting were unsatisfactory. So together we worked out a method of assessing blood clotting in the operating theatre and giving clotting factors before, rather than after, the patient had lost vast quantities of blood.

Thus, within a relatively short period of time we were able to remove the anxieties over blood supplied that had hung like the sword of Damocles over the whole liver transplantation programme. Today, we still sometimes require large volumes of blood, up to 50 or even 100 units, but usually we require less than ten units and occasionally none at all.

One special case was a young boy from a family of Jehovah's Witnesses. Oddly, Jehovah's Witnesses permit transplantation of organs, which, of course, always contain blood, but do not permit in any way a blood transfusion — not even taking blood from the patient, storing it and giving it back. This presents a certain challenge for the doctor, and in fact the parents who brought their boy to see me said that they had been to all the liver transplant units in Britain, and nobody would do a liver transplant unless they had permission to use blood. I explained that to start such a procedure, which nearly always requires some blood, and see a young boy die just because a few units of blood were needed and couldn't be given, would have a disastrous effect on the nursing and medical team: it was not reasonable to expect them to proceed in this way. Nevertheless, I said, I would try not to use blood but I wouldn't start the operation unless I had their permission to give blood *in extremis*.

The boy's parents reluctantly agreed to this and the anaesthetist was extremely careful, as were we on the surgical side. Blood substitutes were given, and at the end of the operation the patient was in good condition with a new liver, and no blood had been used. The parents were delighted and every year I receive a card telling me that the boy is doing well and they are so grateful that no blood was needed.

Liver transplantation moved from an experimental and highly dangerous form of therapy to a respectable and much sought after treatment following the introduction of cyclosporin as an immunosuppressant in clinical organ grafting. The handful of centres doing liver transplantation in the 1970s included Denver, Cambridge/Kings, Hanover and Paris. The overall results were poor and the number of patients transplanted were small. All this changed following the introduction of cyclosporin. The better results being largely attributable to safer immunosuppression but also to improvements in technique, better selection of cases, and a body of experience established in centres worldwide. The introduction of a pumped veno/venous bypass to stabilize the patient during the annepatic phase, greatly facilitated the training of surgeons. Previously speed was important in managing the patients during and after hepatectomy until the new liver was re-vascularized. Veno/venous bypass, although adding to the time and costs of the operation removed the temporal pressure. Most surgeons training in liver grafting first learned the anatomy by joining and eventually leading the donor teams. They also learned the importance of caring for the liver with great gentleness and cooling it rapidly. The introduction of the University of Wisconsin infusion fluid prolonged the permissible cold ischaemia time and also helped with the planning and logistics of the operation.[18] Although veno/venous bypass is often unnecessary in adults and never used in children, it does still have a place in training surgeons and also in the management of patients with severe renal or cardiovascular associated diseases.

The establishment of liver transplant centres has varied in different countries in the Western world. In the United Kingdom, almost all the operations are performed under the auspices of the Health Service, and since the government finances the procedure, the U.K. Department of Health has been able to control the number of centres and audited their performance. There is now a shortage of donors worldwide for all organ transplants. Until recently, this was not particularly acute for livers in the U.K. where the instance of liver disease is less than most other countries. Special techniques have been developed to manage paediatric recipients by use of a reduced size or cut-down liver, often the left lateral segment of an adult liver is sufficient for a small child, and then split-liver techniques allowing two recipients to benefit from one adult liver. This technique always involves a compromise and it is usual practice to given the best vascular arrangements for the child recipient so that for the adult the right half of the liver is not only smaller than the normal adult liver but also may require extra techniques to re-vascularize.

In Japan where there was no law on brain death until very recently, liver transplants from living related donors became very popular and the Japanese are extremely skilled in this technique. The most tricky part of the operation, the micro-vascular anastamoses are done in some Japanese centres by technicians who may not be medically qualified using operating microscopes. They come in at the crucial moment and then leave when they have done a beautiful technical performance. Living related liver transplantation was first attempted in Sao Paolo Brazil by Silvano Raia and then by Russel Strong in Brisbane and it is now gaining popularity woldwide due to the donor shortage.

One worrying consideration in liver transplantation is the recurrence of the patient's original disease, whether it is infective, malignant, alcoholic or auto-immune. Biliary atresia and metabolic defects such as oxalosis cannot recur in the transplant, but the recurrence rate of some diseases is high, hepatitis C always and hepatomas have a very high recurrence rate. With a limited and expensive resource of liver transplantation to be considered in the context of the rest of health care, it is not surprising that the governments responsible for this facility are very concerned with the outcome of the operation and also the proficiency and excellence of the transplant team.

Multi-Visceral Transplants

Multi-visceral transplants in which the liver is a component organ have figured experimentally and clinically over the years with the introduction of cyclosporin there have been a number of attempts at intestinal and liver transplants, when the patient has had liver disease in addition to the need for small bowel replacement. There is some evidence, although not universally accepted, that the liver gives protection of the intestine from rejection, the intestine being notoriously aggressively rejected when transplanted alone. More and more organ have been transplanted. In Cambridge, we have performed six-organ transplants on two patients, both of whom have survived, one for more than five years.[19] Multiple organ transplants and re-transplantation of a patient whose initial transplant has failed, has raised the ethical dilemma of should a patient have extra transplants depriving other patients of the possibility of the life-saving organ? These matters are not easily resolved but it is very difficult to demand that no action is taken when a patient who has been looked after sometimes for years by a transplant team, needs another organ to replace the failed graft.

Looking to the Future

I think it is likely that the competence of liver transplant teams and the outcome of liver transplantation in different categories of disease will become increasingly important considerations for those providing the funds for liver transplantation. There will be further development of split liver techniques with the logistics developed for sharing the two halves of the liver in an efficient manner, and trust between different centres receiving half a liver that has been prepared in another centre. There will undoubtedly be more living-related liver transplants but inevitably there will be a mortality rate for the donors which will have to be fully explained and appreciated by all concerned. Meanwhile, the operation has achieved respectability and is now being performed worldwide. Overall results are improving and this will continue. Perhaps one day livers from animals will be the source of organ donation, but the liver is probably one of the most worrying organs to consider from a physiological point of view, even if rejection could be prevented. All the proteins produced by an animal liver would be different in structure from respective human proteins, although we know that some animal proteins, for example insulin from the pig, function perfectly well in man despite being chemically slightly different.

Looking back, it is extraordinary to consider how formidable and dreadful were the early attempts at liver transplantation. It has now become a routine, and almost boring, though still a challenging operation. Fewer people want to be involved in the donation procedure involving night trips to unfamiliar hospitals where one's presence is often resented by the local doctors and nurses.

As Surgery and Medicine become more and more technological, I think there will be strong argument to plan organ donation as a technical exercise to be performed by specially trained surgical technologists for the specific operative procedure involving removal of organs and preservation of organs for transplantation.

For the patients, liver transplantation is a terrifying and dreadful experience which now fortunately is usually successful. The distinguished painter John Bellany depicted his personal image of liver transplantation in 60 large watercolours executed in three weeks whilst a patient at Addenbrooke's 11 years ago. These are extremely powerful images of how the patient feels. He gave me some tutorials in painting and one of them was to paint him. Instead of the heroic St. Sebastian-like figure being tortured by the surgeons and nurse I showed him as an extremely weak patient just about able to sit in a chair. It struck me that liver transplantation had never been a subject for artists because it didn't exist, and the perception of the operation from the point of view of the surgeon might be something worth pursuing. For surgeons in the future performing liver transplants as a routine procedure, it might help them understand how difficult it was in the early days when we were still struggling to get through the operation with a live patient.

References

1. Welch CS (1955). A note on transplantation of the whole liver in dogs. *Transpl Bull* **2**, 54.
2. Cannon JA (1956). Brief report. *Transpl Bull* **3**, 7.
3. Moore FD, Smith LL, Burnap TK, *et al.* (1959). One stage homotransplantation of the liver following total hepatectomy in dogs. *Transpl Bull* **6**, 103–110.
4. Starzl TE, Bernhard VM, Benvenuto R, Cortes N (1959). A new method for one-stage hepatectomy in dogs. *Surgery* **46**, 880–886.
5. Starzl TE, Marchioro TL, Porter KA, *et al.* (1965). Factors determining short- and long-term survival after orthotopic liver homotransplantation in the dog. *Surgery* **58**, 131–155.
6. Starzl TE, Marchioro TL, von Kaulla KN, *et al.* (1963). Homotransplantation of the liver in humans. *Surg. Gyecol Obstet* **117**, 659–676.
7. Starzl TE, Groth CG, Brettschneider L, Penn I, *et al.* (1968). Orthotopic homotransplantation of the human liver. *Ann Surg* **168**, 392–415.
8. Calne RY, Williams R (1968). Liver transplantation in man. I Observations on technique and organisation in five cases. *Br Med J* 4, 535–550.
9. Calne RY, White HJO, Yoffa DE, *et al.* (1967). Prolonged survival of liver transplants in the pig. *Br Med J* **4**, 645–648.
10. Calne RY, *et al.* (1969). Induction of immunological tolerance by porcine liver allografts. *Nature* **223**, 5205.
11. Calne RY, *et al.* (1969). Strange English pigs. *Lancet* **2**, 940–941.
12. Calne RY (January 8, 1994). Organ graft tolerance: the liver effect. *Lancet* **343**, 67–68.
13. Mazaroegos GV, Reyes J, Marino IR, *et al.* (1997). Weaning of immunosuppression in liver transplant recipients. *Transplantation* **63**, 243–249.
14. Wang C, *et al.* (1999). Antigen load and liver allograft survival. *Transpl Proc* **31**, 455.

15. Sriwatanawongsa V, Davies HFF, Calne RY (1995). The essential roles of parenchymal tissues and passenger leukocytes in the tolerance induced by liver grafting in rats. *Nature Med* **1**, 428–432.
16. Pollard S, Davies HFF, Calne RY (1990). Peri-operative appearance of serum class 1 antigen during liver transplantation. *Transplantation* **49**(3), 659–660.
17. Shaw BW Jr, Martin DJ, Marquez JM, *et al.* (1984). Venous bypass in clinical liver transplantation. *Ann Surg* **200**, 542–534.
18. Jamieson NV, Sundberg R, Lindell S, *et al.* (1988). Preservation of the canine liver for 24–48 hours using simple cold storage with UW solution. *Transplantation* **46**, 517–525.
19. Chan WK, Friend PJ, Jamieson NV, Calne NY (1995). Multivisceral grafting for Gardner's syndrome. *J Wound Care* **4**, 214–216.

VI

THE HISTORY OF PANCREAS TRANSPLANTATION

David ER Sutherland
Carl G Groth

Introduction

Clinical pancreas transplantation began in 1966.[1] It has been used primarily to treat diabetes mellitus.

Diabetes mellitus is a heterogenous disease. For the version classified as Type 1, the cause is total or near-total autoimmune destruction of the insulin-secreting β-cells scattered throughout the pancreas within the islets of Langerhans (autoimmune diabetes). Type 2 diabetes (non-autoimmune) is associated with insulin resistance in individuals whose β-cell mass or secretory capability is inadequate to meet the metabolic demand (Type 2 diabetes). Even though β-cells comprise only 1% of the pancreas, nearly all pancreas transplants have been performed simply to restore or augment this cell type in the afflicted individual, thereby eliminating the need for exogenous insulin and establishing a euglycemic state. The vast majority (95%) of pancreas transplants have been in diabetic recipients classified as Type 1,[2] but the procedure is clearly effective in those classified as Type 2.[3]

Pancreas transplantation has occasionally been used to treat exocrine deficiency,[4–6] but the need for acinar cell restoration appears to be less than that for diabetes. Oral pancreatic enzyme preparations are usually effective at aiding digestion and there are no toxic doses.

For diabetes mellitus, however, exogenous insulin treatment is nearly always sub-optimal, and unless normoglycemia is constant (nearly impossible to achieve), secondary complications may occur.[7] Those affecting the eyes, nerves and kidneys can be particularly devastating, leading to blindness, amputations, and need for renal replacement therapy. Attempts to achieve euglycemia with exogenous insulin requires a near dialysis-like life, with multiple glucose determinations during the day in order to adjust the dose of insulin injections or the rate of delivery through a pump. Thus, pancreas or islet transplantation is a logical therapy of diabetes. Early application is mainly inhibited by the need for immunosuppression to prevent rejection or (for Type 1 diabetes) recurrence of autoimmune destruction of β-cells in the newly transplanted pancreas and, at least for a pancreas transplant, the risk of surgical complications. Certainly for the recipient, islet transplantation is much simpler, but the development and integration of reliable techniques for isolation of a sufficient number of viable human islets to consistently induce insulin independence with immunosuppressive regimes made as non-diabetogenic as possible consumed more than three decades. Meanwhile, immediately vascularized pancreas grafts for β-cell replacement needed only to adapt the vascular surgery techniques and immunosuppressive modalities used for other organ transplants.

The idea of pancreas transplantation as a treatment for diabetes has been present ever since the disease was recognized as a deficiency of insulin secretion in the pancreas. In the same decade (1920s) that insulin was isolated and parental administration for treatment of the disease became widespread, vascularization of pancreas allografts was also done in dogs.[8]

In the next few decades, various investigators worked out some of the surgical details of pancreas transplantation in animal models,[8–15] thus setting the stage for clinical application.[16]

Clinical Chronology

The first pancreas allograft in a human was performed by Drs. William D. Kelly and Richard C. Lillihei on December 16, 1966 at the University of Minnesota. They transplanted a duct ligated segmental graft simultaneous with a kidney in a uremic diabetic recipient who immediately became insulin independent (Fig. 1). The graft functioned until the recipient died two months later of sepsis related to surgical complications.[1]

The seminars[17, 18] organized to forward pancreas transplantation in the decade or two that followed the first case focused largely on surgical techniques. Although critical, evolution from the pioneering era where only an occasional pancreas transplant recipient achieved long-term insulin independence[19] to the current highly successful application,[20] depended just as much on the development of new immunosuppressants and the evolution of modern anti-rejection protocols that enveloped the organ transplant field in general.[21–23] Symposiums on pancreas transplantation in the mid-1980s reflected the broad interests, surgical, immunological, metabolic, of a maturing discipline.[24–26]

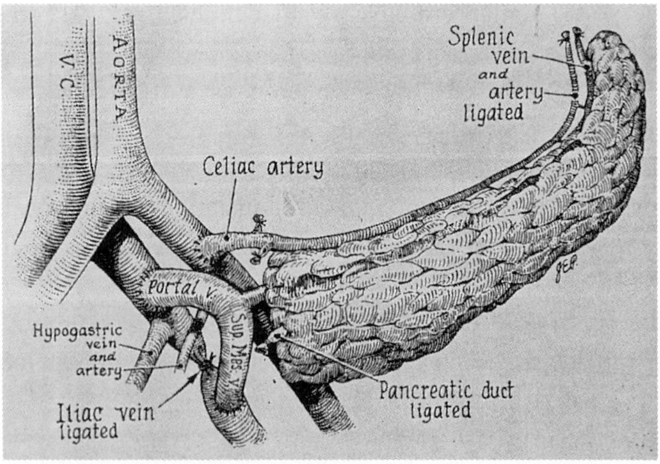

Fig. 1 Technique for segmental transplantation used by Kelly and Lillehei in the first human pancreas transplant.[1]

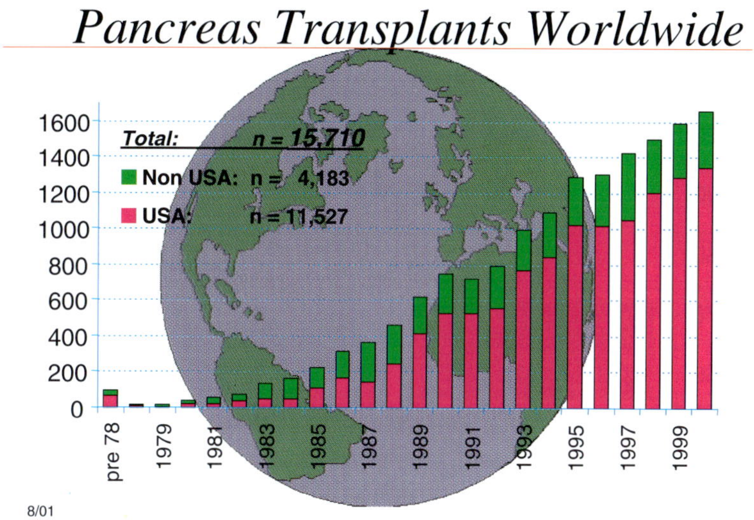

Fig. 2 Number of pancreas transplants tabulated by the International Pancreas Transplant Registry (IPTR) from 1966 to 2000.

The International Pancreas Transplant Registry (IPTR) was formed at the Lyon meeting[17] in 1980.[19] The progressive improvement in pancreas transplant outcomes overtime has been well documented by the IPTR[27] through annual reports.[20] By the year 2000, more than 15,000 pancreas transplants had been reported to the IPTR (80% from the United States), with more than 1000 annually since the mid-1990s (Fig. 2). During the first decade of activity, the one-year graft survival rate (insulin independence) was only 5%, but by 1980 it was up to 20% and by the mid-1980s over 70% for simultaneous pancreas and kidney (SPK) transplants, and approximately 50% for solitary pancreas after kidney (PAK) or pancreas transplants alone (PTA). By the mid-1990s, one-year pancreas graft survival rates were over 80% in the SPK, and over 70% in the PAK and PTA categories (Fig. 3).

Although pancreas transplantation can be done at any stage of diabetes, most centers largely limited its application to diabetic renal allograft recipients who were obligated to immunosuppression in any case. The early experience at Stockholm showed that PTAs in non-uremic

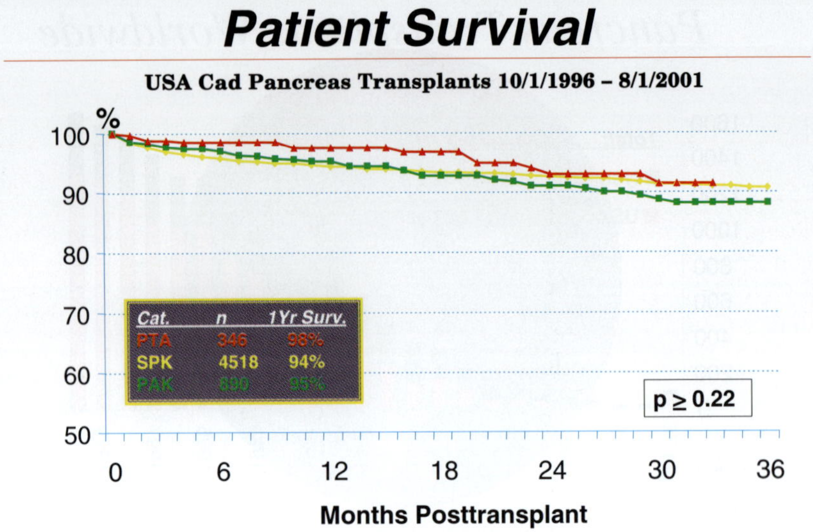

(A)

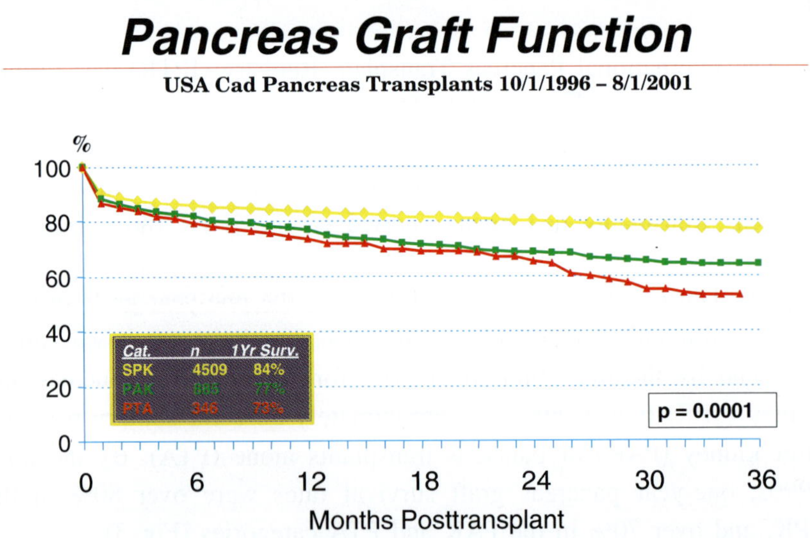

(B)

Fig. 3 Patient (**A**) and pancreas graft (**B**) survival rates from 1996 to 2001 U.S. cadaver SPK, PAK and PTA transplants as reported to UNOS and the IPTR.[2]

diabetic recipients had a substantially higher rejection rate than SPK in uremic recipients.[28] Nevertheless, a few programs persisted in the application of pancreas transplantation in non-uremic diabetic patients, particularly the University of Minnesota,[29] and the differences in outcome between the categories progressively narrowed.[2]

The history of the pancreas transplantation is described by individual contributions to program development; the evolution, physiology, and refinements of new surgical techniques; anti-rejection strategies and methods to diagnose rejection episodes; and studies of metabolism and the effect on diabetic secondary complications.

Surgical Techniques

The pancreas can be transplanted as a whole organ (with or without the duodenum and with or without the need for arterial reconstruction), or as a left (living or cadaver donor) segment (body and tail) on a vascular pedicle of the splenic artery and vein, or as a right segment (head and uncinate process) on a vascular pedicle of the pancreatico-duodenal arteries off the superior mesenteric and hepatic arteries and portal vein; with the duct occluded or exocrine secretions drained to a hollow viscus (gut or urinary system) or other site; and with the graft venous effluent drained into the systemic or portal system of the recipient. Virtually every combination of variations on the theme has been employed, including transplanting the right segment (head and uncinate process) and left segment (tail and body) for two different recipients from the same donor.[30]

Following the first pancreas transplant (segmental) with Kelly,[31] Lillihei and associates[16, 32] went on to perform an additional 13 whole pancreas transplants between 1966 and 1973 (the first five with a cutaneous graft duodenostomy; the next seven with internal exocrine drainage via a Roux-en-Y duodenal jejunostomy (Fig. 4); the last with only the Papilla of Vater retained for anastamosis to the recipient bowel). Of the first 14 cases at the University of Minnesota, the initial 11 were in uremic patients (ten with a simultaneous kidney transplant).

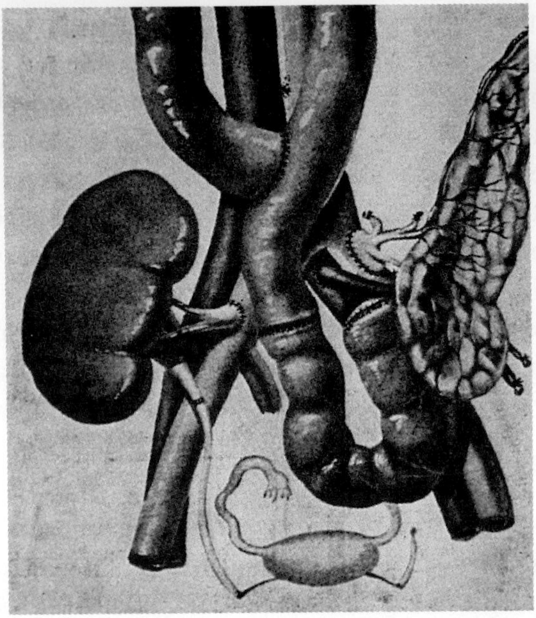

Fig. 4 Technique of whole pancreatico-duodenal transplant with enteric drainage via a Roux en Y limb of recipient bowel, as devised by Lillehei in the initial Minnesota series.[16]

The subsequent three were in non-uremic patients. The surgical complication rate was high, but one recipient in the series remain insulin independent for over a year until dying with a functioning graft after returning to dialysis because of renal allograft thrombosis secondary to arterial stenosis. This was the longest functioning pancreas graft in the world until, a series of ten cases of non-synchronous segmental pancreas and kidney transplants with graft-to-recipient ducto-ureterostomy by Gliedman *et al.* at Montefiore Hospital in New York begun in the early 1970s, produced a recipient whose new pancreas functioned for five years.[33]

A small number of surgeons[34–41] scattered around the world performed a few whole or segmental pancreas transplants in the late 1960s and early 1970s with most failing for technical reasons or rejecting within a year.[42,43] By the mid-1970s, only two centres were systematically offering pancreas transplantation to diabetic patients.[44,45] Because many of the early failures seemed related to complications of the graft duodenum, both groups did segmental grafts (body and tail). In Stockholm, Groth *et al.*[44–47] evolved enteric drainage for duct management, while in Lyon, Dubernard *et al.*[45,48] obliterated the exocrine pancreas by duct injection of a sclerosing synthetic polymer. The latter technique was very safe, although long term-function was still rare.

Meanwhile, a clinical islet transplant program was initialized by the University of Minnesota in 1974.[49] Although clinical islet autografts were successful,[50] none of the initial allograft islet recipients became insulin independent.[51] Ultimately islet allografts were done successfully in a few patients,[52] but the initial experience prompted resumption of pancreas transplants at the University of Minnesota in 1978 with a new surgical team.[53] The new Minnesota series began with open-duct segmental grafts drained into the peritoneal cavity.[54] The very first recipient of an open duct graft (pancreas after kidney, PAK) achieved long-term insulin independence (for 17 years until dying with a functioning graft after being thrown off a horse[55]). However, the open duct technique was technically successful only half the time,[56] and the University of Minnesota program went on to use all of the duct management techniques that were extant or subsequently evolved (duct injection, enteric drainage and urinary drainage) for both segmental and whole organ grafts.[57]

In the early 1980s, several new programs emerged and forwarded the field. Sollinger *et al.*[58] at the University of Wisconsin modified the urinary drainage technique with direct anastamosis of the pancreatic duct of a whole pancreas graft to the bladder. Starzl *et al.*[59] at the University of Pittsburgh resurrected the whole pancreatico-duodenal technique of Lillihei *et al.*[16] for enteric drainage (ED). Shortly thereafter, Ngiem and Corry at the University of Iowa described retention of the duodenum segment for anastamosis to the bladder[60] (Fig. 5). Bladder drainage (BD) via the graft duodenum was quickly adapted by most U.S. centers and was the dominant surgical technique for

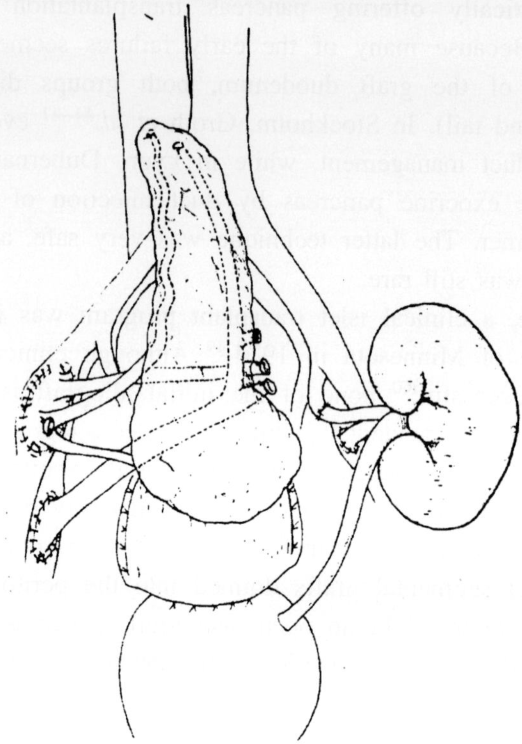

Fig. 5 Technique of whole pancreatico-duodenal transplantation with bladder drainage via a duodenocystostomy as described by Ngheim and Corry.[60] This was a modification of the whole pancreas technique with duct cystostomy described by Sollinger *et al.*,[58] but safer and easier and was adapted by nearly all programmes shortly after it was described.

management of the pancreas graft exocrine secretions for more than a decade.[20] BD was popular because even if it did not have a lower complication rate than enteric drainage, the acute morbidity or severity of the complication was less. In addition, for solitary pancreas transplants, a decrease in urine amylase was shown to be an early marker for rejection that preceded hyperglycemia by several days.[61] Earlier, Gliedman *et al.*[37] had hinted that urine amylase actually correlated with the presence or absence of rejection in urinary drained grafts.

For SPK transplants, BD for urine amylase monitoring was not so important, since the elevation of serum creatinine is a marker for rejection in the kidney that usually preceded exocrine or endocrine dysfunction as a sign of concomitant rejection in the pancreas from the same donor, allowing for early treatment and reversal.[62]

Chronic complications of BD of pancreatico-duodenal transplants (infections, hematuria, acidosis and dehydration) lead to the need for enteric conversion in about 10% of cases.[20] The first conversion was reported by Munda *et al.*[63] from the University of Cincinnati in 1987.

Because of the chronic complications of BD and the lack of need for urinary amylase monitoring in SPK transplants, ED began to be used by more and more centers as the principal drainage technique[64, 65] According to the Registry,[2] in 1993 more than 90% of pancreas transplants in the U.S. were BD; in 1999 more than 70% were ED.

Although segmental pancreas transplantation did not completely disappear, nearly all groups preferred whole pancreatico-duodenal grafts after the mid-1980s. Indeed, the fact that a segmental graft could be done at all led to the first use of living donors (LDs) for pancreas transplantation at the University of Minnesota in 1979.[66] Over the next 20 years, more than 100 LD segmental pancreas transplants were done at the University of Minnesota, nearly more than 10% of the transplants at this institution.[6] Initially, living donor segmental pancreas transplantation was done as a solitary procedure,[67] but in 1994, the University of Minnesota reported the first simultaneous segmental pancreas and kidney transplants with both organs from a living donor.[67] Living donor segmental pancreas transplants have been reported from other institutions,[68] and this option is likely to become more common as waiting times for cadaver pancreas grafts increase even further.

Besides the choice of where to drain (or not drain) the pancreas graft exocrine secretions, there is also a choice as to the site transplantation and of venous drainage (systemic or portal). Most pancreas transplants are placed in the pelvis with vascular anastamosis to the recipient iliac artery and vein. With the expansion of all organ transplants, techniques were developed to preserve the arterial vessel to both the pancreas and liver when removed and transplanted separately. In the 1980s, several groups reported separately,[69–71] the technique of retaining the entire hepatic artery to the liver by using a Y-graft of the donor common, external and internal iliac artery to join the splenic and superior mesenteric arteries feeding a whole pancreas graft.

Drainage of segmental pancreas transplant venous effluent into the recipient's portal system was done at a few centers in the 1980s. The first case was reported by Calne *et al.*[72] from Cambridge. He anastomosed the graft splenic vein to the recipient's splenic vein transabdominally and performed a ducto-gastrostomy for exocrine drainage

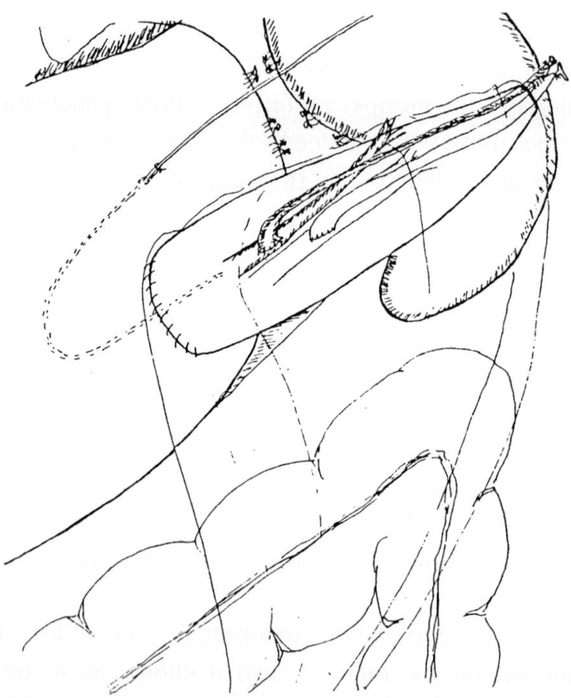

Fig. 6 First report of portal drainage of a pancreas graft by Calne *et al.*,[72] using the recipients splenic vein for sustaining of a segmental graft splenic vein with duct gastrostomy.

(Fig. 6). In Barcelona, Gil-Vernet *et al.*[73] also used the recipient's splenic vein, but via a retroperitoneal approach with a ducto uterectomy for exocrine drainage. In Stockholm, the portal system was accessed via the superior mesenteric vein with enteric drainage.[74, 75] At the University of Minnesota,[76] the inferior mesenteric vein was used and combined with either duct injection or enteric drainage.

The first whole pancreatico-duodenal transplant with portal drainage was reported from Vienna by Muhlbacher *et al.*[77] They anastomosed the distal splenic vein at the tip of the graft tail to the recipient's portal

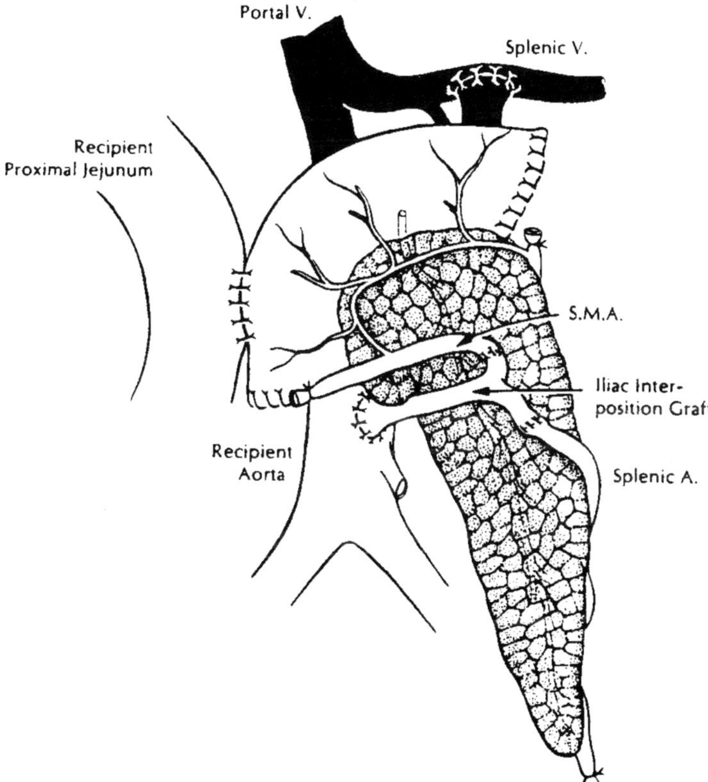

Fig. 7 Technique of whole pancreatico-duodenal enteric drained pancreas transplant and portal drainage as devised by Rosenlof *et al.*[124] at the University of Virginia in 1990.

vein, and performed bladder drainage via a duodenocystostomy. This is the only report of drainage of a pancreas graft venous effluent directly into the recipient's portal vein and not a tributary. In 1992, two groups reported a series of whole pancreatico-duodenal grafts with enteric drainage and portal drainage at the junction of the superior mesenteric and splenic veins.[78, 79] Rosenlof *et al.*[78] at the University of Virginia, used the splenic vein (Fig. 7) and Shokouh-Amri *et al.*[79] at the University of Tennessee at Memphis used the superior mesenteric vein. Except for the venous anastomosis differing by a centimetre or two, the operations described were nearly identical. In the 1990s, several groups adapted this technique as their routine.[64, 80–83] Systemic insulin levels in pancreas transplant recipients are lower with portal venous drainage, and it is, thus, more physiological.[84] A retrospective analysis at one institution suggested that portal drainage might be associated with a lower rejection rate than for systemically drained grafts,[82] but a prospective, randomized study at another institution found no difference.[85]

To summarize the surgical evolution of pancreas transplantation, it went from predominantly whole organ transplantation in the late 1960s and early 1970s to predominately segmental transplantation in the late 1970s and early 1980s and then back to whole organ transplantation. Enteric drainage (ED), duct injection and bladder drainage (BD) were successively dominant from the early 1970s to the mid-1980s in all recipients categories. For solitary pancreas transplants, BD is still the most frequent, but ED has now re-emerged as dominant for SPK transplants.[2] Indeed, the whole pancreatico-duodenal transplant technique with systemic venous and enteric exocrine drainage devised early on by Lillehei (Fig. 2) is nearly identical to that currently used by many surgeons (in 2000, about 40% duplicated Lillehei by using a Roux-en-Y limb of recipient bowel[2]). This shift is explained by the fact that although the immediate technical complications rate is nearly the same for ED and BD drainage, the chronic complications rate is higher for BD. Early diagnosis and treatment of pancreas graft rejection (before hyperglycemia), however, remains a problem for solitary pancreas transplants (PAK and PTA), while for SPK transplants from the same donor, an elevation in serum creatinine is a surrogate marker for rejection likely involving both organs. For solitary pancreas transplants with bladder drainage, rejection will always eventually lead to a decline in urinary amylase (UA). Although not entirely specific, UA is extremely sensitive, while an elevation in serum pancreatic enzymes (also not specific) does not always occur (lower sensitivity).[61] Thus, urinary drainage for pancreas transplantation has persisted more than 30 years after its introduction by Gliedman *et al.*[33] using the ureter and nearly 20 years after direct drainage to the bladder by Sollinger *et al.*[58]

Antirejection Treatment

The immunosuppressive regimens derived for pancreas allograft recipients are similar to those for other solid organ recipients. From the 1960s to the early 1980s, the only universally available maintenance immunosuppressants were Azathioprine (Aza) and steroids. In 1980, the one-year overall pancreas graft survival rate was 20%.[19] Following the introduction of cyclosporine for general use in the mid-1980s, one-year graft survival rates reached approximately 75% for SPK and approximately 50% for PAK and PTA cases.[86] Tacrolimus (Tac) and Mycophenolate Mofetil (MMF) came into general use in the mid-1990s,[22, 23] and current one-year pancreas graft survival rates are ≥ 80% in all categories.[20] It should be noted that the transplant surgeons instrumental in introducing Csa, Tac and MMF, Calne,[21] Starzl,[22] and Sollinger,[23] respectively, were also pioneers and innovators in the surgical techniques for pancreas transplantation.[58, 59, 72]

Calcineurin inhibitors suppress insulin secretion and action, and are diabetogenic in some patients. The diabetogenic effect of cyclosporine was first reported from its use in pancreas transplant recipients in Stockholm in 1983.[87] Tacrolimus, was also shown to have this property shortly after its initial clinical use.[22] Nevertheless, the incidence of drug-induced hyperglycemia has not been any higher in pancreas transplant recipients than other organ allograft recipients, and long-term insulin independence rates, are highest in pancreas recipients given one or the other of these drugs.[20]

The diagnosis and treatment of rejection episodes is based on the same principle as for other organ allografts. Graft dysfunction leads to a tentative diagnosis, a confirmative biopsy is done if possible, and if positive, immunosuppression is temporarily increased. Pancreatic graft biopsies were first done by laparotomy at the University of Minnesota,[88] then transcystoscopic for bladder-drained grafts at the Mayo Clinic,[89] and then percutaneously by Allen *et al.*[90] in Australia and Gaber *et al.*[91] in Memphis.

Impact of Pancreas Transplantation on the Understanding of Pathogenesis of Diabetes Mellitus and Its Associated Complications

The introduction of exogenous insulin therapy for the treatment of diabetes in the 1920s, allowed diabetic patients to survive to develop the complications that affect the eyes, nerves, kidneys and other tissues. Well before the Diabetes Control and Complications Trial (DCCT) results published in 1993[92] showed that the lower the glycosylated hemoglobin levels (a reflection of the average blood sugar levels) were in the study patients, the lower the probability that retinopathy, neuropathy, or nephropathy would develop. However, in the 1980s, well before the DCCT study was published, observations in pancreas transplant recipients had already settled the debate on etiology of the diabetic complications.[93–97]

Studies at the University of Minnesota in the 1970s showed that normal kidneys transplanted into diabetic recipients developed diabetic nephropathy.[98] In 1985, Bohman *et al.*[93] in Stockholm, showed that in recipients of simultaneous kidney and pancreas transplants with long-term function lesions of diabetic nephropathy did not occur in a transplanted kidney. Shortly thereafter, Bilous *et al.*[97] at the University of Minnesota, showed that a successful pancreas transplant (PAK) could prevent, reverse or stabilize lesions in a previously transplanted kidney. Wilczek *et al.* later confirmed these observations in long-term follow-up studies of diabetic renal and pancreas allograft recipients.[99] Finally, Fioretto and Mauer *et al.*,[100] at the University of Minnesota, showed that lesions of diabetic nephropathy could be reversed in the native kidneys of non-uremic recipients of pancreas transplants alone, but it took more than five years.

In regard to neuropathy, Solders *et al.*[94, 101] at Stockholm, in studies of recipients of combined pancreas and kidney transplants (SPK), and Kennedy and associates[95, 102] at the University of Minnesota, in studies of recipients of successful solitary pancreas transplants, showed neuropathy improved as opposed to its inexorable progression in those with no transplant or with a failed graft. The Minnesota group also showed that neuropathic patients had improved survival probability after successful pancreas transplantation.[103]

Finally, Ramsay *et al.*,[96] also at the University of Minnesota, showed that in patients with advanced retinopathy, retinal events occurred in up to 30% during the first three years post-successful pancreas transplant. Thereafter, they remained stable, while those with failed grafts continued to have retinal events. No patients without disease at baseline developed retinopathy as long as the pancreatic graft functioned.

Thus, the scientific principle that the diabetic lesions of the eye, nerve and kidneys are linked to disordered metabolism was fully established from the observations in pancreas transplant recipients well before the DCCT was completed.

The pathogenesis and pathophysiology of Type 1 and Type 2 diabetes mellitus has also been enlightened by observation in pancreas transplant recipients. By early 1980s, consistent evidence implicated autoimmunity as responsible for the destruction of β-cells in patients with Type 1 diabetes.[104] That this must be the case was finally shown by observations in diabetic recipients of segmental pancreas transplants from their non-diabetic identical twin counterparts carried out at the University of Minnesota in the 1980s.[105] The first three recipients in the identical twin series were done without prophylactic immunosuppression (since isografts cannot be rejected) and all three twin recipients developed biopsy confirmed autoimmune isletitis (no rejection) and recurrence of diabetes mellitus within a few months of the transplant.[106] In contrast, immunosuppressed recipients of identical twin segmental grafts did not develop recurrence of disease;[107] indeed, two identical twin recipients have been insulin independent for ten and 14 years, respectively.[55]

Although unusual, islet directed autoimmunity could break through immunosuppression capable of preventing allograft rejection. In 1996, Tyden *et al.* in Stockholm documented selective β-cell destruction in two cadaver pancreas transplants.[108] However, immunosuppression given to prevent allograft rejection is nearly always sufficient to suppress autoimmunity,[107] given the rarity of recurrent disease.[109]

In regards to Type 2 diabetes, a disorder characterized by insulin resistance, pancreas dysfunction may be equally important. Light *et al.*[3] at Washington Transplant Center has shown that pancreas transplant recipients who are insulin dependent but C-peptide positive pre-transplant (a marker for Type 2 diabetes and presence of endogenous β-cells) achieve insulin independence post-transplant with the same frequency as do Type 1 diabetic recipients.

In summary, observations in pancreas transplant recipients have definitely proved that Type 1 diabetes is an autoimmune disease, that the diabetic complications which may occur in other organs are secondary to disorder in metabolism, and that Type 2 diabetes is as much a disorder of the pancreas as of insulin sensitivity.

Evolution of Recipient Selection and Program Development

In addition to the observations on secondary complications, studies in the 1980s documented an improvement in day-to-day quality of life of a diabetic patient by the insulin independence induced by a functioning pancreas graft.[110, 111] Follow-up studies confirmed this benefit,[112, 113] providing an impetus to consider pancreas (or islet) transplantation at any stage of diabetes.

The most physiological treatment of diabetes mellitus is transplantation of β-cells, ideally as a free graft of isolated islets. The optimism that islet transplantation could be rapidly developed to achieve this simplicity probably delayed the development of clinical pancreas transplantation. However, develop it did. In turn, the surprising success of pancreas transplantation (after a pessimistic start) in part has probably inhibited the development of clinical islet transplantation.

Immunological tolerance for clinical application has not yet been developed, so immunosuppression is currently needed for both pancreas and islet transplant recipients. Thus many programs still limit pancreas or islet transplants to uremic diabetic recipients of renal allografts who are thus obligated to immunosuppression. As the 20th century ended, approximately 1500 pancreas transplants were being performed annually worldwide (the majority in the U.S.), of which more than 80% were SPK.[20] During the 1980s, solitary pancreas transplants were routine only at the University of Minnesota.[114] In the 1990s, the number of programs promoting solitary pancreas transplants has expanded[115, 116] and the results in the modern immunosuppressive era have been good.[20] In the year 2000, nearly 20% of pancreas transplants were solitary.[2] Pancreas (or islet) transplantation in non-uremic patients should predominate and the foundation for its application has been laid.

In regard to islet transplantation, since 1989, several groups have succeeded in establishing insulin independence in occasional diabetic recipient by intraportal islet transplantation.[52, 117–119] Consistent islet allograft success has recently been achieved by using multiple donors and a steroid-free immunosuppression regimen at the University of Alberta in Edmonton.[120] Thus, engrafting an adequate β-cell mass appears to be the critical factor for clinical islet transplant to succeed. The challenge now is to achieve consistent success with a single donor. It appears that may be possible by using large donors for small recipients with low insulin requirements, as has been done at the University of Minnesota.[121] The fact that auto islet transplantation after total pancreatectomy can sustain insulin independence,[122] as shown by the very first case in the late 1970s,[50] is further indication that single donor islet allografts will succeed.

For the immediate future, it is likely that islet and pancreas transplantation will co-exist, with islet transplantation reserved for diabetic individuals with a low insulin requirement, while pancreas transplant may be preferable for those with a high insulin requirement or insulin resistance such as those with Type 2 diabetes. Pancreas transplantation will also be preferred in individuals diabetic as a result of total pancreatectomy where enteric drainage could be used to correct exocrine deficiency, as was first done in the 1980s.[123] Indeed, if an unlimited supply of β-cells for transplantation could be obtained (xenografts or human cell lines), the future of pancreas transplantation could primarily be to correct exocrine deficiency, for which the groundwork has already been laid.[5, 123, 124]

References

1. Kelly WD, Lillehei RC, Merkel FK (1967). Allotransplantation of the pancreas and duodenum along with the kidney in diabetic nephropathy. *Surgery* **61**, 827–835.
2. Gruessner AC, Sutherland DER (2001/2002). Report of the international pancreas transplant registry. In Cecka JM, Terasaki PI (eds.) *Clin Transpl*, in press.
3. Light JA, Sasaki TM, Currier CB, Barhyte DY (2001). Successful long-term kidney-pancreas transplants regardless of C-peptide status or race. *Transplantation* **71**, 152–154.
4. Gruessner RWG, Manivel DC, Dunn DL, Sutherland DER (1991). Pancreaticoduodenal transplantation with enteric drainage following native total pancreatectomy for chronic pancreatitis: a case report. *Pancreas* **6**, 479–488.
5. Stern RC, Mayes JT, Weber FL, Jr., Blades EW, Schulak JA (1994). Restoration of exocrine pancreatic function following pancreas-liver-kidney transplantation in a cystic fibrosis patient. *Clin Transpl* **8**, 1–4.

6. Sutherland DE, Gruessner RWG, Dunn DL, Matas AJ, Humar A, Kandaswamy R, Mauer SM, Kennedy WR, Goetz FC, Robertson RP, Gruessner AC, Najarian JS (2001). Lessons learned from more than 1000 pancreas transplants at a single institution. *Ann Surg* **233**, 463–501.
7. Lum CT, Sutherland DER, Yasmineh WG, Fryd DS, Howard RJ, Najarian JS (1979). Adenosine deaminase activity in cytomegalovirus-related graft and patient loss. *Transpl Proc* **2**, 83–88.
8. Gayet RC, Guillamie M (1927). La regulation de la secretion interne pancreatique par un processus humoral, demonstree par des transplantations de pancreas. Experiences sue des animaux normaux. *C R Soc Biol* **97**, 1613–1614.
9. Brooks JR, Gifford GH (1959). Pancreatic homotransplantation. *Transpl Bull* **6**, 100–103.
10. DeJode LR, Howard JM (1962). Studie in pancreaticoduodenal homo-transplantation. *Surg Gynecol Obstet* **114**, 553–558.
11. Bergan J, Hoehn JG, Porter N (1965). Total pancreatic allografts in pancreatectomized dogs. *Arch Surg* **90**, 521–526.
12. Largiader F, Lyons GW, Hidalgo F, Dietzman RH, Lillehei RC (1967). Orthotopic allotransplantation of the pancreas. *Am J Surg* **113**, 70–78.
13. Merkel FK, Kelly WD, Goetz FC, Maney J (1968). Irradiated hetertopic segmental canine pancreatic allografts. *Surgery* **63**, 291–297.
14. Idezuki Y, Feemster JA, Dietzman RH, Lillehei RC (1968). Experimental pancreaticoduodenal preservation and transplantation. *Surg Gynecol Obstet* **126**, 1002–1014.
15. Sutherland DER (1981). Pancreas and islet transplantation. I. Experimental studies. *Diabetologia* **20**, 161–185.
16. Lillehei RC, Simmons RL, Najarian JS, Weil R, Uchida H, Ruiz JO, Kjellstrand CM, Goetz FC (1970). Pancreatico-duodenal allotransplantation: experimental and clinical experience. *Ann Surg* **172**, 405–436.

17. Dubernard JM, Traeger J (1980). Pancreas and islet transpantation workshop. *Transpl Proc* **12**, 1–2.
18. Land W, Landgraf R (1983). Segmental pancreatic transplantation international workshop. *Horm Metab Res* **13**, 1–104.
19. Sutherland DER (1980). International human pancreas and islet transplant registry. *Transpl Proc* **12**(4) (Suppl. 2), 229–236.
20. Gruessner A, Sutherland DER (2000). Analyses of pancreas transplant outcomes for United States cases reported to the United Network for Organ Sharing (UNOS) and non-US cases reported to the International Pancreas Transplant Registry. In Cecka JM, Terasaki PI (eds.). *Clinical Transplant — 1999*. Los Angeles: UCLA Immunogenetics Center, pp. 51–69.
21. Calne RY, Rolles K, White DJ, Thiru S, Evans DB, McMaster P, Dunn DC, Craddock GN, Henderson RG, Aziz S, Lewis P (1979). Cyclosporin A initially as the only immunosuppressant in 34 recipients of cadaveric organs: 32 kidneys, 2 pancreases, and 2 livers. *Lancet* **2**, 1033–1036.
22. Starzl TE, Todo S, Fung J, Demetris AJ, Venkataramman R, Jain A (1989). FK 506 for liver, kidney, and pancreas transplantation. *Lancet* **2**, 1000–1004.
23. Rayhill SC, Kirk AD, Odorico JS, Heisey DM, Cangro CB, Pirsch JD, D'Alessandro AM, Knechtle SJ, Sollinger HW (1995). Simultaneous pancreas-kidney transplantation at the University of Wisconsin. *Clin Transpl* 261–269.
24. van Schilfgaarde R, Persijn GG, Sutherland DER (1984). International symposium on organ transplantation in diabetics. *Transpl Proc* **16**, 565–566.
25. Land W, Landgraf R (1987). The world experience in clinical pancreas transplantation. *Transpl Proc* **19**(Suppl. 4), 1–2.
26. Groth CG, Tyden G, Bolinder J (1989). First international congress on pancreatic and islet transplantation. *Diabetes* **38**(Suppl. 1), 1–335.
27. Sutherland DER, Moudry KC. Pancreas Transplant Registry: History and Analysis of Cases 1966 to October 1986. *Pancreas* **2**, 473–488.

28. Groth CG, Lundgren G, Ostman J, Gunnarsson R (1980). Experience with nine segmental pancreatic transplantations in preuremic diabetic patients in Stockholm. *Transpl Proc* **12**, 68–71.
29. Sutherland DER (1986). Pancreas transplantation in non-uremic diabetic patients. *Transpl Proc* **18**, 1747–1749.
30. Sutherland DER, Morel P, Gruessner RWG (1990). Transplantation of two diabetic patients with one divided cadaver donor pancreas. *Transpl Proc* **22**(2), 585.
31. Kelly WD, Lillehei RC, Merkel FK (1967). Allotransplantation of the pancreas and duodenum along with the kidney in diabetic nephropathy. *Surgery* **61**, 827–835.
32. Lillehei RC, Ruiz JO, Aquino C, Goetz FC (1976). Transplantation of the pancreas. *Acta Endocrinol* **83**(Suppl. 205), 303–320.
33. Gliedman ML, Tellis VA, Soberman R (1978). Long-term effects of pancreatic transplant function in patients with advanced juvenile onset diabetes. *Diabetes Care* **1**, 1–9.
34. Bortagaray MC, Zelasco JF, Bava A (1970). Homotransplante parcial del pancreas en el diabetico. *Prensa Med Argent* **57**, 220–224.
35. Teixeira E, Monteiro G, De Cenzo M, Teixeira A, Bergan JJ (1970). Transplantation of the isolated pancreas: report on the first human case. *Bull Soc Int Chir* **29**, 337–344.
36. Merkel FK, Poticha SM, Nudelman EJ, Colwell JA, Bergan JJ (1971). Pancreas transplantation in man. I. The donor and recipient operation. *Arch Surg* **103**, 205–210.
37. Gliedman ML, Gold M, Whittaker J, Rifkin H, Soberman R, Freed S, Tellis V, Veith FJ (1973). Clinical segmental pancreatic transplantation with ureter-pancreatic duct anastomosis for exocrine drainage. *Surgery* **74**, 171–180.
38. Connolly JE, Martin DC, Steinberg T, Gwinup G, Gazzaniga AB, Bartlett RH (1973). Clinical experience with pancreaticoduodenal transplantation. *Arch Surg* **106**, 489–494.
39. Merkel FK, Ryan WG, Armbruster K, Seim S, Ing TS (1973). Pancreatic transplantation for diabetes mellitus. *IMJ Ill Med J* **144**, 477–479.

40. Largiader F, Uhlschmid G, Binswanger U, Zaruba K (1975). Pancreas rejection in combined pancreaticoduodenal and renal allotransplantation in man. *Transplantation* **19**, 185–187.
41. Brynger H, Blohme G, Claes G, Gustafsson A, Gelin LE (1975). Transplantation of a duct-ligated pancreatic allograft to a diabetic patient. *Scand J Urol Nephrol* **29**(Suppl.), 59–62.
42. Sutherland DER (1981). Pancreas and islet transplantation. II. Clinical trials. *Diabetologia* **20**, 435–450.
43. Sutherland DER, Rynasiewicz JJ, Najarian JS (1981). Current status of pancreas and islet transplantation. In Brownlee M (ed.) *Handbook of Diabetes Mellitus*. New York: Garland Publishing, Inc., pp. 273–416.
44. Groth CG, Lundgren G, Arner P, Collste H, Hardstedt C, Lewander R, Ostman J (1976). Rejection of isolated pancreatic allografts in patients with with diabetes. *Surg Gynecol Obstet* **143**, 933–940.
45. Dubernard JM, Traeger J, Neyra P, Touraine JL, Traudiant D, Blanc-Brunat N (1978). A new method of preparation of segmental pancreatic grafts for transplantation: trials in dogs and in man. *Surgery* **84**, 633–640.
46. Groth CG, Lundgren G, Gunnarsson R, Hardstedt C, Ostman J (1980). Segmental pancreatic transplantation with special reference to the use of ileal exocrine diversion and to the hemodynamics of the graft. *Transpl Proc* **12**(Suppl. 2), 62–67.
47. Groth CG, Collste H, Lundgren G (1982). Successful outcome of segmental human pancreatic transplantation with enteric exocrine diversion after modifications in technique. *Lancet* **2**, 522–524.
48. Traeger J, Dubernard JM, Touraine JL, Neyra P, Malik MC, Pelissard C, Ruitton A (1979). Pancreatic transplantation in man: a new method of pancreas preparation and results on diabetes correction. *Transpl Proc* **11**, 331–335.
49. Najarian JS, Sutherland DER, Matas AJ, Steffes MW, Simmons RL, Goetz FC (1977). Human islet transplantation: a preliminary report. *Transpl Proc* **9**, 233–236.

50. Sutherland DER, Matas AJ, Najarian JS (1978). Pancreatic islet cell transplantation. *Surg Clin No Am* **58**, 365–382.
51. Sutherland DER, Matas AJ, Goetz FC, Najarian JS (1980). Transplantation of dispersed pancreatic islet tissue in humans: autografts and allografts. *Diabetes* **29**, 3–44.
52. Gores PF, Najarian JS, Stephanian E, Lloveras JJ, Kelley SL, Sutherland DER (1993). Insulin independence in type I diabetes after transplantation of unpurified islets from a single donor using 15-deoxyspergualin. *Lancet* **341**, 19–21.
53. Sutherland DER, Goetz FC, Najarian JS (1979). Intraperitoneal transplantation of immediately vascularized segmental pancreatic grafts without duct ligation: a clinical trial. *Transplantation* **28**, 485–491.
54. Sutherland DER, Baumgartner D, Najarian JS (1980). Free intraperitoneal drainage of segmental pancreas grafts: clinical and experimental observations on technical aspects. *Transpl Proc* **12**, 26–32.
55. Sutherland DE, Gruessner RWG, Dunn DC, Matas AJ, Humar A, Kandaswamy R, Mauer SM, Kennedy WR, Goetz FC, Robertson RP, Gruessner AC, Najarian JS (2000). More than 1000 pancreas transplants at a single institution. *Ann. Surg*. Ref Type, in press.
56. Sutherland DER, Goetz FC, Rynasiewicz JJ, Baumgartner D, White DC, Elick BA, Najarian JS (1981). Segmental pancreas transplantation from living related and cadaver donors: a clinical experience. *Surgery* **90**, 159–169.
57. Sutherland DE, Gores PF, Farney AC, Wahoff DC, Matas AJ, Dunn DL, Gruessner RW, Najarian JS (1993). Evolution of kidney, pancreas, and islet transplantation for patients with diabetes at the University of Minnesota. *Am J Surg* **166**, 456–491.
58. Sollinger HW, Cook K, Kamps D (1984). Clinical and experimental experience with pancreaticocystostomy for exocrine pancreatic drainage in pancreas transplantation. *Transpl Proc* **16**, 749–751.

59. Starzl TE, Iwatsuki S, Shaw BW, Jr., Greene DA, Van Thiel DH, Nalesnik MA, Nusbacher J, Diliz-Pere H, Hakala TR (1984). Pancreaticoduodenal transplantation in humans. *Surg Gynecol Obstet* **159**, 265–272.
60. Nghiem DD, Corry RJ (1987). Technique of simultaneous pancreaticoduodenal transplantation with urinary drainage of pancreatic secretion. *Am J Surg* **153**, 405–406.
61. Prieto M, Sutherland DER, Fernandez-Cruz L, Heil JE, Najarian JS (1987). Experimental and clinical experience with urine amylase monitoring for early diagnosis of rejection in pancreas transplantation. *Transplantation* **43**, 71–79.
62. Gruessner RWG, Dunn DL, Tzardis PJ, Tomadze G, Moudry-Munns KC, Matas AJ, Najarian JS, Sutherland DER (1990). Simultaneous pancreas and kidney transplants versus single kidney transplants and previous kidney transplants in uremic patients and single pancreas transplants in nonuremic diabetic patients: comparison of rejection, morbidity, and long-term outcome. *Transpl Proc* **22**(2), 622–623.
63. Tom WW, Munda R, First MR, Alexander JW (1987). Autodigestion of the glans penis and urethra by activated transplant pancreatic exocrine enzymes. *Surgery* **102**, 99–101.
64. Gaber AO, Shokouh-Amiri MH, Hathaway DK, Hammontree L, Kitabchi AE, Gaber LW, Saad MF, Britt LG (1995). Results of pancreas transplantation with portal venous and enteric drainage. *Ann Surg* **221**, 613–622.
65. Bartlett ST, Schweitzer EJ, Johnson LB, *et al.* (1996). Equivalent success of simultaneous pancreas kidney and solitary pancreas transplantation. P prospective trial of tacrolimus immunosuppression with percutaneous biopsy. *Ann Surg* **224**, 440–449.
66. Sutherland DER, Goetz FC, Najarian JS (1980). Living-related donor segmental pancreatectomy for transplantation. *Transpl Proc* **12**, 19–25.
67. Gruessner RWG, Sutherland DER (1996). Simultaneous kidney and segmental pancreas transplants from living related donors—the first two successful cases. *Transplantation* **61**, 1265–1268.

68. Benedetti E, Dunn T, Massad MG, Raofi V, Bartholomew A, Gruessner RW, Brecklin C (1999). Successful living related simultaneous pancreas-kidney transplant between identical twins. *Transplantation* **67**, 915–918.
69. Sutherland DER, Moudry KC, Najarian JS (1987). Pancreas transplantation. In Cerilli J (ed.) *Organ Transplantation and Replacement*. Philadelphia: J.B. Lippincott Company, pp. 535–574.
70. Marsh CL, Perkins JD, Sutherland DER, Corry RJ, Sterioff S (1989). Combined hepatic and pancreaticoduodenal procurement for transplantation. *Surg Gynecol Obstet* **168**, 254–258.
71. Delmonico FL, Jenkins RL, Auchincloss H, Jr., Etienne TJ, Russell PS, Monaco AB, Cosimi AB (1989). Procurement of a whole pancreas and liver from the same cadaveric donor [see comments]. *Surgery* **105**, 718–723.
72. Calne RY (1984). Paratopic segmental pancreas grafting: a technique with portal venous drainage. *Lancet* **1**, 595–597.
73. Gil-Vernet JM, Fernandez-Cruz L, Caralps A, Andreu J, Figuerola D (1985). Whole organ and pancreaticoureterostomy in clinical pancreas transplantation. *Transpl Proc* **17**, 2019–2022.
74. Tyden G, Lundgren G, Ostman J, Gunnarsson R, Groth CG (1984). Grafted pancreas with portal venous drainage. *Lancet* **1**, 964.
75. Tydén G, Wilczek H, Lundgren G, Ostman J, Gunnarsson R, Jaremko G, Groth C (1985). Experience with 21 intraperitoneal segmental pancreatic transplantations with enteric or gastric exocrine diversion in humans. *Transpl Proc* **17**, 331–335.
76. Sutherland DER, Goetz FC, Moudry KC, Abouna GM, Najarian JS (1987). Use of recipient mesenteric vessels for revascularization of segmental pancreas grafts: technical and metabolic considerations. *Transpl Proc* **19**, 2300–2304.
77. Muhlbacher F, Gnant MF, Auinger M, Steininger R, Klauser R, Prager R, Karnel F (1990). Pancreatic venous drainage to the portal vein: a new method in human pancreas transplantation. *Transpl Proc* **22**, 636–637.

78. Rosenlof LK, Earnhardt RC, Pruett TL, Stevenson WC, Douglas MT, Cornett GC, Hanks JB (1992). Pancreas transplantation. An initial experience with systemic and portal drainage of pancreatic allografts. *Ann Surg* **215**, 586–595.
79. Shokouh-Amiri MH, Gaber AO, Gaber LW, Jensen SL, Hughes TA, Elmer D, Britt LG (1992). Pancreas transplantation with portal venous drainage and enteric exocrine diversion: a new technique. *Transpl Proc* **24**, 776–777.
80. Reddy KS, Stratta RJ, Shokouh-Amiri MH, Alloway R, Egidi MF, Gaber AO (1999). Surgical complications after pancreas transplantation with portal-enteric drainage. *J Am Coll Surg* **189**, 305–313.
81. Bruce DS, Newell KA, Woodle ES, Cronin DC, Grewal HP, Millis JM, Ruebe M, Josephson MA, Thistlethwaite JR, Jr (1998). Synchronous pancreas-kidney transplantation with portal venous and enteric exocrine drainage: outcome in 70 consecutive cases. *Transpl Proc* **30**, 270–271.
82. Philosophe B, Farney AC, Schweitzer E, Colonna J, Jarrell BE, Krishnamurthi V, Wiland A, Bartlett S (2000). The superiority of portal venous drainage over systemic venous drainage in pancreas transplantation. *Ann Surg*. Ref. Type, in press.
83. Cattral MS, Bigam DL, Hemming AW, Carpentier A, Greig PD, Wright E, Cole E, Donat D, Lewis GF (2000). Portal venous and enteric exocrine drainage versus systemic venous and bladder exocrine drainage of pancreas grafts: clinical outcome of 40 consecutive transplant recipients. *Ann Surg* **232**, 688–695.
84. Diem P, Abid M, Redmon JB, Sutherland DE, Robertson RP (1990). Systemic venous drainage of pancreas allografts as independent cause of hyperinsulinemia in Type 1 diabetic recipients. *Diabetes* **39**, 534–540.
85. Petruzzo P, DaSilva M, Feitosa LC (2000). Simultaneous pancreas-kidney transplantation: portal versus systemic venous drainage of the pancreas allografts. *Clin Transpl* **14**, 287–291.

86. Sutherland DER, Moudry-Munns KC (1989). International Pancreas Transplant Registry Report. In Terasaki PI (ed.) Clinical Transplants 1988. California: UCLA Tissue Typing Laboratory, pp. 53–64.
87. Gunnarsson R, Klintmalm G, Lundgren G, Wilczek H, Ostman J, Groth CG (1983). Deterioration in glucose metabolism in pancreatic transplant recipients given cyclosporin [letter]. *Lancet* **2**, 571–572.
88. Sutherland DER, Casanova D, Sibley RK (1987). Role of pancreas graft biopsies in the diagnosis and treatment of rejection after pancreas transplantation. *Transpl Proc* **19**, 2329–2331.
89. Perkins JD, Munn SR, Marsh CL, Barr D, Engen DE, Carpenter HA (1990). Safety and efficacy of cystoscopically directed biopsy in pancreas transplantation. *Transpl Proc* **22**, 665–666.
90. Allen RD, Wilson TG, Grierson JM, Greenberg ML, Earl MJ, Nankivel BJ, Pearl TAC (1991). Percutaneous biopsy of bladder-drained pancreas transplants. *Transplantation* **51**, 1213–1216.
91. Gaber AO, Gaber L, Shokouh-Amiri MH, Hathaway D (1992). Percutaneous biopsy of pancreas transplants. *Transplantation* **54**, 548–550.
92. DCCT Research Group (1993). Diabetes control and complications trial (DCCT): the effect of intensive diabetes treatment in long term complications in IDDM. *New Engl J Med* **329**, 977–986.
93. Bohman SO, Tyden G, Wilczek H, Lundgren G, Jaremko G, Gunnarsson R, Ostman J, Groth CG (1985). Prevention of kidney graft diabetic nephropathy by pancreas transplantation in man. *Diabetes* **34**, 306–308.
94. Solders G, Wilczek H, Gunnarsson R, Tyden G, Persson A, Groth CG (1987). Effects of combined pancreatic and renal transplantation on diabetic neuropathy: a two-year follow-up study. *Lancet* **2**, 1232–1235.
95. van der Vliet JA, Navarro X, Kennedy WR, Goetz FC, Najarian JS, Sutherland DER (1988). The effect of pancreas transplantation on diabetic polyneuropathy. *Transplantation* **45**, 368–370.

96. Ramsay RC, Goetz FC, Sutherland DER, Mauer SM, Robinson LL, Cantrill HL, Knobloch WH, Najarian JS (1988). Progression of diabetic retinopathy after pancreas transplantation for insulin-dependent diabetes mellitus. *New Engl J Med* **318**, 208–214.
97. Bilous RW, Mauer SM, Sutherland DER, Najarian JS, Goetz FC, Steffes MW (1989). The effects of pancreas transplantation on the glomerular structure of renal allografts in patients with insulin-dependent diabetes. *New Engl J Med* **321**, 80–85.
98. Mauer SM, Barbosa J, Vernier RL, Kjellstrand CM, Buselmeier TJ, Simmons RL, Najarian JS, Goetz FC (1976). Development of diabetic vascular lesions in normal kidneys transplanted into patients with diabetes mellitus. *N Engl J Med* **295**, 916–920.
99. Wilczek H, Jaremko G, Tydén G, Groth CG (1993). A pancreatic graft protects a simultaneously transplanted kidney from developing diabetic nephropathy: a 1–6 year follow-up study. *Transpl Proc* **25**, 1314–1315.
100. Fioretto P, Steffes MW, Sutherland DER, Goetz FC, Mauer SM (1998). Reversal of lesions of diabetic nephropathy after pancreas transplantation. *New Engl J Med* **339**, 69–75.
101. Solders G, Tyden G, Persson A, Groth CG (1992). Improvement of nerve conduction in diabetic neuropathy: a follow-up study 4 years after combined pancreatic and renal transplantation. *Diabetes* **41**, 946.
102. Kennedy WR, Navarro X, Goetz FC, Sutherland DER, Najarian JS (1990). Effects of pancreatic transplantation on diabetic neuropathy. *New Engl J Med* **322**, 1031–1037.
103. Navarro X, Kennedy WR, Loewensen RB, Sutherland DER (1990). Influence of pancreas transplantation on cardiorespiratory reflexes, nerve conduction, and mortality in diabetes mellitus. *Diabetes* **39**, 802–806.
104. Eisenbarth GS (1986). Type I diabetes mellitus: a chronic autoimmune disease. *New Engl J Med* **314**, 1360–1368.
105. Sutherland DER, Sibley RK, Xu XZ, Michael AF, Srikanta S, Taub F, Najarian JS, Goetz FC (1984). Twin-to-twin pancreas

transplantation: reversal and re-enactment of the pathogenesis of Type I diabetes. *Trans Assoc Am Physicians* **97**, 80–87.
106. Sibley RK, Sutherland DER, Goetz FC, Michael AF (1985). Recurrent diabetes mellitus in the pancreas iso- and sllograft: a light and electron microscopic and immunohistochemical analysis of four cases. *Lab Invest* **53**, 132–144.
107. Sutherland DER, Goetz FC, Sibley RK (1989). Recurrence of disease in pancreas transplants. *Diabetes* **38**, 85–87.
108. Tyden G, Reinholt FP, Sundkvist G, Bolinder J (1996). Recurrence of autoimmune diabetes mellitus in recipients of cadaveric pancreatic grafts [see comments] [published erratum appears in *N Engl J Med* (December 5) **335**(23), 1778]. *N Engl J Med* **335**, 860–863.
109. Papadimitriou JC, Drachenberg CB, Wiland A, Klassen DK, Fink J, Weir MR, Cangro C, Schweitzer EJ, Bartlett ST (1998). Histologic grading of acute allograft rejection in pancreas needle biopsy: correlation to serum enzymes, glycemia, and response to immunosuppressive treatment. *Transplantation* **66**, 1741–1745.
110. Nakache R, Tyden G, Groth CG (1989). Quality of life in diabetic patients after combined pancreas-kidney or kidney transplantation. *Diabetes* **38**(Suppl. 1), 40–42.
111. Zehrer CL, Gross CR (1991). Quality of life of pancreas transplant recipients. *Diabetologia* **34** (Suppl. 1), S145–S149.
112. Nakache R, Tyden G, Groth CG (1994). Long-term quality of life in diabetic patients after combined pancreas-kidney transplantation or kidney transplantation. *Transpl Proc* **26**, 510–511.
113. Gross CR, Limwattananon C, Matthees BJ (1998). Quality of life after pancreas transplantation: a review. *Clin Transplant* **12**, 351–361.
114. Sutherland DER, Kendall DM, Moudry KC, Navarro X, Kennedy WR, Ramsay RC, Steffes MW, Mauer SM, Goetz FC, Dunn DL, Najarian JS (1988). Pancreas transplantation in nonuremic, type 1 diabetic recipients. *Surgery* **104**, 453–464.
115. Bartlett ST, Schweitzer EJ, Johnson LB, Kuo PC, Papadimitriou JC, Drachenberg CB, Klassen DK, Hoehn-Saric EW, Weir MR,

Imbembo AL (1996). Equivalent success of simultaneous pancreas kidney and solitary pancreas transplantation. A prospective trial of tacrolimus immunosuppression with percutaneous biopsy. *Ann Surg* **224**, 440–449.

116. Stratta RJ, Weide LG, Sindhi R, Sudan D, Jerius JT, Larsen JL, Cushing K, Grune MT, Radio SJ (1997). Solitary pancreas transplantation. Experience with 62 consecutive cases. *Diabetes Care* **20**, 362–368.
117. Scharp DW, Lacy PE, Santiago JV, McCullough CS, Weide LG, Falqui L, Marchetti P, Gingerich RL, Jaffe AS, Cryer PE (1990). Insulin independence after islet transplantation into Type 1 diabetic patient. *Diabetes* **39**, 515–518.
118. Warnock GL, Kneteman NM, Ryan E (1992). Long-term follow-up after transplantation of insulin-producing pancreat islets into patients with Type 1 (insulin independent) diabetes mellitus. *Diabetologia* **35**, 89–95.
119. Meyer C, Hering BJ, Grossmann R, Brandhorst H, Brandhorst D, Gerich J, Federlin K, Bretzel RG (1998). Improved glucose counterregulation and autonomic symptoms after intraportal islet transplants alone in patients with long-standing Type 1 diabetes mellitus. *Transplantation* **66**, 233–240.
120. Shapiro AMJ, Lakey JRT, Ryan E, Korbutt GS, Toth E, Warnock GL, Kneteman NM, Rajotte RV (2000). Insulin independence after solitary islet transplantation in Type 1 diabetic patients using steroid free immunosuppression. *N Engl J Med* **343**, 230–238.
121. Hering BJ, Kandaswamy R, Harmon JV, Ansite J, Clemmings S, Sakai T, Zabloski J (2001). Insulin independence after single-donor islet transplantation in Type 1 diabetes with hOKT 3-1 (ala-ala), sirolimus, and tacrolimus therapy. *Am J Transpl* **1**, 180.
122. Wahoff DC, Papalois BE, Najarian JS, Kendall DM, Farney AC, Leone JP, Jesserun J, Dunn DL, Robertson RP, Sutherland DER (1995). Autologous islet transplantation to prevent diabetes after pancreatic resection. *Ann Surg* **22**(4), 562–579.

123. Gruessner RWG, Manivel DC, Dunn DL, Sutherland DER (1991). Pancreaticoduodenal transplantation with enteric drainage following native total pancreatectomy for chronic pancreatitis: a case report. *Pancreas* **6**, 479–488.
124. Rosenlof LK, Earnhardt RC, Pruett TL, Stevenson WC, Douglas MT, Cornett GC, Hanks JB (1992). Pancreas transplantation: an initial experience with systemic and portal drainage of pancreatic allografts. *Ann Surg* **215**, 586–595.

VII

A SHORT HISTORY OF THE DEVELOPMENT OF ISLET TRANSPLANTATION

Derek WR Gray

Early Contributory Studies

The early history of transplantation is littered with what now appears as over-optimistic or even laughable attempts at transplantation with insufficient knowledge. However, we must always try to understand the leap of imagination required to even attempt such endeavours. In the case of pancreatic islet transplantation, the first recorded attempt to transplant pancreatic tissue as a free graft was undertaken by Williams, a general practitioner from Bristol, England who in 1892 transplanted two "Brazil nut" sized chunks of sheep pancreas to the abdominal wall of a patient dying of diabetes.[1] This attempt was inspired by the discoveries of von Mering and Minkowski, that extirpation of the pancreas led to diabetes mellitus.[2, 3] Although the attempt looks hopelessly optimistic from today's perspective, indeed insulin had yet to be discovered, it is clear that Williams was the first to suggest that the capacity of pancreatic tissue to reverse diabetes might withstand transplantation as a free graft. No further attempts at transplanting insulin secreting tissue as a free graft were reported until some years after the discovery of insulin, although sporadic attempts at vascularised pancreatic transplantation were published. However, a number of important advances were made which improved techniques or provided data useful for subsequent islet studies. These include the studies of Bensley[4] who described useful ways of identifying islet tissue with various dyes, Clarke[5] who attempted to quantify the volume of human islets for the first time, a task subsequently done most elegantly by Hellman.[6] Dunn[7] first produced experimental diabetes with alloxan, subsequently followed by the excellently documented studies of other diabetogenic agents such as streptozotocin,[8] and Wolffe who described the affinity of dithizone for islet tissue.[9]

The first interest in islet isolation techniques was not for the purpose of transplantation but was driven by a desire to investigate the physiology of islets separated from the exocrine pancreatic tissue. The pioneer in the field was Hellerstrom from Uppsaala, Sweden[10] who developed a micro-dissection technique which was possible in species such as the ob/ob mouse, where the islets were large, numerous and the exocrine tissue relatively transparent. The micro-dissection technique was tedious and provided islets numbered in tens, but the data produced was novel and started an approach to investigating islet physiology and function that continues to this day. It was the desire to increase the number of islets available for such investigations that first motivated Moskalewski[11] to investigate the use of bacterial collagenase to disintegrate the minced pancreas of the guinea pig.

The First Successful Islet Isolation and Transplantation

The transplantation of vascularised organs such as the kidney was gaining much attention by the late 1950s with the phenomenon of allograft rejection documented and under intense investigation. The possibility of transplanting endocrine tissue (parathyroid) as a free graft was first investigated as a scientific endeavour by Russell, who raised the intriguing possibility that such tissue might show reduced immunogenicity.[12] Subsequently, Coupland showed that foetal pancreatic tissue would survive transplantation to the anterior chamber of the eye and underwent fortuitous atrophy of the exocrine tissue when transplanted to this site.[13] The improved yields of islets obtained using the collagenase digestion technique allowed Moskalewski (working with Hellman in Uppsala, Sweden) to report the first study of survival of transplanted isolated islets.[14] However, the real credit for realising the full potential of isolated islet transplantation must go to Lacey in St. Louis[15] who took the collagenase digestion technique reported by Moskalewski and combined it with a pre-digestion process of intraductal distension with mechanical mincing of the pancreatic tissue in order to improve the efficiency of the collagenase. Using this technique, Lacey and co-workers[16] were able to obtain sufficient islets to demonstrate an effect on plasma glucose following isologous islet transplantation to the peritoneal cavity of streptozotocin diabetic rats. By today's standards, these first islet transplants showed minimal function, but nevertheless the results were greeted with much excitement.

Development of Islet Transplantation in Murine Models

Important contributions continued from Lacy's Group in St. Louis in association with Scharp, first developing the use of a density gradient to separate islet from endocrine tissue[17] and demonstrating the metabolic advantage of intraportal islet transplantation.[18] A number of centres worldwide began their interest in islet transplantation at this time, rapidly performing a series of islet transplantation experiments in rodents which formed the basis for our knowledge of the pathophysiology of islet transplantation today. These included the groups of Barker in New York who undertook metabolic studies[19] and the first islet xenograft studies,[20] the Minneapolis group, under the direction of Najarian and Sutherland who first described the regression of diabetic nephropathy following islet transplantation,[21] Federlin's group in Giessen, Germany who documented further the metabolic responsiveness of transplanted isologous islets,[22] and later allogeneic islets.[23, 24] The peculiar vulnerability of islets to allograft rejection were also documented by several groups; notably Lacy and colleagues in St. Louis,[25] Sutherland's group in Minneapolis,[26] Morris's group in

Oxford,[27] and the vulnerability of islets to the humoral rejection process was first documented by Naji and colleagues in Philadelphia.[28] However, it was Kevin Lafferty's original suggestion that passenger leukocytes were the main drive to the alloresponse,[29, 30] that started a drive to identify the cell involved, which culminated in the eventual identification of the dendritic cell (originally identified by Steinman[31]) being shown to be the key mobile antigen presenting cell.[32] These concepts inspired experiments by Lacy to remove passenger leucocytes and prevent rejection,[25, 33] elegantly proving their crucial role.

Effect of Immunosuppression on Rejection of Islets

The effect of the conventional prednisolone and azathioprine immunosuppression available in the late 1970s was minimal in terms of preventing rejection of islets, so alternative immunosuppressive approaches were investigated, the most effective being anti-lymphocyte serum used with bone marrow, as first demonstrated by Monaco's group at Harvard University in Boston.[34] The possibility of preventing islet rejection by introducing a physical membrane barrier was investigated most intensively at this time by Chick, Like and colleagues from Worcester, Massachusetts, U.S., although the basic principles of the approach were actually first raised by Algire in experiments utilising allogeneic and xenogeneic tumour tissue in 1954, and Algire's findings of a different outcome for xenografts versus allografts were ignored for many years, until the finding was re-confirmed and the mechanism of indirect antigen recognition elegantly demonstrated relatively recently by Loudovaris working with Mandel in Melbourne, Australia.[35] In the interim, a rapid expansion of venture capital-funded commercial enterprises was inspired by the report of microencapsulation as an approach to immunoisolation of individual islets, first described by Sun and colleagues from Toronto,[36] but these ventures were not successful in producing a device suitable for clinical use.

Foetal Pancreas Transplantation

Following slightly behind the progress in isolated adult islet transplantation, the use of foetal pancreas as an alternative source of islet tissue was originally suggested by Brown,[37] utilising the original observation by Coupland that foetal pancreas tissue implanted as a free grant under exocrine tissue atrophy.[13] Reversal of diabetes was first demonstrated in rats[37] and later in mice, utilising tissue culture of minced foetal pancreas to produce small particles (termed "proislets") which were simple to culture and transplant, and producing a model which Lafferty used to good effect for immunological investigations of passenger leukocyte depletion.[38] Following the untimely death of Josiah Brown, the investigation of foetal pancreas as an alternative source of tissue for replacement of islet cells was continued by Mullen and colleagues who showed it was possible to reverse experimental diabetes with a single foetal pancreas provided sufficient time was given for the pancreas to develop in a normoglycemic environment.[39] Subsequently, the mouse proislet model was used extensively by Simeonovic to investigate immune rejection and prove that foetal tissue was not immunologically less reactive than adult islet tissue.[40] It can be seen that for a while the progress of research using foetal-derived tissue paralleled the progress in adult islet transplantation. However, there was a major difference in that successful development of repeatable large animal transplantation models was not achieved for foetal islet tissue, and this was the key to the subsequent successful clinical application of isolated adult islet transplantation.

Development of Techniques for Identification, Tissue Culture and Cryopreservation of Islet Tissue and Assessment of Islet Viability and Function

There are a number of ancillary techniques developed over the years that have been important in facilitating advances in islet isolation and transplantation. The ability to easily identify islet tissue *in vitro* is a facility now taken for granted as the use of dithizone allows visualisation of islet tissue causing the zinc-containing islets to stain bright red. The affinity of dithizone for islet tissue was noted many years before it was re-discovered as useful in isolated islet identification.[41, 42] The credit for the re-discovery goes jointly to Latif working with Alejandro in Miami[43] and Hansen working with Christie in Lernmark's group in Sweden.[44] Prior to 1988, the only way of identifying islet tissue was by another dye specific for islet tissue called neutral red, described initially by Bensley[4] and applied to islet isolation by Downing working with Scharp in St. Louis. It had the disadvantage of requiring to be given intravascularly and was therefore of little use for human islets, but did allow progress to be made in isolation of islets from other species. The ability of islets to survive tissue culture and secrete insulin was an integral part of Hellerstrom's groundbreaking experiments on on microdissected islets,[45] and it was Andersson working in the same group in Stockholm and later Uppsaala in Sweden who first defined the tissue culture characteristics of isolated islets,[46] and Gingerich working in Lacy's group in St. Louis demonstrated the utility of measurement of insulin secretion in a perifusion system.[47] These techniques formed the baseline for viability measurements of isolated islets, which were particularly useful for assessing the survival of islet tissue following islet cryopreservation, a technique extensively investigated by Rajotte in Edmonton, Canada,[48] but still lacking a clearcut experimental or clinical application. However, the above techniques for assessing islet viability suffered from the problem of being slow and tedious to perform and only giving an answer several days later. For rapid islet viability assessment, many groups used dye exclusion techniques common to all tissue culture labs but the use of supravital fluorescent dyes requiring active uptake provided more accurate definition of viability and was first introduced and validated for islets by Gray from Morris' group in Oxford, U.K.[49]

Development of Techniques for Islet Isolation from the Pancreas of Larger Animal Species

The first reports of successful islet isolation and transplantation in murine models inspired several attempts to immediately apply the same technique to human pancreas. Despite some initial optimistic reports, it is now clear that the original technique for islet isolation from the rodent pancreas fails completely when applied to the dense pancreas of larger animals. Indeed, instead of isolating purified islets, the most successful approach at the time was to abandon attempts at islet isolation and use a combination of mechanical mincing of the tissue combined with collagenase digestion and then transplant the resulting tissue directly to the splenic pulp. This approach was use by Mirkovitch working in Lausanne, Switzerland, to produce the first repeatable model of large animal islet transplantation as an autotransplantation model utilising total pancreatectomy to produce a reliable model of diabetes and provide the tissue for subsequent autotransplantation.[50] Subsequent studies by the Minneapolis group directed by Sutherland and Najarian confirmed the utility of the total pancreatectomy/unpurified islet transplantation model, investigated the metabolic efficiency (which was low) and attempted vascularised islet allotransplantation,[51] although successful allotransplantation with diabetes reversal was not achieved until later by Kolb in the group led by Largiader in Zurich, Switzerland.[52] However, this approach was much less successful, even dangerous when applied to humans, and no progress was made for several years.

Progress in Preparation of Collagenase

Considering the fact that collagenase is central to the islet isolation process, the amount of research devoted to this field was remarkably little for many years. In part, this reflects the fact that the enzyme was commercially produced and the market size was deemed insufficient to justify major investment. In fact, little progress was made until one company, Boehringer from Indianapolis, U.S. decided to invest considerable sums of money in a programme to develop improved collagenase for human islet isolation, using porcine pancreas as a surrogate. In this endeavour, they were aided by contact with several groups who were regularly isolating islets human pancreas, including those from Giessen, Germany, Miami, U.S., and Leicester, England. However, the most influential advice came from Rajottes' group in Edmonton, Canada.[53] The result was a greatly improved collagenase with high activity for human islet isolation and much less batch-to-batch variation, subsequently sold as "Liberase".

The Development of Techniques for Islet Isolation from the Human Pancreas

We know now that the crucial element of successful islet isolation in larger species is the delivery of collagenase into the exocrine pancreatic duct. This innovation was first described for canine pancreas by Horaguchi and Merrell[54] and developed to a repeatable method by the Miami group led by Mintz.[55] The first successful application of a modified intraductal collagenase technique to the human pancreas was reported by Gray from Morris' group in Oxford, U.K. With the advent of a technique that could produce large numbers of islets from the human pancreas, increasing numbers of laboratories became interested in clinical application. The workload involved in processing a whole human pancreas for islet isolation by the manual method was considerable and laboratories that approached clinical islet transplantation soon found the demands of processing large numbers of donor pancreas to be formidable. Some years earlier, Scharp and colleagues in St. Louis had reported the use of a device they termed a digestion/filtration chamber for use with primate[56] and canine[57] pancreas. However, at that time the advantage of the digestion/filtration chamber method above simpler manual dissociation methods was not clear and it was not adopted outside St. Louis. Ricordi, working with Scharp, applied the digestion filtration concept to the human pancreas and produced a device which substantially reduced the workload involved in processing the human pancreas and also improved the consistency and yield of islet tissue.[58] These benefits were enough to convince most laboratories involved in human islet isolation and the digestion filtration chamber became a standard item of equipment for human islet isolation, as it remains to the present day. Another important advance in improving the practicality of the human islet isolation procedure was introduced by Lake working with London in Bell's group in Leicester, U.K., which was the adaptation of a machine designed originally for separation of cells from bone marrow, namely the COBE 2991 Cell Separator.[59] A further innovation from the same group was the use of continuous density gradients to replace discontinuous gradients.[60] The net effect of these automation procedures was to lower the processing time for a single human pancreas from more than 12 hours to between two to three hours, and to reduce the minimum number of technical staff required by at least one. These practical gains were sufficient to make clinical islet transplantation a realistic undertaking, and several centres took up the challenge.

Further Modifications to the Automated Islet Isolation Technique

Every centre had its own modification of the automated technique described above — some with fairly radical changes to the protocol. The importance of these variations in terms of yield and purity of the islet preparation obtained was difficult to assess because of a lack of standardised assessment criteria, and a meeting which produced a subsequent manuscript that made a valiant attempt to correct this problem deserves mention. However, standardised assessment of islet yields between centres remains a problem today, and therefore only those centres which were prepared to scientifically validate alterations to techniques can be mentioned here. The benchmark study investigating and proving the superiority of the automated method, first using canine pancreas[61] and subsequently human pancreas with a potentially important modification, was introduced by Warnock in Rajottes' group in

Edmonton, Canada. They showed that the use of continuous perfusion of intraductal collagenase gave superior yields of islets when compared to the single injection loading technique,[62] recently re-confirmed by Lakey in the same group.[63] A further important variable has been the composition of the density gradient used to separate islets from exocrine tissue. Many centres still use a variant of the Ficoll gradient originally used by Scharpe in St. Louis. A major improvement in islet yield and viability was reported by the Oxford and Leicester groups in the U.K., namely the use of bovine serum albumin gradients, but unfortunately this modification fell victim to the subsequent Bovine Spongiform Encephalopathy (BSE) virus scare. By changing to media containing the inert dense molecules of the type originally developed for use in angiography, a significant improvement in islet yield and purity (developed by application to porcine islets) was noted by several groups although only recently reported by the Giessen group directed by Federlin and subsequently Bretzel.[64] The ionic and non-ionic content of the solution has also been the subject of considerable investigation, particularly the importance of using impermeant molecules such as those shown to be suitable for preservation of vascularised organs, e.g. the University of Wisconsin solution,[65] which van der Berg, working in Gooszen's laboratory, first showed to be appropriate for islet isolation.[66]

Clinical Islet Transplantation

The ultimate goal of all the aforementioned experiments has been to successfully transplant islets into diabetic patients, producing as close to a true cure by replacement as it may be possible to get. Early attempts to achieve this goal using tissue obtained from techniques designed for rodent pancreas were un-critical in terms of definition of exactly what was being transplanted and the results were very poor. Indeed, clinical islet transplantation more or less ceased (ignoring sporadic reports that were not repeatable) until a repeatable technique which could regularly isolate large numbers of islets from the human pancreas became available, with the aforementioned advances in automation. Several centres vied to be the first to produce an insulin-independent patient using the new islet isolation technique. Technically, the first documented patient to achieve its independence was reported by Scharp from St. Louis,[67] but the duration of insulin independence was brief and the convincing claim to be the first group to successfully transplant human islets into patients using a repeatable technique must go to the Edmonton group in Canada led by Warnock and Kneteman under the direction of Rajotte.[68] Successful islet transplantation with insulin independence was subsequently reported by several groups, notably the Milan group including Secchi,[69] the Giessen group including Hering (initially) and Brendel under the direction of Federlin and then Bretzel,[70] and the Pittsburgh group, directed by Ricordi and sponsored by Starzl.[71] Landmark achievements from these groups included the first truly long-term (more than three years) functioning islet allograft in a Type 1 diabetic patient reported by the Milan group.[72] A further important finding was an improved rate of success when an islet graft from a single donor was combined with a major multi-organ transplant,[71] thus demonstrating that the allograft reaction, which was probably attenuated in these patients, was a significant factor in reducing success in other recipients of islet transplantation without multi-organ grafting. Most reports of successful islet transplantation utilised several donors to obtain sufficient islets in order to obtain insulin independence, and most of the centres took the approach of trying to reduce the number of donors required to one, the Giessen group being the most successful in this aim.[70] However, the importance of being able to achieve insulin independence, albeit using multiple organ donors was highlighted by the extraordinary interest generated from the report by Lakey and Shapiro from Rajotte's group in Edmonton who obtained insulin independence in seven consecutive patients by using islets from multiple donors,[2–4] combined with a novel, steroid sparing immunosuppressive regimen.[73] At the time of writing, the latter approach is under intensive investigation with the aim of reproducing the results and hopefully shedding light on the crucial factor or factors involved.

A Historical View of the Future

At the time of writing, there is great interest in the future of islet transplantation, and the reaction of patients' groups to the news of progress has underlined just how important an advance islet transplantation available to treat all patients would be viewed. Approaches to attain this aim include the use of non-animal species, as donor sources of tissue (xenotransplantation), and varied molecular engineering/stem cell proliferation approaches to produce insulin-secreting cell lines. It is likely the latter approach will prevail in the end, but the interim period may be long enough to provide a substantial role for xenotransplantation. The latter approach is too new to have a significant history, but the xenograft approach has been explored over a sufficient period, and some benchmark studies have been recorded. The first islet xenografts were recorded by Weber in Reemstma's group in New York in rodent recipients,[20] and extensive investigations of immunosuppression and other approaches to prevent xenograft rejection have been undertaken, notably by Simeonovic's group in Canberra, Australia,[74] Ildstad working with Starzl in Pittsburg, and Korsgren working in Groths' group in Stockholm. The application of the various approaches to prevent rejection reported as successful in rodent models has yet to be demonstrated in large animal models, and there is one study (from the Oxford U.K. group) which suggests that islet xenografts into primate recipients may suffer early damage within hours. However, both these fields are still under intensive investigation and much of their history has yet to be written.

References

1. Williams PW (1894). Notes on diabetes treated with extract and by grafts of sheep's pancreas. *Br Med J* **2**, 1303–1304.
2. Mering JV, Minkowski O (1890). Diabetes mellitus nach pankreas-exstirpation. *Arch Exp Pathol Pharmaco* **26**, 371–387.
3. Minkowski O (1892). Weitere Mittheilungen uber den diabetes mellitus nach exstirpation des pankreas. *Berlin Klin Wochenschr* **5**, 90–94.

4. Bensley RR (1911). Studies on the pancreas of the guinea pig. *Am J Anat* **12**, 297–388.
5. Clark E (1913). The number of islands of Langerhans in the human pancreas. *Anat Anz* **43**, 81–94.
6. Hellman B (1959). Actual distribution of the number and volume of the islets of Langerhans in different size classes in non-diabetic humans of varying ages. *Nature* **184**, 1498–1499.
7. Dunn JS, McLetchie NG (1943). Experimental alloxan diabetes in the rat. *Lancet* **2**, 384–387.
8. Junod A, Lambert AE, Orci L, Pictet R, Gonet AE, Renold AE (1967). Studies of the diabetogenic action of streptozotocin. *Proc Soc Exp Biol Med* **126**, 201–205.
9. Wolffe M, Maske H, Stampff B, Baumgarten F (1952). Untersuchungen uber den dithizondiabetes. *Arch Exp Pharmacol* **216**, 440–456.
10. Hellerstrom C (1964). A method for the microdissection of intact pancreatic islets of mammals. *Acta Endocrinol* **45**, 122–132.
11. Moskalewski S (1965). Isolation and culture of the islets of Langerhans of the guinea pig. *Gen Comp Endocrinol* **5**, 342–353.
12. Russell PS, Gittes RE (1959). Parathyroid transplants in rats — a comparison of their survival time with that of skin grafts. *J Exp Med* **109**, 571–588.
13. Coupland RE (1960). The survival and growth of pancreatic tissue in the anterior chamber of the eye of the albino rat. *J Endocrinol* **20**, 69–77.
14. Moskalewski S (1970). Comparison of cultured and transplanted islets of the guinea pig. In Hellman B, Taljedal I (eds.) *The Structure and Metabolism of the Pancreatic Islets: A Centennial of Paul Langerhans*. Oxford: Pergamon Press, pp. 69–72.
15. Lacy PE, Kostianovsky M (1967). Method for the isolation of intact islets of Langerhans from the rat pancreas. *Diabetes* **16**, 35–39.
16. Ballinger WF, Lacy PE (1972). Transplantation of intact pancreatic islets in rats. *Surgery* **72**, 175–186.

17. Scharp DW, Kemp CB, Knight MJ, Ballinger WF, Lacy PE (1973). The use of ficoll in the preparation of viable islets of Langerhans from the rat pancreas. *Transplantation* **16**, 686–689.
18. Kemp CB, Knight MJ, Scharp DW, Ballinger WF, Lacy PE (1973). Effect of transplantation site on the results of pancreatic islet isografts in diabetic rats. *Diabetologia* **9**, 486–491.
19. Ziegler MM, Reckard CR, Barker CF (1974). Long term metabolic and immunological considerations in transplantation of pancreatic islets. *J Surg Res* **16**, 575–581.
20. Weber C, Weil R, McIntosh R, Hogle H, Warden G, Reemtsma K (1975). Xenotransplantation of piscine islets into hyperglycemic rats. *Surgery* **77**, 208–215.
21. Mauer SM, Steffes MW, Sutherland DE, Najarian JS, Michael AF, Brown DM (1975). Studies of the rate of regression of the glomerular lesions in diabetic rats treated with pancreatic islet transplantation. *Diabetes* **24**, 280–285.
22. Slijepcevic M, Helmke K, Federlin K (1975). Islet transplantation in experimental diabetes of the rat. I. Comparative studies: pancreatectomy — streptozotocin. *Horm Metab Res* **7**, 20–25.
23. Slijepcevic M, Helmke K, Federlin K (1975). Islet transplantation in experimental diabetes in the rat. III. Studies in allogeneic streptozotocin-treated rats. *Horm Metab Res* **7**, 456–461.
24. Federlin K, Slijepcevic M, Helmke K (1976). Islet transplantation in experimental diabetes of the rat. IV. The influence of transplantation site and of histocompatability on islet function. *Horm Metab Res* **8**, 97–101.
25. Lacy PE, Davie JM, Finke EH (1979). Prolongation of islet allograft survival following *in vitro* culture (24 degrees C) and a single injection of ALS. *Science* **204**, 312–313.
26. Sutherland DE, Rynasiewicz J, Kawahara K, Gorecki P, Najarian JS (1980). Rejection of islets versus immediately vascularized pancreatic allografts — a quantitative comparison. *J Surg Res* **29**, 240–247.
27. Finch DR, Morris PJ (1976). Passive enhancement of isolated pancreatic islet allografts. *Transplantation* **22**, 508–512.

28. Naji A, Reckard CR, Ziegler MM, Barker CF (1975). Vulnerability of pancreatic islets to immune cells and serum. *Surg Forum* **26**, 459–461.
29. Lafferty KJ, Bootes A, Dart G, Talmage DW (1976). Effect of organ culture on the survival of thyroid allografts in mice. *Transplantation* **22**, 138–149.
30. Lafferty KJ, Woolnough J (1977). The origin and mechanism of the allograft reaction. *Immunol Rev* **35**, 231–262.
31. Steinman RM, Adams JC, Cohn ZA (1975). Identification of a novel cell type in peripheral lymphoid organs of mice. IV. Identification and distribution in mouse spleen. *J Exp Med* **141**, 804–820.
32. Steinman RM, Witmer MD (1978). Lymphoid dendritic cells are potent stimulators of the primary mixed leukocyte reaction in mice. *Proc Natl Acad Sci USA* **75**, 5132–5136.
33. Faustman DL, Steinman RM, Gebel HM, Hauptfeld V, Davie JM, Lacy PE (1984). Prevention of rejection of murine islet allografts by pretreatment with anti-dendritic cell antibody. *Proc Natl Acad Sci USA* **81**, 3864–3868.
34. Panijayanond P, Monaco AP (1974). Enhancement of pancreatic islet allograft survival with ALS and donor bone marrow. *Surg Forum* **25**, 379–381.
35. Loudovaris T, Mandel TE, Charlton B (1996). CD4+ T cell mediated destruction of xenografts within cell-impermeable membranes in the absence of CD8+ T cells and B cells. *Transplantation* **61**, 1678–1684.
36. Sun AM, Shea GMO, Goosen MF (1984). Injectable micro-encapsulated islet cells as a bioartificial pancreas. *Appl Biochem Biotechnol* **10**, 87–99.
37. Brown J, Molnar IG, Clark W, Mullen Y (1974). Control of experimental diabetes mellitus in rats by transplantation of foetal pancreases. *Science* **184**, 1377–1379.
38. Bowen KM, Andrus L, Lafferty KJ (1979). Survival of pancreatic islet allografts. *Lancet* **2**, 585–586.

39. Mullen YS, Clark WR, Molnar IG, Brown J (1977). Complete reversal of experimental diabetes mellitus in rats by a single foetal pancreas. *Science* **195**, 68–70.
40. Simeonovic CJ, Lafferty KJ (1982). Immunogenicity of isolated foetal mouse proislets. *Aust J Exp Biol Med Sci* **60**, 391–395.
41. McNary WF (1954). Zinc-dithizone reaction of pancreatic islets. *J Histochem Cytochem* **2**, 185–193.
42. Maske H (1957). Interaction between insulin and zinc in the islets of Langerhans. *Diabetes* **6**, 335–341.
43. Latif ZA, Noel J, Alejandro R (1988). A simple method of staining fresh and cultured islets. *Transplantation* **45**, 827–830.
44. Hansen WA, Christie MR, Kahn R, Norgaard A, Abel I, Petersen AM, Jorgensen DW, Baekkeskov S, Nielsen JH, Lernmark A, Egeberg J, Olesen HR, Grainger T, Kristensen JK, Brynitz S, Bilde T (1989). Supravital dithizone staining in the isolation of human and rat pancreatic islets. *Diabetes Res* **10**, 53–57.
45. Hellerstrom C, Nilsson B, Sandler S, Andersson A, Agren A, Petersson B (1979). Isolation and preservation *in vitro* of human pancreatic islets intended for transplantation. *Adv Exp Med Biol* **119**, 465–471.
46. Andersson A (1976). Tissue culture of isolated pancreatic islets. *Acta Endocrinol* **205**(Suppl.), 283–294.
47. Gingerich RL, Aronoff SL, Kipnis DM, Lacy PE (1979). Insulin and glucagon secretion from rat islets maintained in a tissue culture-perifusion system. *Diabetes* **28**, 276–281.
48. Rajotte RV, Stewart HL, Voss WA, Shnitka TK (1977). Viability studies on frozen-thawed rat islets of Langerhans. *Cryobiology* **14**, 116–120.
49. Gray DW, Morris PJ (1987). The use of fluorescein diacetate and ethidium bromide as a viability stain for isolated islets of Langerhans. *Stain Technol* **62**, 373–381.
50. Mirkovitch V, Campiche M (1977). Absence of diabetes in dogs after total pancreatectomy and intrasplenic autotransplantation of pancreatic tissue. *Transpl Proc* **9**, 321–323.

51. Kretschmer GJ, Sutherland DE, Matas AJ, Payne WD, Najarian JS (1978). Autotransplantation of pancreatic fragments to the portal vein and spleen of totally pancreatectomized dogs: a comparative evaluation. *Ann Surg* **187**, 79–86.
52. Kolb E, Urfer K, Largiader F (1979). Early rejection of allotransplanted pancreatic islets in the dog. *Transplant Proc* **11**, 543–548.
53. Lakey JR, Cavanagh TJ, Zieger MA, Wright M (1998). Evaluation of a purified enzyme blend for the recovery and function of canine pancreatic islets. *Cell Transpl* **7**, 365–372.
54. Horaguchi A, Merrell RC (1981). Preparation of viable islet cells from dogs by a new method. *Diabetes* **30**, 455–458.
55. Noel J, Rabinovitch A, Olson L, Kyriakides G, Miller J, Mintz DH (1982). A method for large-scale high-yield isolation of canine pancreatic islets of Langerhans. *Metabolism* **31**, 184–187.
56. Scharp DW, Murphy JJ, Newton WT, Ballinger WF, Lacy PE (1975). Transplantation of islets of Langerhans in diabetic rhesus monkeys. *Surgery* **77**, 100–105.
57. Downing R, Scharp DW, Ballinger WF (1980). An improved technique for the isolation and identification of mammalian islets of Langerhans. *Transplantation* **29**, 79–83.
58. Ricordi C, Lacy PE, Finke EH, Olack BJ, Scharp DW (1988). Automated method for isolation of human pancreatic islets. *Diabetes* **37**, 413–420.
59. Lake SP, Bassett PD, Larkins A, Revell J, Walczak K, Chamberlain J, Rumford GM, London NJ, Veitch PS, Bell PR, James RF (1989). Large-scale purification of human islets utilizing discontinuous albumin gradient on IBM 2991 cell separator. *Diabetes* **38**(Suppl 1), 143–145.
60. Robertson GS, Chadwick DR, Contractor H, James RF, Bell PR, London NJ (1993). The use of continuous density gradients for the assessment of islet and exocrine tissue densities and islet purification. *Acta Diabetol* **30**, 175–180.
61. Ao Z, Lakey JR, Rajotte RV, Warnock GL (1992). Collagenase digestion of canine pancreas by gentle automated dissociation in

combination with ductal perfusion optimizes mass recovery of islets. *Transpl Proc* **24**, 2787.

62. Warnock GL, Ellis D, Rajotte RV, Dawidson I, Baekkeskov S, Egebjerg J (1988). Studies of the isolation and viability of human islets of Langerhans. *Transplantation* **45**, 957–963.
63. Lakey JR, Warnock GL, Shapiro AM, Korbutt GS, Ao Z, Kneteman NM, Rajotte RV (1999). Intraductal collagenase delivery into the human pancreas using syringe loading or controlled perfusion. *Cell Transpl* **8**, 285–292.
64. Lau D, Hering BJ, El-Ouaghlidi A, Jahr H, Brandhorst H, Brandhorst D, Vietzke R, Federlin K, Bretzel RG (1999). Isokinetic gradient centrifugation prolongs survival of pig islets xenografted into mice. *J Mol Med* **77**, 175–177.
65. Wahlberg JA, Southard JH, Belzer FO (1986). Development of a cold storage solution for pancreas preservation. *Cryobiology* **23**, 477–482.
66. van der Burg MP, Guicherit OR, Frolich M, Prins FA, Bruijn JA, Gooszen HG (1994). Cell preservation in University of Wisconsin solution during isolation of canine islets of Langerhans. *Cell Transpl* **3**, 315–324.
67. Scharp DW, Lacy PE, Santiago JV, McCullough CS, Weide LG, Falqui L, Marchetti P, Gingerich RL, Jaffe AS, Cryer PE, *et al.* (1990). Insulin independence after islet transplantation into type I diabetic patient. *Diabetes* **39**, 515–518.
68. Warnock GL, Kneteman NM, Ryan E, Seelis REA, Rabinovitch A, Rajotte RV (1991). Normoglycemia after transplantation of freshly isolated and cryopreserved pancreatic islets in Type 1 (insulin-dependent) diabetes mellitus. *Diabetologia* **34**, 55–58.
69. Secchi A, Socci C, Maffi P, Taglietti MV, Falqui L, Bertuzzi F, De Nittis P, Piemonti L, Scopsi L, Di Carlo V, Pozza G (1997). Islet transplantation in IDDM patients. *Diabetologia* **40**, 225–231.
70. Bretzel RG, Brandhorst D, Brandhorst H, Eckhard M, Ernst W, Friemann S, Rau W, Weimar B, Rauber K, Hering BJ, Brendel MD (1999). Improved survival of intraportal pancreatic islet cell

allografts in patients with type-1 diabetes mellitus by refined peritransplant management. *J Mol Med* **77**, 140–143.

71. Ricordi C, Tzakis AG, Carroll PB, Zeng YJ, Rilo HL, Alejandro R, Shapiro A, Fung JJ, Demetris AJ, Mintz DH, *et al.* (1992). Human islet isolation and allotransplantation in 22 consecutive cases. *Transplantation* **53**, 407–414.
72. Davalli AM, Maffi P, Socci C, Sanvito F, Freschi M, Bertuzzi F, Falqui L, Di Carlo V, Pozza G, Secchi A (2000). Insights from a successful case of intrahepatic islet transplantation into a type 1 diabetic patient (in Process Citation). *J Clin Endocrinol Metab* **85**, 3847–3852.
73. Shapiro AM, Lakey JR, Ryan EA, Korbutt GS, Toth E, Warnock GL, Kneteman NM, Rajotte RV (2000). Islet transplantation in seven patients with type 1 diabetes mellitus using a glucocorticoid-free immunosuppressive regimen (see comments). *N Engl J Med* **343**, 230–238.
74. Simeonovic CJ, Ceredig R, Wilson JD (1990). Effect of GK1.5 monoclonal antiody dosage on survival of pig proislet xenografts in CD4+ T cell-depleted mice. *Transplantation* **49**, 849–856.

VIII

THE HISTORY OF INTESTINAL TRANSPLANTATION

Kareem Abu-Elmagd

The Historical Perspective

Although the origin of bowel transplantation can be traced back to experiments by Alexis Carrel at the beginning of the 20th century, the modern era was heralded by the publication of three articles in 1959–1960. The first two by Richard Lillehei of the University of Minnesota described the transplantation in modified dogs of intestine alone. Autografts and allografts that had been preserved by immersion in iced saline were implanted in the normal anatomic position.[1,2] The most important observation was that autografts provided good long-term nutritional support.

The following year, transplantation in the orthotopic position was reported of multivisceral canine allografts that included the liver and pancreas plus the entire sub-diaphragmatic gastrointestinal tract.[3] A key observation was that the intestine and other non-hepatic organs appeared to have minimal rejection when accompanied by the donor liver.[4] In an important technical difference from the intestine-alone study, the multivisceral allograft was cooled by intravascular infusion with chilled lactated Ringer's solution, an approach destined for application to kidney allografts[5] and eventually to all kinds of organ procurement.[6]

Other aspects of the intestine-alone and multivisceral procedures were much the same as those in clinical use today, but the practical application of such operations was forestalled for another three decades because of multiple problems or concerns that were not successfully addressed until the last 15 years. Thus, it is not possible to make a clear distinction between the history of intestinal transplantation and the evolving "current status" of this field. However, the principles have been clarified in the recent period, and can be placed in an historical context.

Eras of immunosuppression

Azathioprine

When immunosuppression became available, the attention of those interested in clinical intestinal transplantation was focused almost exclusively on the control of rejection. Early attempts with azathioprine to achieve long-term survival in clinically relevant animal models were uniformly unsuccessful.[7, 8] Nevertheless, at least seven well-documented clinical attempts[9] were made under azathioprine-based immunosuppression, five in the United States and one each in France and Brazil, with the apparent premise that the accomplishment already achieved with kidney transplantation might be possible with bowel. The longest period of nutritional function was two months, in a recipient of a Human Leucocyte Antigen (HLA) identical segmental graft from a sister.[10] The patient died of sepsis after 5 1/2 months, and at autopsy the intestinal allograft was intact but atrophic.

Cyclosporine

With the arrival of this more potent immunosuppressant, good results in rodent small bowel transplant models were obtained in several laboratories beginning in the mid-1980s.[9, 11] Full nutritional support from canine intestinal allografts for as long as one year was achieved in a few experiments,[12, 13] but these were exceptions in a sea of failures. The difficulties with large animals continued until 1988 when Grant *et al.*[14] reported a landmark study in pigs in which extraordinary efforts were made to circumvent the unpredictable enteric absorption of cyclosporine by giving uninterrupted intravenous infusion. The majority of the animals survived for more than 100 days, prompting a small number of clinical trials, from which a handful of hopeful observations emerged.[15]

Survival had not yet been recorded in any animal model of multivisceral transplantation up to this time. However, the demonstration in 1962 that the intestinal component of multivisceral allografts was "privileged" in comparison to the intestine-alone allograft[4] and the further observation that the liver protected co-transplanted other tissues and organs[16, 17] justified a multivisceral transplantation in Pittsburgh in November 1987.[18] The child recipient, who had suffered a mid-gut volvulus perinatally, survived for more than six months, and was sustained with enteral feeding until death from a B-cell lymphoma [post-transplant lymphoproliferative disorder (PTLD)].

Using variants of the multivisceral operation, two subsequent patients treated in London, Ontario, by Grant, Wall and McAllister *et al.*[19, 20] achieved long-term rehabilitation (five or more years) before death from multiple causes (Table 1). Except for a patient of Margreiter,[21] none of the other recipients in the second wave of attempts at multivisceral transplantation had prolonged survival.[15] Reports of the Pittsburgh

Table 1 Survival pre-1990 after intestinal transplantation*

City	Month/Year	Type of Graft	Survival	Ref.
Pittsburgh	Nov '87	Multivisceral	6.5 months	18
†Kiel	Aug '88	Intestine	61 months	24
W. Ontario	Nov '88	Liver/Intestine	66 months	19
Paris	Mar '89	Intestine	12 2/3 years	25
W. Ontario	Nov '89	Multivisceral	58 months	20
Innsbruck	Dec '89	Multivisceral	7.5 months	21

*Length of ultimate survival in the early cases verified through December 1, 2001.
†Jejuno-ilial segment from live-related donor.

case,[18] and of a Chicago multivisceral trial by Williams *et al.*[22] in the *Journal of the American Medical Association* were accompanied by a critical editorial.[23]

During the same controversial period, Deltz of Kiel, Germany performed a segmental small bowel transplantation from a related donor.[24] The recipient survived for five years, on enteral alimentation for most of this time. Finally, in 1989, Goulet *et al.*[25] of Paris transplanted a near total cadaveric small bowel into a child who is still alive nearly 13 years later. These first six examples of prolonged human intestinal allograft function (summarized in Table 1) re-ignited interest in the field. Although the intestine was no longer "a forbidden organ", further progress awaited better immunosuppression.

Tacrolimus

In 1989, Murase,[26] Hoffman,[27] and Lee[28] and their associates established the superior efficacy of tacrolimus (then FK 506) relative to cyclosporine in preventing rejection of isolated rat intestinal allografts. In addition, Murase *et al.* reported permanent survival of multivisceral allografts in the rat model,[26] and confirmed unequivocally the protective effect of the liver on co-transplanted abdominal viscera.[29] Tacrolimus already had been demonstrated to rescue about 80% of human liver allografts from intractable rejection under cyclosporine-based therapy.[30] A specific further advantage was that oral absorption of tacrolimus was less influenced by intestinal dysfunction than cyclosporine.[31]

After beginning a clinical trial in 1990,[31] five examples of long-surviving human intestinal recipients were reported by Todo *et al.*[32] from Pittsburgh in 1992: four with liver-intestinal or multivisceral grafts, and one with an isolated small bowel. By 1993, the intestine-alone experience in Pittsburgh had increased to 15 cases with survival of 12 of the recipients for 1.5 to 19 months.[33] Both of the early reports placed less emphasis on the high rate of survival than on the difficulty of clinical care and the need for improved management, including better monitoring of rejection. The same concerns were expressed well into the 1990's.[34–36]

Since 1990, most of the programs of intestinal transplantation have depended on tacrolimus immunosuppression, combined with other drugs. The additional components of graft and host pre-conditioning strategies, and the administration of donor bone marrow augmentation are better discussed in the context of current practice than in an historical context.

The perceived threat of graft-versus-host-disease

In 1959, three prerequisites were defined by Billingham[37] for the development of graft-versus-host disease (GVHD): (1) the presence of mature immunologically competent cells in a graft, (2) sufficient time for these cells to react before they are rejected by the host, and (3) important histocompatibility antigens in the recipient that are lacking in the transplant. Because of the large lymphoid component of the intestine, the possible role of GVHD was raised in explaining the histopathologic findings in canine multivisceral[4] and intestinal allografts[7, 8] and in human allografts.[9]

Subsequently, GVHD and its consequences were delineated clearly by Monchik and Russell[38] in genetically controlled rat experiments in which semi-allogeneic small bowel transplantation was performed from parent strain to F_1 hybrid offspring. Recipients in such models are incapable of rejecting the grafts but are subject to their attack. The rat observations emphasized the peril of a host-graft immunologic imbalance if this favored the graft. However, the use of these models, and of rat strain combinations that are "GVHD prone" (e.g. Lewis to Brown Norway)[39] had the effect of systemically overstating the clinical threat of GVHD.[9]

The resulting anxiety was apparent at the first international small bowel transplant symposium convened in London, September 1988. One of the main themes at this meeting was how to eliminate the lymphoid tissue in intestinal grafts by irradiation, anti-lymphocyte globulin (ALG), or even by surgical lymphadenectomy. Concern about GVHD has been largely discounted by experiments that have shown the GVHD complications to be rat model-specific,[39–41] and especially by a now large clinical experience in which GVHD has been an unusual complication (5%) after both intestine-alone and multivisceral transplantation.[36] Moreover, it has since been learned that the elimination of the passenger leukocytes from the intestine predisposes the graft to chronic rejection.[42]

The double immune reaction

In Kirkman's grimly pessimistic, but exceptionally astute, review of intestinal transplantation in 1984[9] and in the review of the human experience in 1991,[15] there were extensive discussions of the dual immune reaction that occurs after intestinal transplantation: host versus graft (HVG) and graft versus host (GVH). The discussion presaged the paradigm of transplant rejection, engraftment, and tolerance that emerged after the discovery in 1992 of microchimerism in long surviving human recipients of kidney, liver, and other kinds of allografts.[43–47] In the resulting new view of organ engraftment, the seminal mechanism of engraftment is reciprocal clonal activation of the contemporaneous HVG and GVH reactions, followed by variable degrees of mutual clonal exhaustion and deletion[43,45] (Fig. 1). Maintenance of the variable clonal exhaustion-deletion induced at the outset is dependent on residual low level (i.e. micro-) chimerism, the perpetuation of which is dependent on a collaborating mechanism of immune ignorance.[48,49]

The distinction between the two co-existing immune responses is difficult to make after most kinds of organ transplantation. However, separate delineation of the components of the double alloimmune response in GVHD-prone intestinal transplantation models over the last decades[39–42,50] has exposed principles of transplantation immunology that can be generalized to the application of all organs.[48,49] Importantly, it already was strongly suspected by 1990 that the double immune

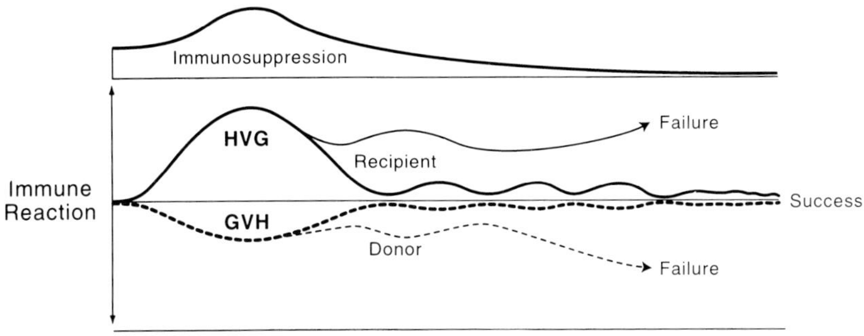

Fig. 1 Contemporaneous host-versus-graft (HVG) and graft-versus-host (GVH) responses following transplantation. Failure is defined as the inability to control one or sometimes both of the reactions. Acute reciprocal clonal exhaustion after successful transplantation is subsequently maintained by chimerism dependent low grade stimulation of one or both leukocyte populations, which may wax and wane. With intestinal, liver-intestine, and multivisceral transplantation, the risk of clinical GVHD is minimal providing the recipient immune system is intact (i.e. not cytoablated). Moreover, host pre-conditioning with ALG or other anti-lymphoid antibody preparations has not been associated with clinical GVHD (from Starzl TE and Zinkernagel RM (2001). *Nature Rev Immunol* **1**, 233–239, used by permission).

reaction in recipients whose immune response had *not been weakened by cytoablation* would be reciprocally deleted.[15] This conviction was the "green light" that led to the institution of the large-scale intestinal transplant program at the University of Pittsburgh.[31]

Subsequent research in chimerism has provided unparalleled insight into the meaning of tolerance generally and allotolerance specifically[48] and the development of treatment principles for the induction of transplantation tolerance.[49] In essence, these principles consist of manipulating one or the other of the countervailing immune reactions to achieve the necessary balance that may lead to reciprocal antigen-specific deletion (Fig. 1). The strategies under the rubric of "immune modulation", are currently undergoing clinical trial with all of the major organs including the intestine,[34, 36, 51] and are part of the still evolving "state of the art" of intestinal and all other kinds of transplantation.

Portal versus caval intestinal venous drainage

Although it has been established that the liver graft is protective of or tolerogenic for concomitantly transplanted organs including the intestine,[4, 29] an historical controversy was whether allograft antigen delivery into the portal vein, instead of the systemic venous system, ameliorated the vigour of rejection. In a study by Murase *et al.*[52] the fate of the isolated small bowel allograft was determined with portal venous versus systemic drainage. Both acute rejection and the distortion of lymphoid architecture associated with chronic rejection were observed in the two cohorts with equal severity and frequency.

However, the metabolic welfare of the rat intestinal recipient differed according to the route of the intestinal venous outflow. Rats with portal drainage of their isolated intestinal grafts had better weight gain than after caval drainage and were healthier,[52] consistent with independent observations by Schraut *et al.*[53] This was not surprising. It has been shown in non-transplant models that optimum liver structure and function depend on hormones (especially insulin), nutrients, and other substances found in high concentration in splanchnic venous blood.[54] Depriving the liver of this kind of blood causes hepatic atrophy, deterioration of the hepatocyte organelles and subtle loss of hepatic function.

The physiologic consequences of diverting splanchnic venous blood from the liver are less extreme in humans than in animals.[54] After clinical intestinal transplantation, it has been necessary for technical expediency in some cases to drain intestinal venous return into the host inferior vena cava. Serious morbidity has not been reported.[51]

Current Status

Transplantation operations

The three kinds of allografts are shown in Fig. 2: isolated intestinal, combined intestine and liver, and multivisceral. Patients who still have good liver function, or reversible hepatic dysfunction, are candidates

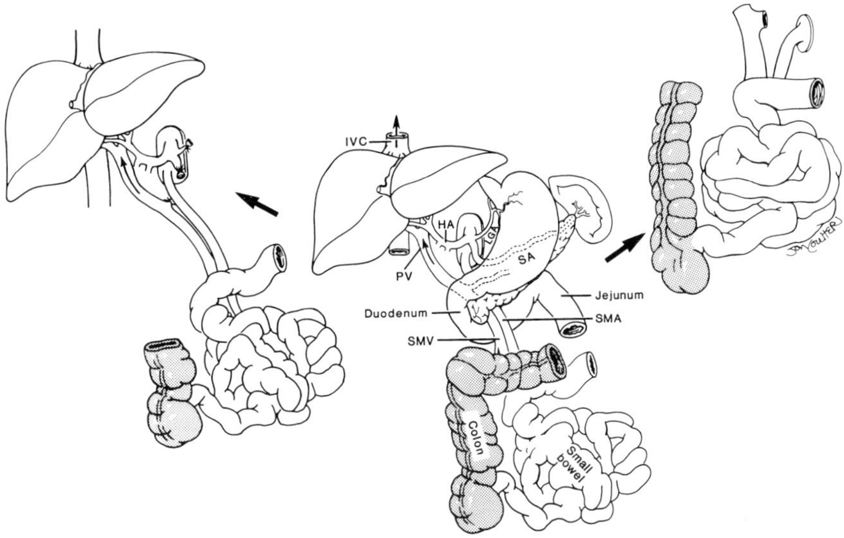

Fig. 2 Three kinds of visceral allografts: intestine alone (right), liver and intestine (left), and multivisceral (middle). In the Pittsburgh experience, colonic segments (shaded) were included in recipients scattered through the three cohorts, a practice that has been discontinued (see text). IVC: inferior vena cava; PV: portal vein; HA: hepatic artery; SMA: superior mesenteric artery; SMV: superior mesenteric vein; SA: splenic artery; and LGA: left gastric artery (from Todo S *et al.* (1995). *Ann Surg* **222**, 270–282, used by permission).

for intestine only. Those who have liver-based inborn errors responsible for intestinal disease and or severe total parenteral nutrition (TPN) related hepatic dysfunction receive liver and intestine. Abdominal multivisceral grafts are reserved for patients who have extensive abnormalities of the gastrointestinal tract caused by absorptive, motility, or vascular disorders. The principles[15] and various modifications of the three generic procedures have been described extensively.[51]

In all three kinds of recipients, grafts have been used that included large bowel. Use of the colon was prompted by the high post-operative stomal output and frequent re-admissions for treatment of dehydration that were frequently observed early in our experience.[34] Because inclusion of a colonic segment in 29 Pittsburgh recipients appeared to be a significant risk factor with all three procedures (especially in adults), we remove the colon from the allograft in most cases. A possible explanation for the added morbidity and mortality of the colon was a greater post-operative increase in recipient plasma tumor necrosis factor-α (and by implication, endotoxemia) compared to patients whose grafts did not contain colon.[34]

The world experience

Between April 1985 and May 2001, 651 patients worldwide received 696 intestinal transplants at 55 centers,[55] 180 (28%) of which were performed at the University of Pittsburgh. The number of centers and the procedures have increased each year. The kind of allografts used were isolated intestine (42%), combined liver-intestine (43%), and multivisceral (14%). In children (61% of the recipients), the underlying disease was gastroschisis (21%), volvulus (18%), dysmotility syndromes (18%), necrotizing enterocolitis (12%), intestinal atresia (7%), and microvillus inclusion (6%). For adults (39% of total cases), the indications for transplantation were intestinal vascular occlusive disease (22%), Crohn's disease (13%), gastrointestinal neoplasms (13%), and trauma (12%). The re-transplantation rate was 7% among children and 5% for adults.

According to the data contributed to the International Intestinal Transplant Registry and summarized in Stockholm at the 7th International Small Bowel Transplant Symposium (September 2001), a total of 335 intestinal recipients were alive worldwide. As of May 2001, the overall survival was 51%.[55] The leading causes of death have been sepsis (n = 166), rejection (n = 33), technical complications (n = 31), lymphoma (n = 26), and respiratory failure (n = 22). The actuarial one and five-year survival rates for grafts transplanted after February 1995, were 61% and 42% at the two follow-up times. Interestingly, there was no difference in survival after intestine alone, liver-intestine, and multivisceral transplantation.

Based upon the 1999[56] report and the subsequent account presented in Stockholm in September 2001 from the International Intestinal

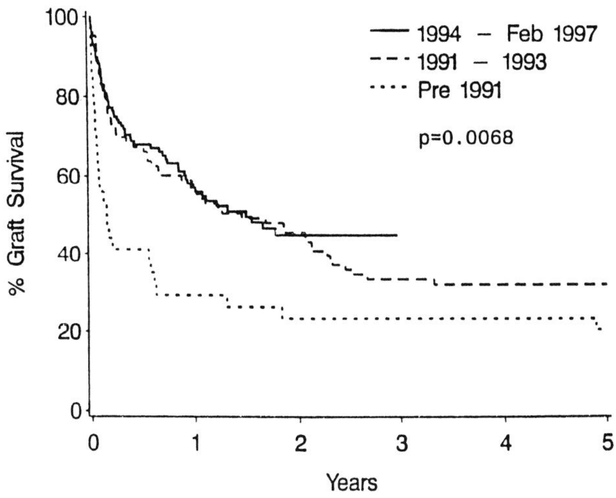

Fig. 3 The cumulative graft survival of the world experience according to the era of transplantation (from Grant D (1999). *Transplantation* **67**, 1061–1064, used by permission).

Transplant Registry,[55] the worldwide survival outcome with intestinal transplantation has been influenced by two important factors. First, survival has significantly improved over time (Fig. 3). Second, both patient and graft survival outcomes were significantly better at centers that have performed more than a total of ten intestinal transplants. These two variables are surrogate markers for the cumulative refinement in surgical techniques and post-operative management (i.e. the learning curve).

For patients who survive the perilous early post-transplant period, the long-term rehabilitation with all three kinds of the intestinal transplant procedure is similar to that achieved with other types of thoraco-abdominal organ transplantation. Of 288 worldwide survivors who lived for at least six post-operative months, 70% have maintained complete enteric nutritional autonomy. Importantly, 85% of these recipients achieved a modified Karnofsky performance score of 90–100%.[55,56] These therapeutic indices are even higher at centers with a large experience.

The Pittsburgh experience

At the University of Pittsburgh, a total of 165 transplants were given to 155 consecutive recipients between May 1990, and February 2001. The actuarial patient survival for the total population was 75% at one year, 54% at five years, and 42% at ten years (Fig. 4), with achievement of full nutritional autonomy in more than 90% of survivors.[51] Beyond five years, recipients of liver-intestinal grafts have had the best prognosis. Despite the numerous gratifying successes generated during the early part of this experience, however, we self-imposed a moratorium on further case accrual throughout most of 1994. The decision was made because we had been unable despite much experience to reduce the mortality rate and difficulty of post-operative management, or to eliminate the need for lengthy and frequent hospitalization.[34, 36]

Multivariate analysis of the cases up to 1994,[34] and subsequently,[36] demonstrated multiple high risk factors: multiple previous operations, lengthy operations, long cold ischemia time, inclusion of the donor

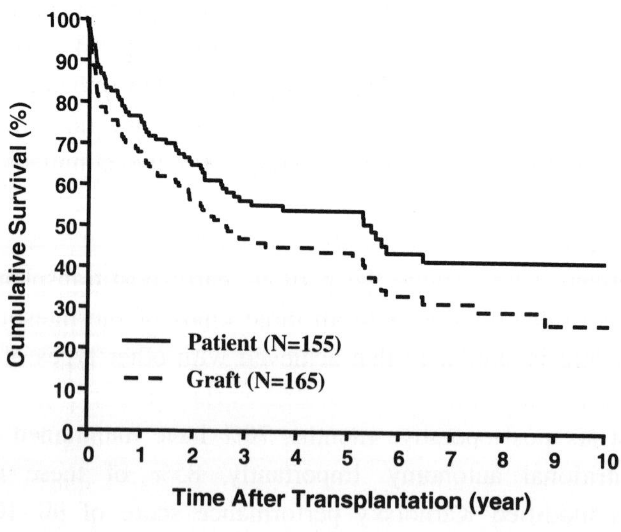

Fig. 4 Kaplan-Meier patient and graft survival curves for the Pittsburgh population (from Abu-Elmagd K *et al.* (2001). *Ann Surg* **234**, 404–417, used by permission).

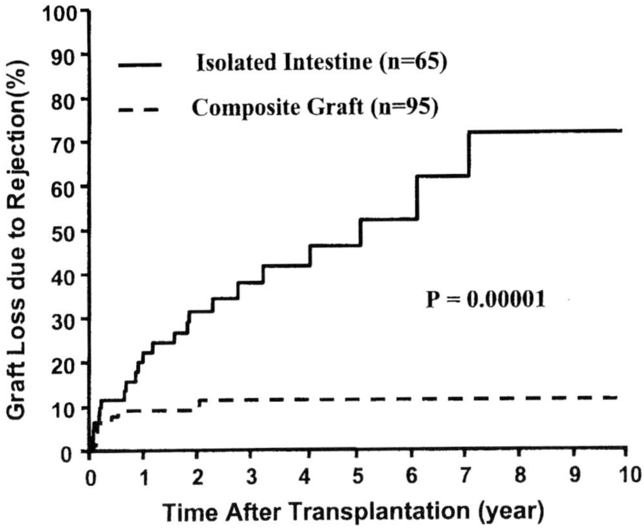

Fig. 5 Cumulative risk of graft loss from rejection in the intestine-only and composite visceral grafts that contained liver (from Abu-Elmagd K *et al.* (2001). *Ann Surg* **234**, 404–417, used by permission).

colon with the allograft, development of rejection, sustained high blood levels of tacrolimus, excessive prednisone doses, the development of Epstein-Barr Virus (EBV)-associated B-cell lymphomas, and complications of Cytomegalovirus (CMV) infection. Correction of the non-immunologic risk factors (e.g. avoidance of predictably difficult operations, reduction of the donor pool), would have precluded treatment for many, if not most, of the candidates who needed it most. Consequently, the prime objective when the program was re-opened was to develop strategies that would reduce the need for heavy chronic immunosuppression that was at the root of many of the risk factors. Particular attention was directed to recipients of intestine alone in whom the cumulative risk of graft loss due to acute/chronic rejection was significantly greater than in recipients of composite grafts that contained liver (Fig. 5).

The efforts were weighted at first to the addition of short-term adjunct immunosuppressants to the continuous treatment with tacrolimus and prednisone: daclizumab (Zenapax® is a humanized IgG1 monoclonal antibody directed at the α-subunit of the human interleukin-2 receptors), sirolimus, cyclophosphamide, OKT_3, or a potent antilymphoid globulin (ALG, i.e. thymoglobulin®). With the possible exception of daclizumab, the additional drugs did not dramatically improve either survival or the ease of management. Consequently, we looked to strategies beyond more potent immunosuppressive drug regimens as a means of improving the outlook for intestinal recipients.

The first initiative was to infuse adjunct donor bone marrow cells at the time of intestinal transplantation.[34, 36] In well-controlled rat studies, Murase *et al.*[42] have shown that chronic rejection of intestinal allografts can be prevented by this expedient. Reflecting the historically-based anxiety about a threat of GVHD, and citing a study of adjunct donor bone marrow infusion in irradiated pigs,[57] Gruessner predicted that the incidence of death from rejection, infection, GVHD, or a combination of these three immunologic events would be seen in clinical cases, and further that the incidence of B-cell lymphoma would be increased.[58] Because human intestine recipients are not cytoablated, none of these dire consequences were observed.[36, 51] In fact, the results were improved with all three kinds of intestinal transplant procedures.[36] However, the morbidity and mortality remained too high for the widespread application of intestinal transplantation. Nevertheless, establishment of the safety of the adjunct bone marrow infusion opened the way to its further use in immune modulation strategies.

In selected inbred rat strain combinations (e.g. Brown Norway and Lewis), it has been shown by Murase *et al.*[41] that the quantity and lineage profiles of the passenger leukocytes contained in different kinds of organ grafts strongly influences the relative strengths of the interactive HVG and GVH reactions that are depicted in Fig. 1, the quantity and lineages of the chimerism found in the recipients, graft survival, and graft function. As a practical consideration, the passenger leukocytes of the liver and bone marrow cells, both of which include large numbers of immature leukocytes and cells of myeloid origin, have been shown to be highly tolerogenic with a lower risk of GVHD than the intestinal passenger leukocytes which are rich in mature lymphocytes.[41] While depletion of the passenger leukocytes from the intestine with irradiation or anti-lymphoid antibodies eliminates essentially all risk of GVHD, loss of these passenger leukocytes predisposes the allograft to chronic rejection.[42]

It was shown in rat experiments, however, that intestinal graft irradiation and replacement of the cytoablated intestinal leukocytes with a peri-operative infusion of donor bone marrow cells was safe and highly tolerogenic.[42] This strategy has been extensively tested in human intestinal recipients under high and low dose immunosuppression. After 7.5 Gy bowel irradiation, 2.4–9.0 $\times$ 10^8/kg donor bone marrow cells were infused as "substitute passenger leukocytes". The high dose immunosuppressive protocol consisted of post-operative daclizumab induction and a tacrolimus-prednisone regimen; sirolimus, mycophenolate mofetil, azathioprine, or OKT_3 were added as contingency adjuncts. Survival was 89% or higher after intestinal (n = 16), liver-intestinal (n = 3), or multivisceral transplantation (n = 5) was achieved in patients treated during 2000–2001. When *ex vivo* bowel irradiation was combined with adjunct donor bone marrow infusion under the high dose immunosuppression (n = 19), the incidence of rejection was reduced from 67% to 16% (Table 2).

Adjunct donor bone marrow infusion has also been administered to intestinal recipients under a low dose immunosuppressive protocol in which the recipients were pre-conditioned with approximately 5 mg/kg thymoglobulin® and treated post-operatively with tacrolimus only. This modification of immunosuppression was based on the recent elucidation of the mechanisms of allograft acceptance: donor-specific clonal exhaustion deletion plus immune ignorance.[48] Two treatment principles emerged from this insight into mechanisms of allo-engraftment and acquired tolerance.[49] The first principle is that effective host pre-treatment before arrival of the allograft (i.e. in our protocol with thymoglobulin®) reduces the donor-specific response enough to be

Table 2 Intestinal allograft irradiation and recipient pre conditioning at the University of Pittsburgh

Dose of Immunosuppression	Irradiation Plus Bone Marrow Infusion	No. of Patients	Graft Survival	Rejection
High dose immunosuppression	Yes	18	16 (89%)	3 (17%)
(April 2000–June 2001)	No	6	6 (100%)	4 (67%)
Low dose immunosuppression with pre-conditioning	Yes	9	8 (89%)	1 (10%)
(July–September 2001)	No	1	1 (10%)	1 (100%)

more readily deleted with arrival of migratory antigen (the donor leukocytes). Second, the use of low dose tacrolimus monotherapy allows the immune activation → exhaustion → deletion to proceed rather than subverting this seminal mechanism with over-immunosuppression. This more tolerogenic strategy has been used for ten recipients (eight intestine, two liver-intestine), eight of whom received irradiated bowel and adjunct bone marrow cells (Table 2).

Although bowel irradiation plus adjunct bone marrow cells was associated with a reduced rate of rejection under both high and low dose immunosuppression, equivalent graft survival rates (89–100%) were obtained in all four cohorts (Table 2). The single lost graft in the pre-conditioned low dose immunosuppression group had severe ischemia-reperfusion damage after transplantation against a positive B-cell cross-match, and was removed with patient survival. Opportunistic infections in the first few months were strongly associated with additional treatment for rejection. None of the pre-conditioned recipients who have subsequently undergone dose reduction of maintenance tacrolimus has experienced graft rejection. With the unprecedented early success of combined pre-treatment/low dose immunosuppression, we are cautiously proceeding with complete weaning of some of these patients from tacrolimus monotherapy. Whether or not complete weaning can be accomplished, the stage has been set for major further improvements in the field.

Future Prospects

With the last 12 years, the transplantation of the intestine-alone or as a component of multivisceral grafts has progressed from a completely experimental procedure(s) to one that appears certain to supplant chronic total parenteral nutritional (TPN) support as the preferred treatment of intestinal failure. Based upon Medicare data, the average cost of TPN in 1992 was more than US$150,000 per patient per year not including the cost of frequent hospitalization, medical equipment and nursing care.[59] With the current average cost of the intestine-only transplant, the procedure becomes cost-effective by the second year after surgery.[60] The cost-effectiveness of the liver-intestine and multivisceral procedures is immeasurable because of the lack of an alternative life-saving therapy.

The emergence of intestinal transplantation has depended on the development of more potent immunosuppression, and particularly the advent of tacrolimus. The progress culminated in the qualification of these procedures for funding as a service by the United States Health Care Financing Administration (HCFA) as of October 4, 2000. The further maturation of the field will almost certainly be contingent on better timing and more discriminating doses of immunosuppressants. This trend has been driven by elucidation during the 1990s of the mechanisms of allo-engraftment and acquired tolerance. The results with the modified approach to immunosuppression already suggests that many if not most of the vicissitudes born by the first wave of heroic intestinal recipients will be fading memories by the end of the present decade.

Acknowledgments

The author would like to thank Ms. Terry Mangan for preparing the manuscript and the staff of the Intestinal Rehabilitation and Transplant Center at the University of Pittsburgh Medical Center for their team efforts.

References

1. Lillehei RC, Goott B, Miller FA (1959). The physiologic response of the small bowel of the dog to ischemia including prolonged *in vitro* preservation of the bowel with successful replacement and survival. *Ann Surg* **150**, 543–560.
2. Lillehei RC, Goott B, Miller FA (1959). Homografts of the small bowel. *Surg Forum* **10**, 197.
3. Starzl TE, Kaupp HA Jr (1960). Mass homotransplantation of abdominal organs in dogs. *Surg Forum* **11**, 28–30.
4. Starzl TE, Kaupp HA Jr, Brock DR, Butz GW Jr, Linman JW (1962). Homotransplantation of multiple visceral organs. *Am J Surg* **103**, 219–229.
5. Starzl TE (1964). *Experience in Renal Transplantation*. Philadelphia, PA: WB Saunders Company, pp. 1–383.
6. Starzl TE, Hakala TR, Shaw BW Jr, Hardesty RL, Rosenthal TJ, Griffith BP, Iwatsuki S, Bahnson HT (1984). A flexible procedure for multiple cadaveric organ procurement. *Surg Gynecol Obstet* **158**, 223–230.
7. Ruiz JO, Uchida H, Schultz LS, Lillehei RC (1972). Problems in absorption and immunosuppression after intestinal allotransplantation. *Am J Surg* **123**, 297.
8. Cohen Z, MacGregor AB, Moore KTH, Falk RE, Langer B, Cullen JB (1976). Canine small bowel transplantation: a study of the immunologic responses. *Arch Surg* **111**, 248–253.
9. Kirkman RL (1984). Small bowel transplantation. *Transplantation* **37**, 429–433.
10. Fortner JG, Sichuk G, Litwin SD, Beattie EJ (1972). Immunological responses to an intestinal allograft with HLA-identical donor-recipients. *Transplantation* **14**, 531–535.
11. Raiu S, Cavirli M, Didlake RH (1984). Experimental small bowel transplantation utilizing cyclosporine. *Transplantation* **38**, 561–566.

12. Reznick RK, Craddock GN, Langer B, Gilas T, Cullen JB (1982). Structure and function of small bowel allografts in the dog: immunosuppression with cyclosporin A. *Can J Surg* **25**, 51–55.
13. Diliz-Perez HS, McClure J, Bedetti C, Hong H-Q, de Santibanes E, Shaw BW Jr, Van Thiel D, Iwatsuki S, Starzl TE (1984). Successful small bowel allotransplantation in dogs with cyclosporine and prednisone. *Transplantation* **37**, 126–129.
14. Grant D, Duff J, Zhong R, Garcia B, Lipohar C, Keown P, Stiller C (1988). Successful intestinal transplantation in pigs treated with cyclosporine. *Transplantation* **45**, 279–284.
15. Starzl TE, Todo S, Tzakis A, Alessiani M, Casavilla A, Abu-Elmagd K, Fung JJ (1991). The many faces of multivisceral transplantation. *Surg Gynecol Obstet* **172**, 335–344.
16. Calne RY, Sells RA, Pena JR, Davis DR, Millard PR, Herbertson BM, Binns RM, Davies D (1969). *Nature* **223**, 472–476.
17. Kamada N (1985). The immunology of experimental liver transplantation in the rat. *Immunology* **55**, 369–389.
18. Starzl TE, Rowe MI, Todo S, Jaffe R, Tzakis A, Hoffman AL, Esquivel C, Porter KA, Venkataramanan R, Makowka L, Duquesnoy R (1989). Transplantation of multiple abdominal viscera. *JAMA* **261**, 1449–1457.
19. Grant D, Wall W, Mimeault R, Zhong R, Ghent C, Carcia B, Stiller C, Duff J (1990). Successful small-bowel/liver transplantation. *Lancet* **335**, 181–184.
20. McAlister V, Wall W, Ghent C, Zhong R, Duff J, Grant D (1992). Successful small intestine transplantation. *Transpl Proc* **24**, 1236–1237.
21. Margreiter R, Konigsrainer A, Schmid T, Koller J, Kornberger R, Oberhuber G, Furtwangler W (1992). Successful multivisceral transplantation. *Transpl Proc* **24**, 1226–1227.
22. Williams JW, Sankary HN, Foster PF, Lowe J, Goldman GM (1989). Splanchnic transplantation: an approach to the infant dependent on parenteral nutrition who develops irreversible liver disease. *JAMA* **261**, 1458–1462.

23. Moore FD (1990). The Desperate Case: CARE (Costs, Applicability, Research, Ethics). *JAMA* **261**, 1483–1484.
24. Deltz E, Schroeder P, Gebhardt H, Gundlach M, Timmerman W, Engemann R, Leimenstoll G, Hansman ML, Westphal E, Hamelmann H (1989). Successful clinical small bowel transplantation: Report of a case. *Clin Transpl* **3**, 89–91.
25. Goulet O, Revillon Y, Brousse N, Jan D, Canion D, Rambaud C, Cerf-Bensussan N, Buisson C, Hubert P, de Potter S, Mougenot JF, Fischer A, Ricour C (1992). Successful small bowel transplantation in an infant. *Transplantation* **53**, 940–943.
26. Murase N, Kim D, Todo S, Cramer DV, Fung J, Starzl TE (1990). Induction of liver, heart and multivisceral graft acceptance with a short course of FK 506. *Transpl Proc* **22**, 74–75.
27. Hoffman A, Makowka L, Cai X, Cramer DV, Pascualone A, Todo S, Starzl TE (1990). The effect of FK 506 on small intestine allotransplantation in the rat. *Transpl Proc* **22**, 76–77.
28. Lee K, Stangl MJ, Todo S, Langrehr JM, Starzl TE, Schraut WH (1990). Successful orthotopic small bowel transplantation with short term FK 506 immunosuppressive therapy. *Transpl Proc* **22**, 78–79.
29. Murase N, Demetris AJ, Matsuzaki T, Yagihasi A, Todo S, Fung J, Starzl TE (1991). Long survival in rats after multivisceral versus isolated small bowel allotransplantation under FK 506. *Surgery* **110**, 87–98.
30. Starzl TE, Todo S, Fung J, Demetris AJ, Venkataramanan R, Jain A (1989). FK 506 for human liver, kidney and pancreas transplantation. *Lancet* **2**, 1000–1004.
31. Starzl TE, Abu-Elmagd K, Tzakis A, Fung JJ, Porter KA, Todo S (1991). Selected topics on FK 506: with special references to rescue of extrahepatic whole organ grafts, transplantation of "forbidden organs", side effects, mechanisms, and practical pharmacokinetics. *Transpl Proc* **23**, 914–919.
32. Todo S, Tzakis AG, Abu-Elmagd K, Reyes J, Fung JJ, Casavilla A, Nakamura K, Yagihashi A, Jain A, Murase N, Iwaki Y, Demetris AJ,

Van Thiel D, Starzl TE (1992). Cadaveric small bowel and small bowel-liver transplantation in humans. *Transplantation* **53**, 369–376.

33. Todo S, Tzakis A, Reyes J, Abu-Elmagd K, Furukawa H, Nour B, Casavilla A, Nakamura K, Fung J, Demetris AJ, Starzl T (1994). Small intestinal transplantation in humans with or without colon. *Transplantation* **57**, 840–848.
34. Todo S, Reyes J, Furukawa H, Abu-Elmagd K, Lee RG, Tzakis A, Rao AS, Starzl TE (1995). Outcome analysis of 71 clinical intestinal transplantations. *Ann Surg* **222**, 270–282.
35. Abu-Elmagd K, Todo S, Tzakis A, Reyes J, Nour B, Furukawa H, Fung JJ, Starzl TE (1994). Three years clinical experience with intestinal transplantation. *J Am Coll Surg* **179**, 385–400.
36. Abu-Elmagd K, Reyes J, Todo S, *et al.* (1998) Clinical intestinal transplantation: new perspectives and immunologic considerations. *J Am Coll Surg* **186**, 512–527.
37. Billingham RE (1959). Reactions of grafts against their hosts: transplantation immunity works both ways — hosts destroy grafts and grafts may harm hosts. *Science* **130**, 947.
38. Monchik GJ, Russell PS (1971). Transplantation of small bowel in the rat: technical and immunological considerations. *Surgery* **70**, 693–702.
39. Tanabe M, Murase N, Demetris AJ, Nakamura K, Furuya T, Fujisaki S, Todo S, Starzl TE (1993). Graft-versus-host disease in fully allogeneic small bowel transplantation: incidence of the disease and strain combinations. *Transpl Proc* **25**, 1214–1215.
40. Murase N, Demetris AJ, Woo J, Tanabe M, Furuya T, Todo S, Starzl TE (1993). Graft versus host disease (GVHD) after BN to LEW compared to LEW to BN rat intestinal transplantation under FK 506. *Transplantation* **55**, 1–7.
41. Murase N, Starzl TE, Tanabe M, Fujisaki S, Miyazawa H, Qing Ye, Delaney CP, Fung JJ, Demetris AJ (1995). Variable chimerism, graft versus host disease, and tolerance after different kinds of cell and solid organ transplantation from Lewis to Brown-Norway rats. *Transplantation* **60**, 158–171.

42. Murase N, Ye Q, Nalesnik MA, Demetris AJ, Abu-Elmagd K, Reyes J, Ichikawa N, Okuda T, Starzl TE (2000). Immunomodulation for intestinal transplantation by allograft irradiation, adjunct donor bone marrow infusion, or both. *Transplantation* **70**, 1632–1641.
43. Starzl TE, Demetris AJ, Murase N, Ildstad S, Ricordi C, Trucco M (1992). Cell migration, chimerism, and graft acceptance. *Lancet* **339**, 1579–1582.
44. Starzl TE, Demetris AJ, Trucco M, Ramos H, Zeevi A, Rudert WA, Kocoua M, Ricordi C, Ildstad S, Murase N (1992). Systemic chimerism in human female recipients of male livers. *Lancet* **340**, 876–877.
45. Starzl TE, Demetris AJ, Trucco M, Murase N, Ricordi C, Ildstad S, Ramos H, Todo S, Tzakis A, Fung JJ, Nalesnik M, Rudert WA, Kocova M (1993). Cell migration and chimerism after whole organ transplantation: the basis of graft acceptance. *Hepatology* **17**, 1127–1152.
46. Starzl TE, Demetris AJ, Trucco M, Ricordi S, Ildstad S, Terasaki P, Murase N, Kendall RS, Kocova M, Rudert WA, Zeevi A, van Thiel D (1993). Chimerism after liver transplantation for type IV glycogen storage disease and Type I Gaucher's disease. *New Engl J Med* **328**, 745–749.
47. Starzl TE, Demetris AJ, Trucco M, Zeevi A, Ramos H, Terasaki P, Rudert WA, Kocova M, Ricordi C, Ildstad S, Murase N (1993). Chimerism and donor specific nonreactivity 27 to 29 years after kidney allotransplantation. *Transplantation* **55**, 1272–1277.
48. Starzl TE, Zinkernagel RM (1998). Antigen localization and migration in immunity and tolerance. *N Engl J Med* **339**, 1905–1913.
49. Starzl TE, Zinkernagel RM (2001). Regulation of transplant rejection, engraftment, and tolerance by antigen migration and localization: an historical perspective. *Nature Rev Immunol* **1**, 233–239.

50. Nakao A, Nalesnik MA, Ishikawa T, Azhipa O, Demetris AJ, Starzl TE, Murase N (2002). Stable macrochimerism and tolerance in tacrolimus-treated rat recipients of lymphoid-depleted intestinal allografts and "add back" donor bone marrow cells. *Transplantation*, in press.
51. Abu-Elmagd K, Reyes J, Bond G, Mazariegos G, Wu T, Murase N, Sindhi R, Martin D, Colangelo J, Zak M, Janson D, Ezzelarab M, Dvorchik I, Parizhskaya M, Deutsch M, Demetris A, Fung J, Starzl TE (2001). Clinical intestinal transportation: a decade of a single center experience. *Ann Surg* **234**, 404–417.
52. Murase N, Demetris AJ, Furuya T, Todo S, Fung JJ, Starzl TE (1992). Comparison of the small intestines after multivisceral transplantation and the small intestine transplanted with portal versus caval drainage. *Transpl Proc* **24**, 1143–1144.
53. Schraut WH, Rosemurgy AS, Riddell RM (1983). Prolongation of intestinal allograft survival without immunosuppressive drug therapy. *J Surg Res* **34**, 597–607.
54. Starzl TE, Porter KA, Francavilla A (1983). The Eck fistula in animals and humans. *Curr Probl Surg* **20**, 687–752.
55. Grant D (2001). 2001 Report of the International Intestinal Transplant Registry. *7th International Small Bowel Transplant Symposium, September 12–15, 2001, Stockholm.*
56. Grant D (1999). Intestinal transplantation: 1997 Report of the International Registry. *Transplantation* **67**, 1061–1064.
57. Gruessner RW, Uckun FM, Pirenne J, Nakhleh RE, Benedetti E, Bekersky I, Troppmann C, Gruessner AC (1997). Recipient preconditioning and donor-specific bone marrow infusion in a pig model of total bowel transplantation. *Transplantation* **63**, 12–20.
58. Gruessner RWG (1998). Invited commentary for Reference 36.
59. Howard L, Hassan N (1998). Home parenteral nutrition. *Clin Nutrit* **27**, 481–512.
60. Abu-Elmagd K, Reyes J, Fung JJ, Mazariegos G, Bueno J, Janov C, Colangelo J, Rao A, Demetris A, Starzl TE (1999). Evolution of clinical intestinal transplantation: improved outcome and cost effectiveness. *Transpl Proc* **31**, 582–584.

IX

THE HISTORY OF HEART TRANSPLANTATION AND HEART VALVE TRANSPLANTATION

Timothy J Kroshus
Vibhu R Kshettry

The possibility of human heart transplantation was first entertained in the early 1900s, starting with a heterotopic heart transplant performed by Alexis Carrel and Charles Guthrie in 1905 at the University of Chicago.[1] These experiments were performed to establish the techniques of vascular anastomoses. This first reported heart transplant was performed by transplanting a smaller dog heart into the neck of a larger dog. This transplanted heart survived about two hours.[1] Alexis Carrel continued his pioneering efforts by performing the first heart-lung transplant in 1906, and reported his findings in a manuscript read before the Johns Hopkins Hospital Medical Society on April 23, 1906.[2] This heart-lung transplant was also a heterotopic transplant performed in the neck of a large cat.

The next published studies in cardiac transplantation were by Mann in 1933.[3] He studied the technical aspects of heterotopic transplantation into the carotid-jugular circulation. Arterial inflow was established via the aorta and coronary arteries and venous outflow via the coronary vein and pulmonary artery. Mann reported an eight-day graft survival using this model, and made several of the first observations of allograft rejection.[3] Further work using a modification of this model, the Marcus I technique, was published in 1951 by Marcus *et al.* from the Chicago Medical School.[4] A further modification which enabled a pumping function of both ventricles was reported in 1953.[5] This model was adopted by several other investigators and allowed further technical and immunological studies of the transplanted heart.[6,7] Downie *et al.* reported 23 of 30 successful experiments with ischaemic times ranging from 30–45 minutes with a maximal survival of ten days.[7]

Pioneering work was also being peformed in Russia by Vladimir Petrovich Demikhov during the 1940s and 1950s. He reported in a monograph experiments on the transplantation of a second heart in 1940 and then canine heterotopic cardiac transplants in the inguinal region in 1946.[8] In 1956, he performed several intrathoracic heterotopic cardiac transplants, one of which survived for 32 days. Demikhov reported a series of 250 experiments involving intrathoracic heterotopic transplants, with success ranging from graft survival up to five days with non-sustaining heterotopic heart-lung transplantation and 15.5 hours with a heterotopic heart transplant that functioned to fully support the recipient.

Marcus *et al.* reported in 1953 a heart-lung heterotopic transplant with the heart-lung block being placed into the abdomen.[5] Donor lung ventilation was performed by a separate ventilator, with maximum graft survival of nine hours. Additional studies allowing the heterotopic heart-lung transplant to completely support the recipient for 75 minutes by inflow occlusion of the native heart allowed exploration of the native mitral valve.

Reemtsma *et al.* studied graft survival using immunosuppression in a intrathoracic heart transplant model.[9] Total circulatory support was transferred to the donor heart in one experiment by fibrillating the native heart and allowed circulatory support for up to four hours.[10] Johnson *et al.* from the Karolinska Institute in Stockholm reported intrathoracic heterotopic heart transplants in which the graft functioned as an auxillary left heart pump.[11]

Orthotopic heart transplantation was first performed by Demikhov in the early 1950s. He devised a method of sequential anastomoses to maintain both donor and recipient perfusion during orthotopic heart-lung transplantation without use of cardiopulmonary bypass.[8] In 1953, Neptune *et al.* performed heart-lung transplants using deep hypothermia and demonstrated survival up to six hours in dogs.[12] Transplant studies at this time were limited by organ preservation techniques. This led to studies in maintaining the viability of the donor organs to allow for transfer and subsequent transplantation. Webb *et al.* in 1957 reported on the function of the refrigerated heart.[13] They were able to maintain viability and function of the heart for periods of at least eight hours by refrigeration in nutrient medium.[13]

Futher studies in orthotopic heart-lung transplantation were limited by the absence of functional return of spontaneous respiration in dogs.[14] Investigators later found that denervated lungs caused respiratory compromise in dogs but not in primates.[15,16] Further studies in orthotopic heart-lung transplants were performed in primates and reported by Reitz from Stanford.[17,18]

Technical advances in orthotopic heart transplantation were made by Goldberg *et al.* from the University of Maryland in 1958, decribing orthotopic transplantation with transection of the left auricle to circumvent the anastomosis of the several pulmonary veins.[19] This technique of excision of the recipient heart at the mid-atrial level proved to be a significant technical advance in heart tranplantation. Cass and Brock, working at Guy's Hospital in London, carried the atrial cuff technique one step further by dividing the right atium at mid-level also, creating a cuff encompassing the superior and inferior vena cavae and allowing the multiple pulmonary venous and arterial anastomoses to be encompassed into two large and technically simple atrial anastomoses.[20]

In 1959, Lower and Shumway performed the first succcessful orthotopic heart transplant resulting in full recovery of the animal.[21] They used cardiopulmonary bypass with surface cooling for recipient protection and donor heart protection by profound topical hypothermia induced by rapid immersion in iced saline.[22,23] Using this technique, they were able to perform eight consecutive transplants, with five of the eight surviving six to 21 days with no immunosuppression.[21] Kantrowitz at Maimonides Hospital in Brooklyn, New York performed 37 orthotopic transplants in puppies using the technique described by Lower and Shumway.[24] Despite no immunosuppression, 18 animals survived more than one week, 12 more than two weeks, and one for 213 days.

Following the successful technical advances in orthotopic heart transplantation, new studies were then aimed at immunosuppresion and detection of rejection. In 1965, Lower and Shumway described ECG voltage changes associated with cardiac allograft rejection.[25] Voltage was found to decrease significantly with rejection and was restored by administration of azathioprine and methylprednisolone.

Clinical Heart Transplantation in Humans

The era of clinical human heart transplantation began in 1964 when James D. Hardy at the University of Mississippi performed a chimpanzee-to-human orthotopic heart transplant.[26] The transplant was technically satisfactory, however was not able to maintain the circulatory system of the recipient. In 1966, Lower performed a reverse experiment with removal of a human heart and transplanting it into a primate recipient, proving that the human heart could be removed, transplanted and then successfully resuscitated.[27] On December 3, 1967, Christian N. Barnard performed the first successful human orthotopic heart transplant at Groote Schuur Hospital in Cape Town, South Africa.[28] Post-operative immunosuppression consisted of high-dose prednisone and hydrocortisone, azathioprine, actinomycin C, and local irradiation. The recipient survived for 18 days before succumbing to *Pseudomonas* pneumonia. Three days later, Adrian Kantrowitz, in Brooklyn, New York, performed the second human heart transplant by transplanting the heart of an anencephalic infant into an 18-day-old child with Ebstein's anomaly using deep hypothermia and circulatory arrest.[29] The recipient only survived for 6.5 hours post-operatively. Shumway performed the fourth human heart transplant on January 6, 1968, starting the most successful early series of human cardiac transplantation. The first clinical heart-lung transplant was performed on September 15, 1968 by Denton Cooley, who replaced the heart and lungs of a two-month-old with an atrioventricular canal defect with those of an anencephalic infant donor. This recipient survived for 14 hours before succumbing to respiratory insufficiency.

The field of heart transplantation exploded in 1968, with 102 cardiac transplants performed in 17 countries. The mean survival was only 29 days for the first 100 transplants which dampened enthusiasm for the procedure, and by 1970, most institutions had abandoned the procedure.[30] Experimental work continued at Stanford University which resulted in increasing the one-year survival from 22% in 1968 to 65% by 1978.[31] Research efforts were focused on immunosuppresion, infection, indications and contraindications for the procedure, and donor management. Diagnosis of cardiac allograft rejection was advanced by Caves and Billingham, who developed a technique of transvenous endomyocardial biopsies and describing a histologic grading system.[32]

The development of rabbit antithymocyte globulin enhanced the treatment of rejection and improved survival. Antithymocyte globulin, shown to prolong renal allograft survival, was demonstrated to prolong survival following heart transplantation in dogs at Stanford. Rabbit antithymocyte globulin, developed by Philip Oyer and Charles Bieber, was added to the Stanford clinical regimen in 1973.[33,34] Six doses were given intramuscularly to treat acute rejection.[34] With these improvements in immunosuppression, the one-year survival at Stanford improved from 22% in 1968 (ten transplants), 50% in 1970 (eight transplants), and to 70% in 1976 (21 transplants).

During the early clinical phase of human heart transplantation, the definition of death remained a controversial event. Shumway noted in 1969 that the "death of the donor is a diagnosis which must be made by the neurological and neurosurgical team". Brain death laws were enacted by most states in the United States such as that adopted by the state of California in 1974 based on guidance from the American Bar Association.[35, 36] The statute declared "A person shall be pronounced dead if it is determined by a physician that the person has suffered a total and irreversible cessation of brain function". These laws allowed removal of a beating heart from a ventilated patient. Brain death criteria were formulated by several groups including an *ad hoc* committee of the Harvard Medical School and a National Institutes of Health (NIH) report on cerebral survival.[37, 38] The Harvard committee differed by the requirement of an electroencephalogram. However, others, including the Medical Royal Colleges in the United Kingdom and the NIH in the U.S. rely principally on strict criteria based on structural cerebral injury of known cause, in the absence of conditions or medications that depress cerebral function such as hypothermia or drugs.

During the 1970s, further refinements and improvements in clinical heart transplantation originated principally from the Stanford program. Improvements were made in donor management by recognizing the need to support the potential donor in the intensive care unit with monitoring of central venous and arterial pressures in order to optimize fluid management. Failure to properly manage the donor resulted in impairment of graft function.[39] Refinements in recipient selection criteria also evolved in the 1970s. Patients with positive lymphocyte crossmatch, active infection, insulin-dependent diabetes mellitus, age greater than 50 years, severe renal or hepatic dysfunction, or recent pulmonary infarction were identified as having a significantly poorer outcome after transplantation. The Stanford criteria became accepted worldwide until improvements in immunosuppression and infectious disease management allowed relaxation of recipient selection criteria. It also became apparent in the late 1970s that previous operative treatment did not compromise survival and by 1977, successful cardiac re-transplantation had been achieved.[40, 41]

In the 1970s, corticosteroids and azathioprine were the foundation of clinical immunosuppression after transplantation, with the additional use of antithymocyte or antilymphocyte globulin for the treatment of acute rejection. The introduction of cyclosporine in the 1980s markedly improved the results of cardiac transplantation and led to renewed worldwide enthusiasm for the procedure.[42] This was first used clinically following heart transplantation at Stanford in 1980, and resulted in a one-year survival of 80%. In addition to improved patient survival, the use of cyclosporine resulted in accelerated rehabilitation and reduced hospital stay. This led to a renewed interest in cardiac transplantation with a dramatic expansion of heart transplant centers throughout the world. Since the advent of cyclosporine, the survival of heart transplantation improved to 75–80% at five years and 65–70% at ten years. Further advances in immunosuppresion have been made with newer, more effective drugs aimed at preventing rejection, treating acute rejection episodes and reducing the incidence of opportunistic infections.

Monoclonal antibodies have been developed in recent years, targeted at specific T-lymphocyte markers and receptors using hybridoma techniques.[43] The most widely used monoclonal antibody clinically is OKT3, which was the first monoclonal antibody approved by the Food and Drug Administration for therapeutic use. OKT3 modifies allograft rejection by inhibiting the T-cell antigen recognition site bound to the CD3 antigen.[44] OKT3 was first used in cardiac transplantation by the University of Utah for the successful treatment of refractory rejection.[45] Subsequent use of OKT3 as early prophylactic therapy decreased the incidence of acute rejection and late rejection and reduced the number of patients needing corticosteroid therapy.[46, 47] Additional more selective monoclonal antibodies have been developed as well as newer immunosuppressive drugs such as FK506, Rapamycin, RS61443, and mycophenolate mofetil that promise to potentially reduce acute and chronic rejection as well as opportunistic infections leading to improved survival following transplantation.

In recent years, potential cardiac recipients who cannot be maintained on conventional medical measures have required the placement of a mechanical assist device while they await a suitable donor organ. A number of devices are available, including Thermo-cardiosystems Inc. (TCI), Novocor, Thoratec and Abiomed. Indications for implantation include worsening heart failure refractory to maximal medical management in a patient of suitable size appropriate for the device. These devices have allowed many patients to survive until a suitable donor organ is available, and subsequently have undergone successful transplantation.

Current Results of Heart Transplantation

The Registry of the International Society for Heart and Lung Transplantation recently published their results on thoracic organ transplantation extending through 1999, including heart, heart-lung, single lung and bilateral single/double lung.[48] Through 1999, a total of 55,359 cardiac transplants have been performed and reported to the International Registry. The number of centers performing heart transplantation in 1998 was 213, down from 241 centers in 1993. In addition, the number of heart transplants performed has undergone a steady decline over the past four years with 4466 transplants performed in 1995, 4430 in 1995, 4245 in 1996, 4019 in 1997, and 3646 in 1998. Data from 1999 is incomplete. Indications for adult heart transplantation included coronary artery disease (44.3%), cardiomyopathy (43.7%), valvular heart disease (3.6%), re-transplantation (2.0%), congenital defects (1.5%), and miscellaneous (4.9%).

Total heart transplant actuarial survival over the past 16 years was reported to be 81% at one year. The patient half-life (time to 50% survival) was 9.8 years, and for those surviving past the first year, the half-life was 12.1 years. Results in recent years have improved, with the one- and three-year survival at 85.6% and 79.5%, respectively. Early mortality is most commonly caused by primary graft failure. Cardiac allograft vasculopathy, non-specific graft failure, and malignancy predominate as causes of death after the first year.[48]

History of Heart Valve Transplantation

Early heart valve replacements were performed with crude mechanical valves which were highly thrombogenic. This led to the implantation of homografts, first inserted in 1962 by Donald Ross in London and Brian Barrett-Boyes in Auckland.[49, 50] In the 1960s and early 1970s, homografts were the biological valve of choice. However, there were difficulties in harvesting as well as limited life-span of the homograft. This stimulated further research efforts for alternative bioprostheses, paving the way for the introduction of porcine, and later, pericardial bioprosthetic valves. By 1985, interest was renewed in homografts after showing reliable long-term patient survival[51, 52] and with the demonstration that glutaraldehyde-treated bioprosthetic valves had a limited duration of effectiveness.

Homografts are considered to be either viable or non-viable depending on the methods of harvesting and storage. In reality, none of the stored homografts are viable as none had an alive and functioning vascular endothelium. No differences have been found with regards to primary tissue failure or long-term performance whether the valves were sterilized by antibiotics, ethylene oxide or γ-irradiation, or with storage in a nutrient solution, freeze dried, or cryopreserved.[52]

Use of homograft valves have primarily been for aortic valve replacement, aortic root replacement, or pulmonary valve replacement. Results of aortic homograft replacement have been excellent in experienced hands. Overall actuarial survival 21 years after homograft replacement was 62%, and 67% if excluding non-cardiac related deaths.[52] However, only 9% of homograft valves remained functional by 21 years post-implant, with primary tissue failure as the cause of valve failure.

The use of mitral homografts was introduced into clinical practice in the 1960s. This procedure was soon abandoned due to early failure due to chordal rupture. This technique has recently been re-attempted with more successful results.[53] The tricuspid valve has been shown to be successfully replaced by a mitral valve homograft,[54] followed by successful replacement of the mitral valve by a mitral valve homograft.[55–57] Double mitral and tricuspid valve replacement and partial mitral valve replacement as an extension of mitral valve repair have also been reported. Few numbers of procedures performed make it difficult to draw any firm conclusions from these experiences.

The possibility of using a patient's own pulmonary valve for replacement of the aortic valve was demonstrated experimentally by Lower and Shumway in 1960.[58] Animal studies indicated that the excised pulmonary valve functioned under systemic pressure. In 1982, Gorezynski *et al.* reported no acute failures for pulmonary valve autografts placed in the aortic position.[59] The first patient operated on using this technique in 1967 was alive and well in 1985, 28 years after the procedure.[60] The surgical procedure was initially performed as a sub-coronary technique, but since 1980 has largely been inserted as a free-standing root replacement in order to avoid distortion and allows the valve to grow with the patient. This procedure also requires the placement of a homograft in the pulmonary position, which is at lower pressure than the aortic side. The first clinical series with at least ten years follow-up after pulmonary autograft placement in the aortic position was published in 1994.[61] Survival at ten and 20 years was 86% and 61%, respectively. The main cause of re-operation was aortic regurgitation caused by technical mal-insertion at the original operation. There was a 74% freedom from re-operation or death at 20 years on the pulmonary side. In patients who required replacement of their autografts, all explanted cusps showed viable cells, usually with retention of normal structure and an endothelial lining indicating retained viability.

Summary

Cardiac transplantation has evolved over the past 100 years from an experimental procedure performed in laboratory animals to an effective therapeutic treatment measure for end-stage heart failure. Advancements in technique, preservation, donor and recipient selection, immunosuppression, treatment of post-operative complications and use of mechanical assist devices as bridge therapy have increased the application of heart transplantation to a wider variety of patients. Use of homografts and autografts for valve replacements has resulted in successful treatment for valvular disease. The pulmonary autograft has an established place for the surgery of aortic valve disease, particularly for children, physically active young adults and women desiring children.

References

1. Carrel A, Guthrie CC (1905). The transplantation of veins and organs. *Am Med* **10**, 1101–1102.
2. Carrel A (1907). The surgery of blood vessels, etc. *Bull John Hopkins Hosp* **18**, 18–28.
3. Mann FC, Priestley JT, Markowitz J, *et al.* (1933). Transplantation of the intact mammalian heart. *Arch Surg* **26**, 219–24.
4. Marcus E, Wong SNT, Luisada AA (1951). Homologous heart grafts: transplantation of the heart in dogs. *Surg Forum* **2**, 212–217.
5. Marcus E, Wong SNT, Luisada AA (1953). Homologous heart grafts. I. Technique of interim parabiotic pefusion II. Transplantation of the heart in dogs. *Arch Surg* **66**, 179–191.
6. Cooper DKC (1968). Experimental development of cardiac transplantation. *Br Med J* **4**, 174–181.
7. Downie HG (1953). Homotransplantation of the dog heart. *Arch Surg* **66**, 624–636.
8. Demikhov VP (1962). *Experimental Transplantation of Vital Organs (Haigh B, Trans).* New York: Consultants' Bureau.
9. Reemtsma K (1964). The heart as a test organ in transplantation studies. *Ann NY Acad Sci* **120**, 778–785.
10. McGough EC, Brewer PL, Reemtsma K (1966). The parallel heart: studies of intrathoracic auxiliary cardiac transplants. *Surgery* **60**, 153–158.

11. Johansson L, Soderlund S, William-Olsson G (1967). Left heart bypass by means of a transplanted heart. *Scand J Thorac Cardiovasc Surg* **1**, 23–26.
12. Neptune WB, Cookson BA, Bailey CP, *et al.* (1953). Complete homologous heart transplantation. *Arch Surg* **66**, 174–178.
13. Webb WR, Howard HS (1957). Restoration of function of the refrigerated heart. *Surg Forum* **8**, 302–306.
14. Webb WR, Howard HS (1957). Cardio-pulmonary transplantation. *Surg Forum* **8**, 313–317.
15. Haglin J, Telander RL, Muzzall RE, *et al.* (1963). Comparison of lung autotransplantation in the primate and dog. *Surg Forum* **14**, 196–198.
16. Nakae S, Webb WR, Theodorides T, Sugg WL (1967). Respiratory function following cardiopulmonary denervation in dog, cat, and monkey. *Surg Gynecol Obstet* **125**, 1285–1292.
17. Reitz BA (1990). The history of heart and heart-lung transplantation. In Baumgartner WA, Reitz BA, Achuff SC (eds.) *Heart and Heart-Lung Transplantation.* Philadelphia: WB Saunders, pp. 1–14.
18. Reitz BA (1989). Development of heart-lung transplantation. In Wallwork J (ed.) *Heart and Heart-Lung Transplantation.* Philadelphia: WB Saunders, pp. 437–448.
19. Goldberg M, Berman EF, Akman LC (1958). Homologous transplantation of the canine heart. *J Int Coll Surgeons* **30**, 575–586.
20. Cass MH, Brock R (1959). Heart excision and replacement. *Guy's Hosp Rep* **108**, 285–290.
21. Lower RR, Shumway NE (1960). Studies in orthotopic homotransplantation of the canine heart. *Surg Forum* **11**, 18–19.
22. Lower RR, Stofer RC, Shumway NE (1961). Homovital transplantation of the heart. *J Thorac Cardiovasc Surg* **41**, 196–204.
23. Shumway NE, Lower RR, Stofer RC (1959). Selective hypothermia of the heart in anoxic cardiac arrest. *Surg Gynecol Obstet* **109**, 750–754.

24. Kondo Y, Gradel FO, Chaptal P-A, *et al.* (1965). Immediate and delayed orthotopic homotransplantation of the heart. *J Thorac Cardiovasc Surg* **50**, 781–791.
25. Lower RR, Dong E Jr, Shumway NE (1964). Long-term survival of cardiac homografts. *Surgery* **58**, 110–119.
26. Hardy JD, Chavez CM, Kurrus FD, *et al.* (1964). Heart transplantation in man. *JAMA* **188**, 114–122.
27. Griepp RB, Ergin MA (1984). The history of experimental heart transplantation. *Heart Transpl* **3**, 145.
28. Barnard CN (1967). A human cardiac transplant: an interim report of a successful operation performed at Groote Schuur Hospital, Cape Town. *S Afr Med J* **41**, 1271–1274.
29. Kantrowitz A, Huller JD, Joos H, *et al.* (1968). Transplantation of the heart in an infant and an adult. *Am J Cardiol* **22**, 782–790.
30. Cooley DA, Bloodwell RD, Hallman GL, *et al.* (1969). Organ transplantation for advanced cardiopulmonary disease. *Ann Thor Surg* **8**, 30–46.
31. Griepp RB (1979). A decade of human heart transplantation. *Transpl Proc* **11**, 285–292.
32. Caves PK, Billingham ME, Stinson EB, Shumway NE (1974). Serial transvenous biopsy of the transplanted human heart. Improved management of acute rejection episodes. *Lancet* **1**, 821–826.
33. Griepp RB, Stinson EB, Dong E Jr, *et al.* (1972). The use of antithymocyte globulin in human heart transplantation. *Circulation* **45**(Suppl 1), 147–153.
34. Bieber CP, Griepp RB, Oyer PE, *et al.* (1976). Use of rabbit antithymocyte globulin in cardiac transplantation. *Transplantation* **22**, 478–488.
35. State Laws: California Health and Safety Code, Section 7180-81 (West Suppl 1975).
36. DeMere M. *Report on Definition of Death from Law and Medicine Committee*. American Bar Association, adopted February 1975.
37. Beecher HK, Adams RD, Barger AC, *et al.* (1968). A definition of irreversible coma. *JAMA* **205**, 85.

38. Walker AE (1975). Cerebral death. In Tower DB (ed.) *The Nervous System.* New York: Raven Press, Vol. 2, p. 75.
39. Billingham ME, Baumgartner WA, Watson DC, *et al.* (1980). Distant heart procurement for human transplantation. *Circulation* **62**, 1–11
40. Doug E Jr, Stinson RB, Griepp RB, *et al.* (1973). Cardiac transplantation following failure of previous cardiac surgery. *Surg Forum* **24**, 150.
41. Copeland JG, Griepp RB, Bieber CP, *et al.* (1977). Successful retransplantation of the human heart. *J Thor Cardiovasc Surg* **73**(2), 242–247.
42. Oyer PE, Stinson EB, Jamieson SW, *et al.* (1983). Cyclosporin-A in cardiac allografting: a preliminary experience. *Transpl Proc* **15**, 541–548.
43. Kohler G, Milstein C (1975). Continuous cultures of fused cells secreting antibody of predefined specificity. *Nature* **256**, 495.
44. Bristow MR, Gilbert EM, Renlund DG, *et al.* (1988). Use of OKT3 monoclonal antibody in heart transplantation: review of the initial experience. *J Heart Transpl* **7**, 1.
45. Gilbert EM, DeWitt CS, Eiswirth CC, *et al.* (1987). Treatment of refractory cardiac allograft rejection with OKT3 monoclonal antibody. *Am J Med* **82**, 202.
46. Starnes VA, Oyer PE, Stinson B, *et al.* (1988). Prophylactic OKT3 used as induction therapy for heart transplantation. *Circulation* **78**, 279.
47. Renlund DG, O'Connell JB, Gilbert EM, *et al.* (1987). Feasibility of discontinuation of corticosteroid maintenance therapy for heart transplantation. *J Heart Transpl* **6**, 71.
48. Hosenpund JD, Bennett LE, Keck BM, *et al.* (2000). The Registry of the International Society for Heart and Lung Transplantation: 17th official report — 2000. *J Heart Lung Transpl* **19**, 909–931.
49. Ross DN (1962). Homograft replacement of the aortic valve. *Lancet* **2**, 447.
50. Barratt-Boyes BG (1964). Homograft aortic valve replacement in aortic incompetence and stenosis. *Thorax* **19**, 131–135.

51. Bodnar E, Wain W, Martelli V, Ross DN (1979). Long-term performance of 580 homograft and autograft valves used for aortic valve replacement. *J Thor Cardiovasc Surg* **27**, 31–38.
52. Bodnar E, Parker R, Davies J, *et al.* (1990). Non-viable aortic homografts. In Bodnar E (ed.) *Surgery for Heart Valve Disease*. London: ICR Publishers, pp. 497–504.
53. Revuelta JM, Cagigas JC, Bernal JM, *et al.* (1992). Partial replacement of mitral valve by homograft. An experimental study. *J Thor Cardiovasc Surg* **104**, 1274–1279.
54. Pomar JL, Mestres CA (1993). Tricuspid valve replacement using a mitral homograft. Surgical technique and initial results. *J Heart Valve Dis* **2**, 125–128.
55. Acar C, Farge A, Ramsheyi A, *et al.* (1994). Mitral valve replacement using a cryopreserved mitral homograft. *Ann Thor Surg* **57**, 746–748.
56. Acar C, Gaer J, Chauvand S, Carpentier A (1995). Technique of homograft replacement of the mitral valve. *J Heart Valve Dis* **4**, 31–34.
57. Kumar AS, Trehan H (1994). Homograft mitral valve replacement — a case report. *J Heart Valve Dis* **3**, 473–475.
58. Lower RR, Stofer RC, Shumway NH (1960). A study of pulmonary valve autotransplantation. *Surgery* **48**, 1090–1100.
59. Gorezynski A, Trenker M, Anisurowicz L, *et al.* (1982). Biomechanics of the pulmonary autograft valve in aortic position. *Thorax* **37**, 535–539.
60. Ross DN (1967). Replacement of the aortic and mitral valves with a pulmonary autograft. *Lancet* **2**, 446–447.
61. Chambers JC, Somerville J, Ross DN (1994). Pulmonary autografts: results at 26 years. *J Am Coll Cardiol* **23**, 429A.

X

THE HISTORY OF LUNG TRANSPLANTATION

Timothy J Kroshus
Vibhu R Kshettry

The world's first human lung transplant was performed at the University of Mississippi Medical Center by Dr. James D. Hardy in 1963.[1] This procedure was performed in a 58-year-old prison inmate who was suffering from chronic infection, abscess and atelectasis formation in his left lung. The patient survived for 18 days and maintained excellent arterial oxygen saturations for the first week post-transplant. Immunosuppressive therapy had not been developed and was not used in this first human lung transplant.

Experimental work in lung transplantation was first recorded in 1947 by V.P. Demikhov, a Russian physiologist. His work was published in Russian and was not widely recognized until published in English in 1962 in the book titled *Experimental Transplantation of Vital Organs*.[2] His experiments involved transplanting the right lower lobe in dogs. Others also transplanted the right lower lobe in dogs with survival of up to 12 days for autografts compared to six to eight days for allografts.[3] The first successful whole left lung transplant was performed by Metras in France in 1950.[4] He used an operative approach similar to that used today with performance of a venous anastomosis using a cuff of left atrium.

Functional Aspects of Lung Autotransplantation

Studies at the University of Buffalo in 1951 were conducted by autotransplanting the right lung to study pulmonary physiology.[5] Long-term survival was achieved with demonstration of near normal lung function at 35 months post-transplant. These studies stimulated further work by Neptune at Hahnemann Medical College in Philadelphia using a technique of a left atrial cuff for anastomosis rather than individual pulmonary venous anastomoses.[6] Neptune perfomed 24 left lung allograft tranplants with an average survival of nine days. Four of the dogs received steroids in the form of ACTH which increased the survival to 25 days compared to six days without ACTH. Bronchial anastomotic dehiscence was frequent at time of death. Studies published in 1954 from the University of Kansas of left lung allograft transplants demonstrated increased survival with using litter mates as the donor and recipients, which was the first recorded attempt at living-related lung transplantation. Addition of cortisone at 35 to 45 mg/day also significantly increased survival.

The role of hilar advential structures in pulmonary function was first evaluated by Bogardus *et al.* at the University of Washington.[7] The bronchial circulation was shown to play an important critical role in lung function with failure of autologous lung function with disruption of bronchial and advential circulation. The effect of lung denervation was studied by Faber and Beattie by sequentially dividing and reanastomosing the bronchus, left atrial cuff, and pulmonary artery and then three weeks later, performing contralateral pneumonectomy.[8] After pneumonectomy all animals died with an average survival of five hours. These results were confirmed by other investigators.[9–10]

Effect of Immunosuppression on Lung Graft Survival

In 1961, a series of experiments were published demonstrating prolongation of graft survival with the use of immunosupression. Histologic changes reflective of acute rejection were demonstrated in dog lung allografts at days 4–7 post-transplant.[11] Use of methotrexate demonstrated a striking prolongation of graft survival.[12] However, complications of methotrexate use became apparent, leading to further studies with various combinations of immunosuppressive agents. Blumenstock *et al.* reported results of various combinations of immunosuppression consisting of preoperative blood transfusions from the lung donor, irradiation and methotrexate.[13] One dog survived 300 days and demonstrated function of the transplanted lung consisting of 26% total O_2 consumption and 29% of total alveolar ventilation. Hardy *et al.* performed further experiments with various combinations of immunosuppression with vaiable results.[14] He demonstrated a three- to four-fold increase in survival with use of azathioprine. These results led Hardy to conclude that "drugs currently available for suppression of the immune response leave much to be desired," and that "the experimental progress achieved does not yet justify sacrifice of even a minimal amount of lung function in a patient to permit insertion of a homograft in the treatment of chronic pulmonary insufficiency".

Clinical Lung Transplantation in Humans

The first attempt at human lung transplantation by Hardy was described in the beginning of this chapter.[1] The patient's immunosuppression consisted of daily azathioprine, five courses of cobalt therapy to the thymic region, and prednisone. The patient survived for 18 days and died of renal failure with a functioning lung graft. The next human lung transplant was performed by Magovern and Yates at the University of Pittsburgh in a patient with respiratory failure and ventilator dependence secondary to emphysema.[15] The recipient was able to come off the ventilator for short periods of time for walking on postoperative day 4. However, over the next four days the patient's respiratory status deteriorated with consolidation of the transplanted left lung and died on postoperative day 8. Autopsy showed confluent bronchopneumonia and patent vascular anastomoses.

The world's first lobar lung transplant was performed in 1966 at the Tokyo Medical College in Japan.[16] This was the third attempt at human lung transplantation and was performed in a patient with hemoptysis secondary to bronchiectasis of both lower lobes and the right middle lobe. The donor was a 49-year-old man with carcinoma of the left lung requiring pneumonectomy but appeared to spare the lower lobe. The recipient was treated with thymic radiation and azathioprine preoperatively, and postoperatively received azathioprine and prednisone. The recipient initially did well but required eventual removal of the transplanted lobe on postoperative day 18. Histologic findings in the resected transplanted lobe demonstrated findings consistent with early acute rejection. The patient went on to recover without a tranplanted lung, relieved of his hemoptysis.

By 1970, a review by Wildevuur and Benfield showed a total of 23 human lung transplants having been performed.[17] These consisted of three lobar, 17 single lung, and two heart-lung. Only one patient survived beyond 30 days. The major obstacle at that time appeared to be effective long-term immunosuppression. Over the next eight years, an additional 13 lung transplants were performed. All were unsuccessful. During this time, it became apparent that bronchial anastomotic disruption was the leading cause of death for those patients surviving beyond two weeks.[17, 18] Research efforts began focusing more on airway healing in the subsequent years. Bronchial complications have been attributed to ischemia of the donor bronchus. The bronchial arterial circulation is not re-established during transplantation, and rearterialization via recipient bronchial arteries requires at least one to two weeks after the operation.[19] Therefore, the viability of the donor bronchus is dependent upon retrograde collaterals from the pulmonary artery.[20] Early investigators advocated revascularization of the transplant bronchial arteries.[21, 22] However, this proved to be technically demanding and complicated the procedure. Currently, many simpler operative techniques are used to minimize bronchial anastomotic complications. These include shortening the donor bronchial stump to two or less cartilaginous rings proximal to the origin of the upper lobe,[23] reinforcing the anastomosis with a vascularized tissue pedicle such as the omentum,[24] or intercostal muscle pedicle flap,[25] and using the intussuscepting bronchial anastomosis technique.[26, 27] These maneuvers have helped improve bronchial anastomotic healing. Despite these advances, the continued occurrence of airway anastomosis complications, frequently manifesting weeks to months after transplantation, suggests that there are multiple events involved in the pathogenesis of airway anastomosis problems.

Research efforts directed towards bronchial anastomotic healing were led by the Toronto Lung Transplant Group. These studies were prompted by an early experience with a 19-year-old patient who underwent successful right single lung transplantation but later died at postoperative day 17 due to bronchial anastomotic disruption.[28] Initial studies on the effects of steroids and azathioprine on bronchial anastomotic healing were performed in 1981 by Goldberg *et al.* from the Toronto Lung Transplant Group.[29] They demonstrated markedly decreased bronchial anastomotic breaking strength at day 23 and attributed these finding to steroids and azathioprine. Subsequent studies showed that steroids were responsible for the impaired wound healing and that azathioprine had no effect. The addition of cyclosporin A (CSA) in 1981 did not affect bronchial wound healing.[30]

The role of the bronchial arterial supply in bronchial healing has been debated. A study in 1966 by Stone *et al.* showed re-establishment of bronchial healing and circulation by six weeks after bronchial stripping with division and reanastomosis of the left main stem bronchus in dogs.[31] Mills *et al.* studied the effect of bronchial artery reimplantation and demonstrated a lower incidence of bronchial anastomotic complications with reimplantation.[32] Pearson *et al.* reported complete reconstitution of the bronchial artery circulation by the fourth week post-transplant following right lung hilar stripping and reimplantation.[33] Similar work by Rabinovich with left lower lobe autotransplants found complete re-establishment of the bronchial circulation by 30 days.[34] The results of these studies suggested that the greatest ischemic risk is in the first two weeks after transplantation.

The above studies did not address the potential detrimental effects of immunosuppression on bronchial healing. Siegelman and colleagues obtained bronchial arteriograms after canine lung allotransplants immunosuppressed with azathioprine, prednisone, and antithymocyte serum and found bronchial atery re-establishment by 31 days.[35] However, they found a lack of correlation between the presence of bronchial artery regeneration and the occurrence of bronchial necrosis. These findings suggested a powerful impact of immunosuppression and rejection on bronchial healing. The increased risk of bronchial healing combined with steroids and rejection prompted Lima and colleagues to study the effect of buttressing the bronchial anastomosis with an omental predicle wrap.[36] They found increased neovascularity originating from the omental wrap by day four post-transplant. Use of omental wraps in clinical human lung transplantation was later abandoned due to a high incidence of abdominal complications.[37]

Airway stenosis after lung transplantation has been a significant problem in a number of patients after lung transplantation. Stenosis has been managed with a variety of stents, including silicone rubber stents[38, 39] and expandable metallic stents.[40, 41] Airway stenoses have also been managed surgically with sleeve resection of stenotic segments, bilobectomy and retransplantation.[42] Airway anastomoses with bronchomalacia are also effectively managed with stents.[43] The combined use of bronchodilation with stenting has been successful in management of airway stenoses after lung transplantation.[44] Modifications of bronchial anastomotic technique have improved the incidence of airway complications. Use of a telescoping technique has reduced the incidence of airway complications compared to omental wrapping procedures.[44, 45]

Steroids have been implicated in impaired healing of the bronchial anastomosis.[46] Early experience with clinical lung transplantation suggested that avoidance of steroids in the first week post-transplant resulted in fewer airway complications. Avoiding steroids combined with the use of bronchial omentopexy in canine models demonstrated a marked decrease in the incidence of airway complications.[47, 48] With the introduction of cyclosporine, steroid doses were lowered, and by using cyclosporine in combination with azathioprine, adequate immunosuppression could be achieved with bronchial healing. Further improvements in immunosuppression, preservation and operative techniques have improved bronchial anastomosis healing and steroids are now given early after transplantation.[26, 45]

Introduction of Cyclosporine

The success noted with cyclosporine in human heart transplantation led to further efforts at human heart-lung transplantation. Reitz and Shumway reported the results of three heart-lung transplants treated with cyclosporine.[49] One recipient died at four days, but the other two were alive at eight and ten months after transplant. These results were the first to demonstrate long-term function and survival of lung allogratfs. Subsequent laboratory work in single-lung transplants in dogs using cyclosporine demonstrated long-term survival, function, and more easily managed episodes of acute rejection. This led to a renewed interest in human lung transplantation.

Clinical experiences in New York and Toronto with three patients provided further optimism for lung transplantation.[50] Although all three patients died, they demonstrated prolonged allograft function, lack of irreversible rejection, and healing of the bronchial anastomosis with the use of cyclosporine. The first long-term success of single-lung transplantation was reported in 1986 by the Toronto Lung Transplant Group.[50] Two patients with pulmonary fibrosis underwent single-lung transplantation and were reported alive and well at 14 and 26 months after transplantation. The success achieved with pulmonary fibrosis prompted transplantation for other diseases including cystic fibrosis, bronchiectasis, and obstructive lung disease (emphysema and α-1-antitrypsin deficiency). All these diseases were initially thought to need bilateral lung transplants, which later was required only in those patients with infectious lung diseases.

Initially there was an interest in performing heart-lung transplantation in these patients. It later became apparent that rejection of the heart or lungs could occur simultaneously or independently, making the postoperative monitoring complicated.[51,52] Furthermore, it became apparent that it was unnecessary to transplant a heart when only the lungs needed replacement. It was also observed that severe right heart dysfunction was potentially reversible after bilateral lung transplantation.

In 1986, Dark and colleagues reported on the technique of simultaneous bilateral *en bloc* lung transplantation using a single tracheal anastomosis.[53] This technique was found to be a technically complicated procedure requiring cardiopulmonary bypass. In addition, the tracheal anastomosis suffered from a high rate of complications secondary to ischemia despite omental wrapping, leading to mortality rate of 20–25%. This led to adoption of the technique of bilateral single-lung transplantation, initially described in 1950 by Metras,[4] which is the technique still in use today.

Living-Related Lobar Lung Transplantation

The success of human lung transplantation has improved due to advancements in surgical technique, donor organ preservation, prevention and treatment of rejection, and management of infection. This has resulted in an expansion of recipient selection criteria and increased demand for donor organs. Unfortunately, this also resulted in an increased shortage of organs. Starnes and colleagues have performed living-donor lobar transplantation in patients with cystic fibrosis with a one-year survival of 73.8%.[54] Indications were subsequently expanded to include recipients with other diagnoses. Starnes *et al.* have recently reported the results of eight patients with a variety of diagnoses, including primary pulmonary hypertension, post-chemotherapy pulmonary fibrosis, bronchpulmonary dysplasia, idiopathic pulmonary fibrosis and obliterative bronchiolitis.[55] Overall survival at one year was 75%. These results support an expanded role for living-donor lobar lung transplantation.

Long-Term Complications After Lung Transplantation

Chronic lung dysfunction, thought to be synonomous with chronic rejection, is the most significant long-term complication after lung transplantation.[56, 57] Chronic lung dysfunction was defined by criteria established for patients with bronchiolitis obliterans syndrome (BOS).[58] BOS is presumed to be due to chronic allograft rejection and is characterized clinically by a progressive decline in pulmonary function with the presence or absence of pathologic evidence of obliterative bronchiolitis (OB). OB is a progressive airway disease identified histologically by the presence of small airway inflammation and fibrosis of the lamina propria and lumen and characterized clinically by rapidly progressive airflow obstruction. OB was first reported after heart-lung transplantation in 1985.[59] OB is a major cause of death and disability in long-term survivors of lung transplantation, occurring in up to 50% of patients surviving the first year after transplantation.[60] Augmentation of immunosuppression has been advocated and appears to ameliorate or stabilize the disease in some patients.[61, 62] Despite this, OB often progresses relentlessly, leading to death. OB is frequently unresponsive to medical therapy, with retransplantation as the only therapeutic option. Risk factors noted to have an effect on the development of BOS include cytomegalovirus pneumonitis, acute rejection, human leukocyte antigen mismatches, and absence of donor antigen-specific hyporeactivity.[63]

Current Results of Human Lung Transplantation

The International Society for Heart and Lung Transplantation recently published the reported results for lung transplantation.[64] Through

1999, a total of 6514 single-lung transplants, 4634 bilateral/double-lung transplants and 2698 heart-lung transplants have been performed. The number of centers in the world performing single-lung transplants was 94 and the number of centers performing bilateral/double-lung transplants was 93 in 1999. Despite the use of older donors, the number of lung transplants (single and bilateral/double) performed has remained relatively stable since 1995, with 1353 performed in 1995, 1291 in 1996, 1367 in 1997, 1247 in 1998 and 1057 in 1999 (data for 1999 incomplete). Indications for adult lung transplantation are listed in Table 1.

Table 1 Indications for adult lung transplantation

	Single-lung	Bilateral/double-lung
Emphysema	42.6%	17.5%
Idiopathic pulmonary fibrosis	19.5%	7.1%
α-1 antitrypsin deficiency	10.6%	9.6%
Primary pulmonary hypertension	3.9%	9.2%
Cystic fibrosis	2.4%	31.5%
Re-transplant	2.3%	2.2%
Miscellaneous	18.1%	22.9%

Chronic obstructive pulmonary disease dominates the indications for single-lung transplantation whereas cystic fibrosis is the most common indication for bilateral/double-lung transplantation. In pediatric patients, congenital heart disease, cystic fibrosis and primary pulmonary hypertension are the most common indications. For adult lung transplantation, actuarial survival has consistently improved since 1992. Survival at four years post-transplant is approximately 50% for pediatric and adult lung transplant recipients combined. Continuing research efforts, donor and recipient selection and management, and clinical trials using a variety of new immunosuppressive agents will potentially improve the overall results of lung transplantation.

References

1. Hardy JD, Webb WR, Dalton ML, Walker GR (1963). Lung homotransplantation in man. *JAMA* **186**, 1065–1074.
2. Demikov VP (1962). *Experimental Transplantation of Vital Organs*. New York: Consultants Bureau Enterprises.
3. Staudacher V, Bellinazzo P, Pulin A (1950). Primi rilievi su tentativi di reimpianti autoplastici e di trapianti omoplastici di lobi pulmonari. *Chirugia (Milano)* **5**, 233–237.
4. Metras H (1950). Note preliminaire sur la greffe totale du poumon chez le chien. *C R Acad Sci (Paris)* **231**, 1176–1178.
5. Portin BA, Ramussen GL, Stewart JD, Anderson MN (1960). Physiologic and anatomic studies thirty-five months after successful replantation of the lung. *J Thorac Cardiovasc Surg* **39**, 380–384.
6. Neptune WB, Weller R, Bailey CP (1953). Experimental lung transplantation. *J Thorac Cardiovasc Surg* **26**, 275–289.
7. Bogardus GM (1958). An evaluation in dogs of the relationship of pulmonary, bronchial, and hilar advential circulation to the problem of lung transplantation. *Surgery* **48**, 849–856.
8. Faber LP, Beattie EJ Jr (1958). Respiration following lung denervation. *Surg Forum* **9**, 383–385.
9. Linberg EJ, Demetriades A, Armstrong BW, Konsuwan N (1961). Lung reimplantation in the dog. *JAMA* **178**, 146–147.
10. Alican F, Hardy JD (1962). Lung reimplantation. *JAMA* **183**, 147–151.
11. Blumenstock DA, Kahn DR (1961). Replantation and transplantation of the canine lung. *J Surg Research* **1**, 40–47.
12. Blumenstock DA, Collins JA, Thomas ED, Ferrebee JW (1961). Homotransplantation of the lung in dogs treated with methotrexate. *Surg Forum* **12**, 121–122.
13. Blumenstock DA, Collins JA, Thomas ED, Ferrebee JW (1962). Homotransplants of the lung in dogs. *Surgery* **51**, 541–545.
14. Hardy JD, Eraslan S, Dalton ML Jr, Alican F, Turner MD (1963). Re-implantation and homotransplantation of the lung: laboratory studies and clinical potential. *Ann Surg* **157**, 707–718.

15. Magovern GJ, Yates AJ (1964). Human homotransplantation of left lung: report of a case. *Ann NY Acad Sci* **120**, 710–728.
16. Shinoi K, Hayata Y, Aoki H, *et al.* (1966). Pulmonary lobe homotransplantation in human subjects. *Am J Surg* **111**, 617–728.
17. Wildevuur RH, Benfield O (1970). A review of 23 human lung transplants by 20 surgeons. *Ann Thorac Surg* 489–515.
18. Cooper JD. Lung transplantation (1989). *Ann Thorac Surg* **47**, 28–44.
19. Siegelman SS, Hagstrom JWC, Koerner SK, Veith FJ (1977). Restoration of bronchial artery circulation after canine lung allotransplantation. *J Thorac Cardiovasc Surg* **73**, 792–795.
20. Barman SA, Ardell JL, Parker JC, Perry ML, Taylor AE (1988). Pulmonary and systemic blood flow contributions to upper airway in canine lungs. *Am J Physiol* **255**, H1130–H1135.
21. Mills NL, Boyd AD, Gheranpong C (1970). The significance of bronchial circulation in lung transplantation. *J Thor Cardiovasc Surg* **60**, 866–878.
22. Couraud L, Baudet E, Nashef SAM, *et al.* (1992). Lung transplantation with bronchial revascularisation. *Eur J Cardiothorac Surg* **6**, 490–495.
23. Pinsker KL, Koerner SK, Kamholz SL, *et al.* (1979). Effect of donor bronchial length on healing: a canine model to evaluate bronchial anastomotic problems in lung transplantation. *J Thorac Cardiovasc Surg* **77**, 669–673.
24. Morgan E, Lima O, Goldberg M, *et al.* (1982). Successful revascularization of totally ischemic bronchial autograft with omental pedicle flaps in dogs. *J Thorac Cardiovasc Surg* **84**, 204–210.
25. Fell SC, Mollenkopf FP, Montefusco CM, *et al.* (1985). Revascularization of ischemic bronchial anastomoses by an intercostal pedicle flap. *J Thorac Cardiovasc Surg* **90**, 172–178.
26. Calhoon JH, Grover FL, Gibbons WJ, *et al.* (1991). Single lung transplantation. Alternate indications and technique. *J Thorac Cardiovasc Surg* **101**, 816–825.

27. Veith FJ, Richards K (1970). Improved technique for canine lung transplantation. *Ann Surg* **171**, 553–558.
28. Neelems JMB, Rebuck AS, Cooper JD, *et al.* (1980). Human lung transplantation. *Chest* **78**, 569–573.
29. Lima O, Cooper JD, Peters WJ, *et al.* (1981). Effects of methylprednisolone and azathioprine on bronchial healing following lung autotransplantation. *J Thorac Cardiovasc Surg* **82**, 211–215.
30. Goldberg M, Lima O, Morgan E, *et al.* (1983). A comparison between cyclosporin A and methylprenisolone plus azathioprine on bronchial healing following canine lung autotransplantation. *J Thorac Cardiovasc Surg* **85**, 821–826.
31. Stone RM, Ginssberg RJ, Colapinto RF, Pearson FG (1996). Bronchial artery regeneration after radical hilar stripping. *Surg Forum* **17**, 109–110.
32. Mills NL, Boyd AD, Gheranpong C, Spencer FC (1970). The significance of bronchial circulation in lung transplantation. *J Thorac Cardiovasc Surg* **60**, 866–874.
33. Pearson FG, Golgberg M, Stone RM, Colapinto RF (1970). Bronchial arterial circulation restored after reimplantation of canine lung. *Can J Surg* **13**, 242–250.
34. Rabinovich JJ (1972). Reestablishment of bronchial arteries after experimental lung lobe autotransplantation. *J Thorac Cardiovasc Surg* **64**, 119–126.
35. Siegelman SS, Hagstrom JWC, Koerner SK, Veith FJ (1977). Restoration of bronchial artery circulation after canine lung allotransplantation. *J Thorac Cardiovasc Surg* **73**, 792–795.
36. Lima O, Goldberg M, Peters WJ, *et al.* (1982). Bronchial omentopexy in canine lung transplantation. *J Thorac Cardiovasc Surg* **83**, 418–421.
37. Smith PC, Slaughter MS, Petty MG, *et al.* (1995). Abdominal complications after lung transplantation. *J Heart Lung Transpl* **14**, 44–51.
38. Cooper JD, Pearson FG, Patterson GA, *et al.* (1989). Use of silicone stents in the management of airway problems. *Ann Thorac Surg* **47**, 371–378.

39. Griffith BP, Magee MJ, Gonzalez IF, *et al.* (1994). Anastomotic pitfalls in lung transplantation. *J Thorac Cardiovasc Surg* **107**, 743–754.
40. Higgins R, McNeil K, Dennis C, *et al.* (1994). Airway stenosis after lung transplantation: management with expanding metallic stents. *J Heart Lung Transpl* **13**, 774–778.
41. Nashef SAM, Dromer C, Velly J-F, *et al.* (1992). Expanding wire stents in benign tracheobronchial disease: indications and complications. *Ann Thorac Surg* **54**, 937–940.
42. Schafers H-J, Schafer CM, Zink C, *et al.* (1994). Surgical treatment of airway complications after lung transplantation. *J Thorac Cardiovasc Surg* **107**, 1476–1480.
43. Shennib H, Massard G (1994). Airway complications in lung transplantation. *Ann Thorac Surg* **57**, 506–511.
44. Kshettry VR, Kroshus TJ, Hertz MI, *et al.* (1997). Early and late airway complications after lung transplantation: incidence and management. *Ann Thorac Surg* **63**, 1576–1583.
45. DeHoyos AL, Patterson GA, Maurer JR (1992). Pulmonary transplantation. Early and late results. *J Thorac Cardiovasc Surg* **103**, 295–306.
46. Lima O, Cooper JD, Peters WJ, *et al.* (1981). Effects of methylprednisolone and azathioprine on bronchial healing following lung autotransplantation. *J Thorac Cardiovasc Surg* **82**, 211–215.
47. Goldberg M, Lima O, Morgan, *et al.* (1983). A comparison between cyclosporin A and methylprednisolone plus azathioprine on bronchial healing following canine lung transplantation. *J Thorac Cardiovasc Surg* **85**, 821–826.
48. Lima O, Goldberg M, Peters WJ, *et al.* (1982). Bronchial omentopexy in canine lung transplantation. *J Thorac Cardiovasc Surg* **83**(3), 418–421.
49. Reitz BA, Wallwork JL, Hunt SA, *et al.* (1982). Heart-lung transplantation: successful therapy for patients with pulmonary vascular disease. *N Engl J Med* **306**, 557–564.

50. Toronto Lung Transplant Group (1986). Unilateral lung transplantation for pulmonary fibrosis. *N Engl J Med* **314**, 1140–1145.
51. Griffith BP, Hardesty RL, Trento A, *et al.* (1985). Asynchronous rejection of heart and lungs following cardiopulmonary transplantation. *Ann Thorac Surg* **40**, 488–493.
52. Dawkins KD, Jamieson SW, Hunt SA, *et al.* (1985). Long-term results, hemodynamics, and complications after combined heart-lung transplantation. *Circulation* **71**, 919–926.
53. Dark JH (1986). Experimental *en bloc* double-lung transplantation. *Ann Thorac Surg* **42**, 394–398.
54. Starnes VA, Barr ML, Cohen RG, *et al.* (1994). Lobar transplantation: indications, technique and outcome. *J Thorac Cardiovasc Surg* **108**, 403–411.
55. Starnes VA, Barr ML, Schenkel FA, *et al.* (1997). Experience with living-donor lobar transplantation for indications other than cystic fibrosis. *J Thorac Cardiovasc Surg* **114**, 917–922.
56. Cooper JD, Patterson GA, Trulock EP, *et al.* (1994). Results of single and bilateral lung transplantation in 131 consecutive recipients. *J Thorac Cardiovasc Surg* **107**, 460–471.
57. Sundaresan S, Trulock EP, Mohanakumar T, *et al.* (1995). Prevalence and outcome of bronchiolitis syndrome after transplantation. *Ann Thorac Surg* **60**, 1341–1347.
58. Cooper JD, Billingham M, Egan T, *et al.* (1993). A working formulation for the standardization of nomenclature and clinical staging of chronic dysfunction in lung allografts. *J Heart Lung Transpl* **12**, 713–716.
59. Burke CM, Morris AJ, Dawkins KD, *et al.* (1985). Late airflow obstruction in heart-lung transplant recipients. *Heart Transpl* **4**, 437–440.
60. Burke CM, Theodore J, Baldwin JC, *et al.* (1986). Twenty-eight cases of human heart-lung transplantation. *Lancet* **1**, 517–519.
61. Griffith BP, Paradis IL, Zeevi A, *et al.* (1988). Immunologically mediated disease of the airways after pulmonary transplantation. *Ann Surg* **208**, 371–378.

62. Glanville AR, Baldwin JC, Burke CM, *et al.* (1987). Obliterative bronchiolitis after heart-lung transplantation: apparent arrest by augmented immunosuppression. *Ann Int Med* **107**, 300–304.
63. Kroshus TJ, Kshettry VR, Savik K, *et al.* (1997). Risk factors for the development of bronchiolitis obliterans syndrome after lung transplantation. *J Thorac Cardiovasc Surg* **114**, 195–202.
64. Hosenpud JD, Bennett LE, Keck BM, *et al.* (2000). The registry of the International Society for Heart and Lung Transplantation: seventeenth official report — 2000. *J Heart Lung Transpl* **19**, 909–931.

XI

THE HISTORY OF PEDIATRIC SOLID ORGAN TRANSPLANTATION

Blanche Chavers

Introduction

Solid organ transplantation is the preferred treatment for children with end-stage disease of the kidney, liver and heart because it provides the best potential for normal lifestyles, growth and development. Immunologic research and advances in surgical techniques have led to greater success. Advances in pediatric kidney and liver transplantation have preceded advances in pediatric heart transplantation at most transplant centers. Children with end-stage kidney disease could be maintained on dialysis while awaiting transplantation, but effective bridges to transplantation were not available for end-stage liver and heart patients. Potent immunosuppressive agents needed to advance patient survival in pediatric heart transplantation would not become available until the 1980s.

Surgical techniques for the anastomosis and transplantation of blood vessels were first developed by Alexis Carrel.[1,2] He described circulation in experimental animals after the replantation of a limb and a thyroid gland and after a kidney and a heart transplant. Carrel was awarded the Nobel Prize in 1912 for this accomplishment. A second major advance occurred in the 1940s with experiments conducted by Gibson and Medawar. Their skin grafting experiments in rabbits established the immunologic basis of allograft rejection.[3] Skin grafts would initially take, but then slough at a later time. Histologically, these sloughed grafts showed an inflammatory process. A second graft from the same donor was rejected more rapidly than the first, implying an immune-mediated process. A third major advance was the development of immunosuppressive agents, most of which were initially applied to kidney transplantation.

Kidney Transplantation

The first kidney transplants attempted in humans were heterografts. In 1906, Jaboulay transplanted a kidney from a pig and one from a goat into the antecubital fossa of two separate patients with renal failure but both grafts failed.[4] Attempts by Unger in 1910 using a kidney obtained from an ape and by Neuhof in 1923 using a kidney obtained from a lamb were also unsuccessful.[5,6] Kidney homotransplantation in humans was first reported by Voronoy in 1936.[7] Voronoy performed a cadaver kidney transplant in a patient with renal failure from mercury poisoning, but the patient died two days post-transplant. Unsuccessful attempts were also reported in 1950 and 1951 using grafts from living donors.[8–11]

Dialysis techniques developed in parallel to transplant techniques. Ganter used peritoneal dialysis in 1923 to treat patients with acute renal insufficiency.[12] More than 20 years later, multiple dialysis treatments in patients with renal failure was reported in both adults and children.[13–16] Dialysis not only temporarily improved the patient's status, but also was a bridge, keeping the patient alive until a kidney transplant could be performed.

Allograft survival for more than a few days was reported in the 1950s. Hume and colleagues performed a living donor kidney transplant in 1951. The donor was a two-year-old child whose kidney was removed during an elective operation for treatment of hydrocephalus.[17] The kidney was transplanted into the left thigh of an 18-year-old female with renal failure from chronic glomerulonephritis. It functioned initially, then became anuric after 24 hours. But 12 days later, it began to excrete urine again. After the 26th day post-transplant, urine output began to diminish and ceased altogether 11 days later. Another of the nine kidneys transplanted by Hume and colleagues functioned for 180 days.

The first pediatric kidney transplant was performed by Michon *et al.* in France in December 1952.[18, 19] The patient was a 16-year-old boy who underwent nephrectomy of his native right kidney after a traumatic injury. After the nephrectomy, it was discovered that the boy had congenital absence of his left kidney. The boy's mother was ABO-compatible and donated a normal kidney. It began functioning on the day of transplant, producing three liters of urine during the first 24 hours. But it failed on the 22nd day post-transplant. The boy later died. Histologic evaluation of the transplanted kidney showed allograft rejection.

The first successful kidney transplant was performed by Joseph Murray *et al.* at the Peter Bent Brigham Hospital in Boston, Massachusetts, on December 23, 1954.[20] The kidney was placed retroperitoneally, instead of into the thigh, and anastomosed to the iliac blood vessels. The ureter was implanted into the bladder. This was the first of seven pioneering transplants between identical twins, which Murray described in 1958.[20] Of the seven recipients, two were 14 years old: a girl with renal failure from subacute glomerulonephritis and a boy with congenital megaloureters and chronic pyelonephritis. Both underwent a transplant in 1957. The girl developed recurrence of her original disease in the graft seven weeks post-transplant and died four months post-transplant. The boy's graft never functioned and he died 12 days post-transplant.

Murray *et al.* began a series of six cases in 1958 and 1959 using pre-conditioning with total body irradiation to prevent allograft rejection — the first application of this form of immunosuppressive therapy in humans.[21] Using this immunosuppressive regimen, they performed the first successful kidney transplant between non-identical twins in January 1959.[21, 22] The 24-year-old recipient developed acute rejection eight months post-transplant that was successfully reversed with high-dose prednisone and small doses of whole-body irradiation. The recipient died 25 years post-transplant with a functioning graft. Hamburger *et al.* working in France, reported the second successful transplant between non-identical twins after subtotal irradiation.[23] On October 9, 1959, the first successful transplant between pediatric (12-year-old) identical twins was performed by Murray, Goodwin and Hodges at the University of Oregon in Portland.[24] The recipient had renal failure secondary to chronic glomerulonephritis. A biopsy of the transplanted kidney obtained 18 years post-transplant showed normal morphology.[25] Murray's work in the field of transplantation earned him a Nobel Prize in 1990.

Total body irradiation proved to be quite toxic to the recipient, so alternatives to its use were sought in the late 1950s.[21, 26, 27] In 1959, the hematologists Schwartz and Damashek used the immunosuppressive drug 6-mercaptopurine (synthesized by Elion, Hitchings and Vanderwerff of the Burroughs Wellcome research laboratories) to make mature rabbits tolerant to a soluble antigen.[28] Calne *et al.* showed that 6-mercaptopurine could prolong graft survival and modify allograft rejection in dogs.[29–31] Using another drug synthesized by Elion and Hitchings, Calne showed that the imidazole derivative of 6-mercaptopurine, azathioprine, produced slightly better graft survival in dogs compared with 6-mercaptopurine.[32] One dog survived for four months with a functioning graft on a low maintenance dose of azathioprine.

The first human kidney transplant using azathioprine was performed by Murray *et al.* in March 1961.[33] The living donor was a six-year-old girl who had an expendable kidney removed to treat hydrocephalus. The graft functioned initially, but leukopenia resulted in the azathioprine being stopped on the 23rd day post-transplant. Rejection developed and the patient died on the 38th day post-transplant.[33] The first successful human cadaver kidney transplant was performed by Murray *et al.* in April 1962 using azathioprine and actinomycin C.[33, 34] The allograft functioned for more than a year, thus representing the first case of extended survival in which only drugs were used for immunosuppression.

The first pediatric kidney transplant using azathioprine without subtotal irradiation was performed by Goodwin *et al.* in October 1962 at the University of California in Los Angeles.[24] The 17-year-old recipient with chronic renal failure from glomerulonephritis and pyelonephritis received a kidney from a living donor who underwent a unilateral nephrectomy to treat a urinary fistula caused by cancer. The recipient had previously undergone irradiation of her thymus; bilateral nephrectomy and splenectomy were performed at the time of transplant. Goodwin *et al.* also reported in 1962 the first use of high-dose corticosteroids to successfully reverse kidney graft rejection.[35] In February 1963, Murray *et al.* performed a successful human kidney transplant with an unrelated, healthy, adult volunteer donor; azathioprine and azaserine were used for immunosuppression.[34] Later that year, Starzl *et al.* reported successful reversal of kidney allograft rejection by adding high-dose corticosteroids to maintenance azathioprine immunosuppression: the first prolonged allograft survival in living donor recipients.[36] Azathioprine and prednisone became the mainstay of immunosuppression for kidney recipients.

In November 1962, Starzl *et al.* performed the first successful non-twin, healthy, living related donor, pediatric kidney transplant in a 12-year-old boy with a kidney donated by his mother.[37] Immunosuppression consisted of total body irradiation, azathioprine and prednisone. At the time of Starzl's report, graft survival had been sustained for three years. In March 1963, Hume *et al.* successfully transplanted an 11-year-old girl with a kidney donated by her father.[38] In December 1963, Mowbray *et al.* of St. Mary's Hospital in London performed the first successful pediatric cadaver kidney transplant in a 17-year-old boy.[39] Immunosuppression consisted of prednisone and azathioprine; at the time of Mowbray's report, the graft had survived 586 days. In 1964, Starzl described the technique for transabdominal placement of adult kidneys into infants and small children.[40] In 1964, only 33 centers worldwide were performing human kidney transplants: one in

Belgium, two in France, 24 in the United States, four in England, one in Scotland, and one in Canada.[41]

Chronic, intermittent hemodialysis and peritoneal dialysis techniques were adapted for use in children in the 1960s.[37,42–46] Children with terminal uremia were kept alive by chronic dialysis. At the same time, the appropriateness of kidney transplantation for children was being debated in the medical community. There was concern that the "trauma" to the child exceeded any benefit gained by the transplant. There was also concern that children who were successfully transplanted would become dwarfs secondary to long-term corticosteroid immunosuppression. In an editorial in the *Journal of Pediatrics* in 1964, Riley addressed the ethical issue of kidney transplantation in children, stating that "since the discomfort factors have to be balanced against the length of normal life achieved, a dimension still unknown, no final evaluation as to the value of the procedure can currently be made".[47] This theme would continue to be voiced in the 1970s and 1980s.[48,49]

Major developments in transplantation in the late 1960s and early 1970s included the discovery of human histocompatibility antigens,[50] the use of histocompatibility testing to select the best immunologic donors,[51–53] and the screening of potential kidney transplant recipients for pre-formed antibodies against donor tissue.[54] After antilymphocyte serum was introduced as an immunosuppressive agent, Starzl was the first to use it in human kidney transplantation in 1966.[55–57] One of the recipients was a 16-year-old. Antilymphocyte serum was associated with a decreased incidence of acute rejection and a delay in the time to onset of acute rejection.[57] Starzl recommended its use in combination with prednisone and azathioprine.[57]

In the 1970s, excellent outcomes were reported in children undergoing kidney transplantation. In a series of 56 pediatric recipients at the University of Minnesota in Minneapolis, Najarian *et al.* achieved patient and graft survival rates equivalent to those achieved in adults using antilymphocyte globulin, prednisone and azathioprine.[45] Najarian's report stressed the need for careful assessment and treatment of the lower urinary tract pre-transplant in children with obstructive uropathy, the ability to place the graft retroperitoneally in children weighing more than 20 kg, the recurrence of nephrotic syndrome in the allograft of recipients with steroid-resistant nephrotic syndrome, improved patient and graft survival rates with living related (compared with cadaver) donors, the increased incidence of hypertension in children in the early post-transplant period, and the rare occurrence of "catch-up" growth in children post-transplant.[45] Decreased mortality rates and improved graft survival rates were also reported in 25 children by Potter *et al.*[58] and in 19 children by Gonzalez *et al.*[59] Fine *et al.* reported improved patient and graft survival rates in children receiving cadaver kidney transplants compared with earlier years.[60] By May 1973, of 13,000 kidney transplants reported to the American College of Surgeons/National Institutes of Health Transplant Registry, 946 (7.3%) had been performed in patients under 16.[61] At the end of 1973, 149 children under 15 with functioning grafts had been reported to the European Dialysis and Transplant Association Registry.[62] Now that children could be technically transplanted, concerns began to emerge regarding causes of graft loss (death, rejection and recurrence of original disease), the effect of steroids on the physical appearance of children, and the growth and sexual maturation of children post-transplant.[62–66]

In 1976, J.F. Borel of the Biological and Medical Research Division of Sandoz Laboratories in Basel, Switzerland reported his discovery of a new immunosuppressive agent, cyclosporine, which inhibited cell-mediated cytotoxicity *in vitro*[67] Cyclosporine, a fungal peptide, prolonged graft survival in experimental animals more effectively than other available agents.[68, 69] In 1978 and 1979, cyclosporine was reported to improve graft survival in human cadaver kidney transplant recipients.[70, 71] Later studies supported its effectiveness as a powerful immunosuppressant (10% to 20% improvement in kidney graft survival rates) but also revealed its potential to cause nephrotoxicity in human recipients.[72–74] In 1982, the first use of cyclosporine in pediatric kidney recipients was reported by Starzl *et al.*[75] In 1983, cyclosporine was approved for clinical use in transplantation. Optimism was high that the use of cyclosporine in children would allow reduced steroid dosage and better growth and sexual maturation. Early reports in children were encouraging: graft survival rates ranged from 50% to 100%; however, those follow-up periods were short.[76–84] Cyclosporine use in young children was associated with rapid metabolism of the drug, resulting in subtherapeutic levels and the need for higher doses than required in adults. Hoyer *et al.* recommended cyclosporine dosing based on body surface area, instead of kilogram body weight.[76] Long-term results with cyclosporine also showed improved graft survival rates in children.[85]

In the 1980s, the quality of life for children awaiting kidney transplantation also improved. Both continuous ambulatory and continuous cyclic automated peritoneal dialysis techniques were adapted for use in children.[86–90] Recombinant human erythropoietin was developed and shown to successfully treat the anemia associated with chronic renal failure.[91,92] Available for clinical use in 1989, it decreased the number of blood transfusions administered to patients awaiting kidney transplantation.

In uremic rats, Mehls *et al.* showed that supraphysiologic doses of growth hormone led to substantial improvement in longitudinal growth.[93,94] Likewise, supraphysiologic doses of recombinant human growth hormone in children with chronic renal failure and growth retardation significantly improved their statural growth pre-transplant.[95,96] Another advance in the 1980s pertained to the effect of uremia on the developing brain. In a landmark paper, Rotundo *et al.* reported that infants with chronic renal failure had increased susceptibility to progressive encephalopathy and suggested that early transplantation might protect or stabilize their developing brain.[97] This paper spurred a renewal of efforts to successfully transplant infants, efforts led by John Najarian at the University of Minnesota.[45,65,83,98–102]

Also in the 1980s, techniques to produce monoclonal antibodies against single antigenic determinants were developed.[103] OKT3, a murine monoclonal antibody directed against the CD3 complex on mature human T-lymphocytes, was shown to effectively reverse as well as decrease the incidence of acute rejection in kidney recipients.[104–108] From 1983 to 1986, it was used in pediatric kidney recipients to treat acute rejection resistant to conventional treatment.[109, 110] According to the North American Pediatric Renal Transplant Cooperative Study, 6% of children who underwent a kidney transplant between January 1, 1987 and February 28, 1989 received induction or prophylactic therapy with OKT3.[111] In 1989, the macrolide antibiotic FK506 (tacrolimus) was first used clinically in a pediatric kidney recipient, in combination with steroid therapy.[112] In 1994, it was approved for clinical use by the U.S. Food and Drug Administration (FDA). It has immunosuppressive activity similar to cyclosporine and has been shown to be effective in both adult and pediatric kidney recipients.[112–117]

In the 1990s, several new immunosuppressive agents have been approved for clinical use. Mycophenolate mofetil (Cell Cept), from Roche Laboratories, inhibits the *de novo* pathway of purine synthesis in T- and B-lymphocytes.[118] In 1995, it was approved by the FDA for clinical use in kidney transplantation.[119–123] That same year, the microemulsion formulation of cyclosporine A (Neoral) was approved; it has been shown to be superior to standard cyclosporine A in both absorption and area under the concentration curve.[124] In 1997, the humanized monoclonal antibody daclizumab was approved for prevention of acute rejection in kidney transplantation. This antibody blocks the interleukin-2 receptor expressed on activated T-lymphocytes, thereby providing more selective immunosuppression.[125–127] In a randomized, multicenter trial of kidney recipients, daclizumab was associated with a significant decrease in the incidence of acute rejection during the first six months post-transplant, prolongation of time to the first acute rejection episode, and a decrease in the mean number of rejection episodes per patient (compared with placebo).[128] Randomized trials using daclizumab have not been reported in pediatric kidney recipients.[129, 130]

In 1998, a chimeric anti-human IL-2 receptor monoclonal antibody, basiliximab, was approved for prevention of acute rejection episodes in kidney recipients.[131, 132] No trials in kidney recipients have compared these two monoclonal antibodies (daclizumab and basiliximab). A potential advantage of basiliximab is that only two doses are required for efficacy.

In 1998, thymoglobulin (a rabbit polyclonal antithymocyte globulin), from Sangstat, was approved by the FDA and has been shown to be an effective treatment for acute rejection.[125, 133–135] The efficacy and adverse effects profile of thymoglobulin in pediatric kidney recipients is similar to that in adults.[136, 137] In 1999, sirolimus (rapamycin, Rapamune) was approved for clinical use in kidney transplantation in the US. This potent immunosuppressant inhibits signaling by cytokines and cytokine-induced T-cell proliferation.[138, 139] It is synergistic with cyclosporine, but antagonistic for FK506.[138] Sirolimus has been shown in clinical trials to reduce the incidence of acute rejection episodes, in combination with cyclosporine and steroids.[140, 141] Clinical studies of the efficacy and safety of sirolimus for prevention of acute rejection have not been done in pediatric kidney recipients.

It has now been 40 years since the first successful kidney transplant was performed in a child. Patient and graft survival rates have improved with advances in immunosuppressive therapy. However, we have not overcome the problems of allograft rejection, recurrence of original disease in the allograft, malignancy post-transplant, and the effect of steroids on growth and sexual maturation post-transplant.

Liver Transplantation

In the late 1950s, techniques for whole-organ liver transplantation were developed in animals by Starzl *et al.* and Moore *et al.* working separately.[142, 143] In March 1963, the first human liver transplant was performed by Starzl *et al.* at the University of Colorado. The recipient was a three-year-old child with end-stage liver disease due to congenital biliary atresia who underwent pre-transplant thymectomy and received immunosuppression with azathioprine for 13 days pre-transplant.[144] The child died during the operation of hemorrhage. The next four patients were adults, transplanted between May 1 and October 31, 1963. They survived from 6 1/2 to 22 days post-transplant. Causes of death were pulmonary emboli in three and biliary peritonitis in one.[145] Unsuccessful attempts at human liver transplantation also occurred in Paris and Boston.[37, 146] Attempts at auxiliary liver transplantation were also unsuccessful.[37]

In July 1967, the first successful orthotopic (i.e. the donor liver is placed in the fossa created by removal of the recipient's native liver) liver transplant was performed by Starzl *et al.* The recipient was a 19-month-old girl with end-stage liver failure due to hepatoma. The recipient had no histocompatibility mismatches with her 18-month-old cadaver donor.[147] A splenectomy was performed at the time of transplant. Immunosuppression included azathioprine, prednisone and antilymphoblast globulin. The liver functioned immediately. Although the recipient suffered many post-transplant complications, including tumor recurrence and allograft rejection (treated with graft irradiation), she survived for approximately one year post-transplant before dying of metastatic cancer.[147] Six additional patients (ages 13 months to 16 years) had survived one to six months post-transplant at the time of Starzl's report. He attributed the successes in this series to advances in donor selection, organ preservation and immunosuppression. The 57% mortality rate in the series was largely caused by liver infarction and sepsis.

Unlike maintenance hemodialysis techniques, extracorporeal techniques for long-term maintenance of patients with end-stage liver disease are not available to serve as a bridge to transplantation.[148–150] In 1972, the American College of Surgeons and the National Institutes of Health Organ Transplant Registry published a survey of worldwide transplant activity.[151] This survey showed that 81 orthotopic and 32 heterotopic liver transplants had been performed worldwide by the end of 1969, with 41% of them performed in 1969. An additional 45 liver transplants were performed between January 1, 1970 and December 31, 1971. Biliary atresia was the most common indication for liver transplantation. Only 9% (n = 13, all with orthotopic grafts) of recipients had survived beyond one year, although one survived for three years and four months.[151] The majority of the liver transplants had been performed at two centers: the University of Colorado in Denver (n = 57, 36%) and Cambridge University in England (n = 28, 18%). By 1977, improvement in organ preservation allowed the liver to be maintained in good condition for eight to ten hours after removal from the cadaver donor.[152] By 1979, more than 300 orthotopic liver transplants had been performed, mostly at the University of Colorado and Cambridge University.[153]

Most of the patients transplanted by Starzl *et al.* were children with biliary atresia. Calne *et al.* at Cambridge University transplanted mostly adults, because of concern about the frequent and painful invasive procedures often required post-transplant and because of the undesirable side effects of immunosuppressive drugs that made it difficult for pediatric recipients to integrate into society.[154] Post-transplant immunosuppression at Cambridge University included cyclophosphamide and hydrocortisone for the first three days, followed by azathioprine and prednisolone in low maintenance doses.[154] Starzl's group primarily used antilymphocyte globulin, prednisone and azathioprine.[155] Throughout the 1970s, patient and graft survival rates in liver recipients were poor (compared with kidney recipients). Of 74 patients who underwent liver transplants in England between May 2, 1968 and December 31, 1977, only six (8%) had survived beyond one year post-transplant; however, one survived for over three years.[154] By 1980, one-year graft survival rates had improved to 23.7% for the Cambridge series.[155] Results were somewhat better in Denver, where 51% (n = 86) of the patients transplanted from March 1, 1963 to December 31, 1979 were children and 38% survived beyond one year post-transplant.[155] In the Denver series, patient survival rates were about 10% higher in children than in adults.

In 1978, cyclosporine was first used in liver transplantation by Calne *et al.* as a single agent.[155] In 1980, it was used in combination with steroids in liver transplantation; one-year patient survival rates improved to nearly 70% for both children and adults.[155] Starzl's group also reported that ABO-unmatched livers could be successfully used in patients such as small children, for whom the organ donor supply was more limited.[156] With improved patient and graft survival rates, the number of liver transplant programs increased in North America and Europe. In 1984, reduced-size liver transplantation was reported by Bismuth as an attempt to decrease the number of infants with liver failure who died awaiting transplantation.[157, 158] In 1989, the more potent immunosuppressant FK506 was introduced into liver transplantation by Starzl *et al.*[159] It was initially used to reverse acute rejection in recipients on cyclosporine, then used in combination with low doses of prednisone as primary immunosuppression. Its use has enhanced patient and graft survival rates, reduced the incidence of acute rejection, and allowed for a reduction in prednisone dosage in liver recipients.[160, 161]

Patient and graft survival rates have continued to improve, thanks to advances in surgical and organ preservation techniques as well as in immunosuppressive protocols. One-year patient survival rates now exceed 80% in centers performing larger numbers of pediatric liver transplants.[161–164] Newer developments in pediatric liver transplantation include the use of split-liver and living donor transplants.[165–167]

Heart Transplantation

Techniques developed in dogs in the early 1960s by Lower and Shumway set the stage for clinical heart transplantation in humans.[168, 169] In the experimental model, allograft rejection (a major complication of heart transplantation) could be overcome in the experimental model with potent immunosuppression.[170, 171] In 1964, Hardy *et al.* at the University of Mississippi unsuccessfully transplanted a chimpanzee heart into a man.[172] In December 1967, South African surgeon Christiaan Barnard performed the first orthotopic human heart transplant at the University of Cape Town, South Africa. The recipient was a 54-year-old man with diabetes and a history of recurrent myocardial infarction who survived for 18 days, then died from bacterial pneumonia.[173] A second patient transplanted by Barnard in January 1968 had survived for three months at the time of his report.[174]

Children with end-stage heart disease unamenable to standard medical or surgical procedures were considered appropriate candidates for heart transplantation. Also in December 1967, the first pediatric heart transplant was performed by Kantrowitz. The recipient was a 19-day-old infant with Ebstein's malformation and refractory congestive heart failure.[175] The donor was an anencephalic infant. Hypothermia was induced in both the donor and recipient by immersion in ice water. Extracorporeal perfusion was not used. Hydrocortisone was administered pre-operatively. The recipient died about six hours post-operatively in severe metabolic and respiratory acidosis.[175]

Between January 6, 1968 and October 22, 1968, 62 human heart transplants were attempted by groups in 14 different countries.[176] The largest single series included ten human heart recipients transplanted by Cooley *et al.* at Baylor University in Houston, Texas.[177] In August 1968, Cooley transplanted a five-year-old with endocardial fibroelastosis. Immunosuppression included corticosteroids, azathioprine and antilymphocyte globulin. The child had an uneventful initial recovery, but died eight days post-transplant from allograft rejection.[177, 178] In September 1968, Cooley performed a combined heart-lung transplant in a two-month-old with atrioventricular canal defect and pulmonary hypertension. The infant died 14 hours post-operatively from pneumonia and pulmonary hemorrhage. Fewer than 50% of all patients worldwide who underwent heart transplants in 1968 survived.[176] In an update of the Cooley series, of 16 recipients of 17 allografts only three (18.7%) survived more than 60 days post-transplant.[178] In September 1968, the first successful pediatric heart transplant was performed by DeBakey *et al.* at Baylor University.[179] The recipient was a 16-year-old boy who had an atrioventricular communication defect with deformed mitral and tricuspid valves. He was still alive nine months post-transplant when heart transplant cases worldwide were compiled in a report by Haller and Cerruti.[180]

In December 1968, Kittle *et al.* at the University of Chicago were unsuccessful in their attempt to perform a heart transplant in an eight-day-old infant.[180] In February 1969, a six-year-old transplanted at Cincinnati Children's Hospital survived for 31 days before dying from allograft rejection. By June 22, 1969, 137 heart transplants had been performed in 20 countries worldwide. Survival rates were poor for both adults and children; only 24.5% of recipients were alive six months post-transplant.[180] Acute rejection and sepsis accounted for the large majority of deaths. One study showed that patient survival times were not significantly prolonged in heart recipients compared with potential heart recipients.[181] Because of the poor results, many programs established a moratorium on heart transplantation. Between June 23, 1969, and December 31, 1970, only 30 heart transplants were performed worldwide; 17 additional transplants were performed in 1971.[151]

But the program at Stanford University led by Norman Shumway continued to perform transplants. Between January 1968 and May 1980, seven children (ages 12 to 18 years) underwent heart transplants at Stanford University.[182, 183] In the early 1980s, with the introduction of cyclosporine and its ability to modulate rejection, many centers re-established heart transplant programs or saw improvements in their results.[184] In 1984 alone, 37 children underwent heart transplants worldwide, ten of whom were younger than five years old.[185]

In October 1984, Bailey *et al.* at Loma Linda University in Loma Linda, California transplanted a baboon heart into a 13-day-old female infant with hypoplastic left heart syndrome.[186, 187] Although the procedure was technically successful, she died on the 20th post-operative day.

In November 1984, Cooley *et al.* performed the first successful heart transplant in an infant, using a donor heart from a two-year-old child.[188] The recipient was an eight-month-old female with subendocardial fibroelastosis. Immunosuppression included cyclosporine and steroids. Allograft rejection episodes were detected by endomyocardial biopsy and treated with steroid pulses. In a follow-up report, the recipient was alive six years post-transplant.[189]

As more centers established pediatric heart transplant programs, improved survival rates (ranging from 57% to 86%) were reported in children and infants.[185, 190–192] Between January 1, 1980 and December 31, 1989, according to a report of the International Society for Heart Transplantation Registry, 909 heart transplants were performed worldwide in 874 children under 18 years old by 105 transplant centers.[193] Most of the transplants were performed after January 1, 1984; 58% of the transplants were performed in children ages ten to 18 years old. In 57% of the children, cardiomyopathy was the cause of end-stage heart disease. Re-transplantation occurred in 4% of the patients. Survival rates at one and five years post-transplant were 73.5% and 60%. Transplantation rates in children under one year of age rapidly increased. Early mortality (death in the first 30 days) was higher in the youngest age group (0–4 years old).[193] The major causes of death were cardiac and technical problems, infection and acute rejection. Chronic rejection and infection were the major causes of late mortality. Single-center reports of 84% survival in infants (at four to 40 months post-transplant, with prednisone, cyclosporine and azathioprine immunosuppression) were encouraging.[194]

Heart transplantation has now been successfully performed in children for almost two decades. Current concerns include hyperlipidemia, lethal coronary atherosclerosis, allograft rejection, renal dysfunction caused by immunosuppressive agents, anemia, post-transplant lymphoproliferative disorder and infection.[195–203]

Conclusion

Solid organ transplantation has proved to be a viable therapeutic alternative for children with end-stage kidney, liver and heart disease. Survival rates have improved over the four decades since the first successful human kidney transplant. The improvement in outcomes for children is due to advances in surgical techniques, organ preservation and immunosuppressive therapy. Concerns and future directions for study common to all pediatric solid organ transplantation groups include maximizing growth, development and rehabilitation post-transplant; overcoming allograft rejection; minimizing or preventing complications of immunosuppressive therapy; eliminating non-compliance; and solving the donor organ shortage.[204–207]

Acknowledgement

I thank Sandy Cragg for secretarial assistance, Kari Bunjer and Susan Kupcho for research assistance, and Mary Knatterud for editorial assistance.

References

1. Carrel A (1902). La technique operatoire des anastomoses vasculaires et la transplantation des visceres. *Lyon Med* **98**, 859–864.
2. Carrel A, Guthrie CC (1905). The transplantation of veins and organs. *Am Med* **10**, 1101–1102.
3. Gibson T, Medawar PB (1943). The behavior of skin homografts in man. *J Anat* **77**, 299–310.
4. Jaboulay M (1906). Greffe de reins au pli du coude par soudures arterielles et veineuses. *Lyon Med* **107**, 575–577.
5. Unger E (1910). Nierentransplantation. *Klinische Wochenschrift* **47**, 573–578.
6. Neuhof H (1923). *Transplantation of Tissues.* New York: Appleton and Co.
7. Voronoy Y (1936). Sobre el bloqueo del aparato reticuloendotelial del hombre en algunas formas de intoxicacion por el sublimado y sobre la transplantacion del rinon cadaverico como metodo de tratamiento de la anuria consecutiva a aquella intoxicacion. *Siglo Medico* **97**, 296–298.

8. Lawler RH, West JW, McNulty PH, Clancy EJ, Murphy RP (1950). Homotransplantation of the kidney in the human. *JAMA* **144**, 844–845.
9. Servelle M, Soulie P, Rouguelle J, Delahaye G, Touche M (1951). Greffe d'un rein de supplicie a une malade avec rein unique congenital, atteinte de nephrite chronique hypertensive azotemique. *Bulletins et memoires de la Societe medicale des hospitaux de Paris* **67**, 99–104.
10. Dubost C, Oeconomos N, Vaysse J, Hamburger J, Milliez P, Lebrigand J (1951). Note preliminaire sur l'etude des fonctions renales de reins greffes chez l'homme. *Bulletins et memoires de la Societe medicale des hospitaux de Paris* **67**, 105–106.
11. Kuss R, Teinturier J, Milliez P (1951). Quelques essais de greffe de rein chez l'homme. *Memoires Academie de Chirurgie* **77**, 755–764.
12. Ganter G (1923). Uber die beseitigung giftiger stoffe aus dem blute durch dialyse. *Munchener Medizinische Wochenschrift* **70**, 1478–1480.
13. Fine J, Frank HA, Seligman AM (1946). The treatment of acute renal failure by peritoneal irrigation. *Ann Surg* **124**, 857–878.
14. Murray G, Delorme E, Thomas N (1947). Development of an artificial kidney: experimental and clinical experiences. *Arch Surg* **55**, 505–522.
15. Merrill JP, Thorn GW, Callahan EJ, III, Smith LH, Jr (1950). Use of artificial kidney. I. Technique. *J Clin Invest* **29**, 412–424.
16. Swan H, Gordon HH (1949). Peritoneal lavage in the treatment of anuria in children. *Pediatrics* **4**, 586–595.
17. Hume DM, Merrill JP, Miller BF, Thorn GW (1955). Experiences with renal homotransplantation in the human: report of nine cases. *J Clin Invest* **34**, 327–382.
18. Michon l, Hamburger J, Oeconomos N, Delinotte P, Richet G, Vaysse J, Antoine B (1953). Une tentative de transplantation renale chez l'homme: aspects medicaux et biologique. *Presse Medicale* **61**, 1419–1923.

19. Hamburger J, Crosnier J, Dormont J, Bach J-F (1972). *Theory and Practice*. Baltimore: Williams and Wilkins.
20. Murray JE, Merrill JP, Harrison JH (1958). Kidney transplantation between seven pairs of identical twins. *Ann Surg* **148**, 343–359.
21. Murray JE, Merrill JP, Dammin GJ, Dealy JB, Walter CW, Brooke MS, Wilson RE (1960). Study on transplantation immunity after total body irradiation: clinical and experimental investigation. *Surgery* **48**, 272–284.
22. Merrill JP, Murray JE, Harrison JH, Friedman EA, Dealy JB, Jr, Dammin GJ (1960). Successful homotransplantation of the kidney between nonidentical twins. *New Engl J Med* **262**, 1251–1260.
23. Hamburger J, Vaysse J, Crosnier J, Auvert J, Lalanne C M, Hopper J, Jr (1962). Renal homotransplantation in man after radiation of the recipient: experience with six patients since 1959. *Am J Med* **32**, 854–870.
24. Goodwin WE, Mims MM, Kaufman JJ (1962). Human renal transplantation III: technical problems encountered in six cases of kidney homotransplantation. *Trans Am Assoc Genitourinary Surg* **54**, 116–125.
25. Bohannon LL, Barry JM, Norman DJ, Bennett WM (1988). Renal function 27 years after unilateral nephrectomy for related donor kidney transplantation. *J Urol* **140**, 810–811.
26. Dealy JB, Jr, Dammin GJ, Murray JE, Merrill JP (1960). Total body irradiation in man: tissue patterns observed in attempts to increase the receptivity of renal homografts. *Ann NY Acad Sci* **87**, 572–585.
27. Mathe G, Thomas ED, Ferrebee JW (1959). The restoration of marrow function after lethal irradiation in man: a review. *Transpl Bull* **6**, 407–409.
28. Schwartz R, Dameshek W (1959). Drug-induced immunological tolerance. *Nature* **183**, 1682–1683.

29. Calne RY (1960). The rejection of renal homografts inhibition in dogs by 6-mercaptopurine. *Lancet* **1**, 417–418.
30. Calne RY (1960–1961). Observations on renal homotransplantation. *Br J Surg* **48**, 384–391.
31. Zukoski CF, Lee HM, Hume DM (1960). Prolongation of functional survival of canine renal homografts by 6-mercaptopurine. *Surg Forum* **11**, 470–472.
32. Calne RY (1961). Inhibition of the rejection of renal homografts in dogs by purine analogues. *Transpl Bull* **28**, 445–461.
33. Murray JE, Merrill JP, Dammin GJ, Dealy JB, Jr, Alexandre GW, Harrison JH (1962). Kidney transplantation in modified recipients. *Ann Surg* **156**, 337–355.
34. Murray JE, Merrill JP, Harrison JH, Wilson RE, Dammin GJ (1963). Prolonged survival of human-kidney homografts by immunosuppressive drug therapy. *New Engl J Med* **268**, 1315–1323.
35. Goodwin WE, Kaufman JJ, Mims MM, Turner RD, Glassock R, Goldman R, Maxwell MM (1963). Human renal transplantation. I. Clinical experiences with six cases of renal homotransplantation. *J Urol* **89**, 13–24.
36. Starzl TE, Marchioro TL, Waddell WR (1963). The reversal of rejection in human renal homografts with subsequent development of homograft tolerance. *Surg Gynecol Obstet* **117**, 385–395.
37. Starzl TE, Marchioro TL, Porter KA, Tanous DF, Carey TA (1966). The role of organ transplantation in pediatrics. *Pediatr Clin North Am* **13**, 381–422.
38. Hume DM, Magee JH, Kauffman HM, Jr, Rittenbury MS, Prout GR, Jr (1963). Renal homotransplantation in man in modified recipients. *Ann Surg* **158**, 608–644.
39. Mowbray JF, Cohen SL, Doak PB, Kenyon JR, Owen K, Percival A, Porter KA, Pearth WS (1965). Human cadaveric renal transplantation — report of twenty cases. *Br Med J* **2**, 1387–1394.

40. Starzl TE, Marchioro TL, Morgan WW, Waddell WR (1964). A technique for the use of adult renal homografts in children. *Surg Gynecol Obstet* **119**, 106–108.
41. Murray JE, Gleason R, Bartholomay A (1964). Second report of registry in human kidney transplantation. *Transplantation* **2**, 660–667.
42. Hutchings RH, Hickman R, Scribner BH (1966). Chronic hemodialysis in a pre-adolescent. *Pediatrics* **37**, 68–73.
43. Fine RN, De Palma JR, Lieberman E, Donnell GN, Gordon A, Maxwell MH (1968). Extended hemodialysis in children with chronic renal failure. *J Pediatr* **73**, 706–713.
44. Potter D, Larsen D, Leumann E, Perin D, Simmons J, Piel CF, Holliday MA (1970). Treatment of chronic uremia in childhood. II. Hemodialysis. *Pediatrics* **46**, 678–689.
45. Najarian JS, Simmons RL, Tallent MB, Kjellstrand CM, Buselmeier TJ, Vernier RL, Michael AF (1971). Renal transplantation in infants and children. *Ann Surg* **174**, 583–601.
46. Counts S, Hickman R, Garbaccio A, Tenckhoff H (1973). Chronic home peritoneal dialysis in children. *Am Soc Artificial Internal Organs Trans* **19**, 157–163.
47. Riley, CM (1964). Thoughts about kidney homotransplantation in children. *J Pediatr* **65**, 797–800.
48. Reinhart JB (1970). The doctor's dilemma: whether or not to recommend continuous renal dialysis or renal homotransplantation for the child with end-stage renal disease. *J Pediatr* **77**, 505–507.
49. Kanoti GA (1986). Ethical considerations in solid organ pediatric transplants. *Transpl Proc* **18**, 43–46.
50. Dausset J, Ivanyi P, Ivanyi D (1965). Tissue alloantigens in humans: identification of a complex system (Hu-1). In *Histocompatibility Testing*. Balner H (ed.) Copenhagen: Ejnar Munksgaard, pp. 51–62.
51. Starzl TE, Marchioro, TL, Terasaki PI, Porter KA, Faris TD, Herrmann TJ, Vredevoe DL, Hutt MP, Ogden DA, Waddell WR

(1965). Chronic survival after human renal homotransplantation. Lymphocyte-antigen matching, pathology and influence of thymectomy. *Ann Surg* **162**, 749–787.

52. Terasaki PI, Vredevoe DL, Porter KA, Mickey MR, Marchioro TL, Faris TD, Herrmann TJ, Starzl TE (1966). Serotyping for homotransplantation. V. Evaluation of a matching scheme. *Transplantation* **4**, 688–699.
53. Terasaki PI, Vredevoe DL, Mickey MR, Porter KA, Marchioro TL, Faris TD, Starzl TE (1966). Serotyping for homotransplantation. VII. Selection of kidney donors for thirty-two recipients. *Ann NY Acad Sci* **129**, 500–520.
54. Patel R, Terasaki PI (1969). Significance of the positive crossmatch test in kidney transplantation. *New Engl J Med* **280**, 735–739.
55. Waksman BH, Arbouys S, Arnason BG (1961). The use of specific "lymphocyte" antisera to inhibit hypersensitive reactions of the "delayed" type. *J Exp Med* **114**, 997–1022.
56. Woodruff MFA, Anderson NF (1963). Effect of lymphocyte depletion and thoracic duct fistula and administration of antilymphocytic serum on the survival of skin homografts in rats. *Nature* **200**, 702.
57. Starzl TE, Marchioro TL, Porter KA, Iwasaki Y, Cerilli GJ (1967). The use of heterologous antilymphoid agents in canine renal and liver homotransplantation and in human renal transplantation. *Surg Gynecol Obstetr* **124**, 301–318.
58. Potter DO, Belzer FO, Rames L, Holliday MA, Kountz SL, Najarian JS (1970). The treatment of chronic uremia in childhood. I. Transplantation. *Pediatrics* **45**, 432–443.
59. Gonzalez LL, Martin L, West CD, Spitzer R, McEnery P (1970). Renal homotransplantation in children. *Arch Surg* **101**, 232–240.
60. Fine RN, Korsch BM, Edelbrock HH, Riddell H, Brennan LP, Stiles Q, Grushkin CM, Lieberman E (1971). Cadaveric renal transplantation in children. *Lancet* **1**, 1087–1091.

61. Merkel FK, Ing TS, Ahmadian Y, Lewy P, Ambruster K, Oyama J, Sulieman JS, Belman AB, King LR (1974). Transplantation in and of the young. *J Urol* **111**, 679–686.
62. Scharer K, Parsons FM, Brunner FP, Burck HC, Gurland HJ (1976). Dialysis and renal transplantation of children in Europe, 1973. *Acta Paediatrica Scandanavica* **65**, 187–192.
63. LaPlante MP, Kaufman JJ, Goldman R, Gonick HC, Martin DC, Goodwin WE (1970). Kidney transplantation in children. *Pediatrics* **46**, 665–677.
64. Lilly JR, Giles G, Hurwitz R, Schroter G, Tagaki H, Gray S, Penn I, Halgrimson CG, Starzl TE (1971). Renal homotransplantation in pediatric patients. *Pediatrics* **47**, 548–557.
65. DeShazo CV, Simmons RL, Bernstein DM, DeShazo MM, Wilmert J, Kjellstrand CM, Najarian JS (1974). Results of renal transplantation in 100 children. *Surgery* **76**, 461–468.
66. Grushkin CM, Fine RN (1973). Growth in children following renal transplantation. *Am J Disease Childhood* **125**, 514–516.
67. Borel JF (1976). Comparative study of *in vitro* and *in vivo* drug effects on cell-mediated cytotoxicity. *Immunology* **31**, 631–641.
68. Kostakis AJ, White DJG, Calne RY (1977). Prolongation of rat heart allograft survival by cyclosporin A. *Int Res Commun Syst Med Sci* **5**, 280.
69. Calne RY, White DJG (1977). Cyclosporin A — A powerful immunosuppressant in dogs with renal allografts. *Int Res Commun Syst Med Sci* **5**, 595–596.
70. Calne RY, White DJG, Thiru S, Evans DB, McMaster P, Dunn DC, Craddock GN, Pentlow BD, Rolles K (1978). Cyclosporin A in patients receiving renal allografts from cadaver donors. *Lancet* **2**, 1323–1327.
71. Calne RY, Rolles K, White DJG, Thiru S, Evans DB, McMaster P, Dunn DC, Craddock GN, Henderson RG, Aziz S, Lewis P (1979). Cyclosporin A initially as the only immunosuppressant in 34 recipients of cadaveric organs: 32 kidneys, 2 pancreases, and 2 livers. *Lancet* **2**, 1033–1036.

72. Starzl TE, Klintmalm GBG, Weil R, III, Porter KA, Iwatsuki S, Schroter GPJ, Fernandez-Bueno C, MacHugh N (1981). Cyclosporin A and steroid therapy in sixty-six cadaver kidney recipients. *Surg Gynecol Obstetr* **153**, 486–494.
73. Canadian Multicenter Transplant Group (1983). A randomized clinical trial of cyclosporine in cadaveric renal transplantation. *New Engl J Med* **309**, 809–815.
74. Merion RM, White DJG, Thiru S, Evans DB, Calne RY (1984). Cyclosporine: five years' experience in cadaveric renal transplantation. *New Engl J Med* **310**, 148–154.
75. Starzl TE, Iwatsuki S, Malatack JJ, Zitelli BJ, Gartner JC, Jr, Hakala TR, Rosenthal JT, Shaw BW, Jr (1982). Liver and kidney transplantation in children receiving cyclosporin A and steroids. *J Pediatr* **100**, 681–686.
76. Hoyer PF, Offner G, Wonigeit K, Brodehl J, Pichlmayr R (1984). Dosage of cyclosporin A in children with renal transplants. *Clin Nephrol* **22**, 68–71.
77. Klare B, Walter JV, Hahn H, Emmrich P, Land W (1984). Cyclosporin in renal transplantation in children. *Lancet* **2**, 692.
78. Conley SB, Flechner SM, Rose G, van Buren CT, Brewer ED, Kahan BD (1985). Use of cyclosporine in pediatric renal transplant recipients. *J Pediatr* **106**, 45–49.
79. Flechner SM, Conley SB, van Buren CT, Rose G, Kerman R, Kahan BD (1985). Impact of cyclosporine on renal function and growth in pediatric renal transplant recipients. *Transpl Proc* **17**, 1284–1288.
80. Kahan BD, Kerman RH, Wideman CA, Fletchner SM, Jarowenko M, van Buren CT (1985). Impact of cyclosporine on renal transplant practice at the University of Texas Medical School at Houston. *Am J Kidney Diseases* **5**, 288–295.
81. Ellis D, Avner ED, Rosenthal JT, Taylor RJ, Young LW, Palumbi MA, Hakala TR (1985). Renal function and somatic growth in pediatric cadaveric renal transplantation with cyclosporine-prednisone immunosuppression. *Am J Diseases Children* **139**, 1161–1167.

82. MacDonald AS, Crocker JFS, Belitsky P, Bitter-Suermann H, Lannon SG (1986). Cyclosporine therapy in pediatric renal transplantation. *Transpl Proc* **18**, 153–155.
83. Nevins TE, Knaak M, So SKS, Chavers BM, Melvin T, Simmons RL, Mauer SM, Najarian JS (1986). Preliminary results of low-dose cyclosporin A in pediatric renal transplantation. *Int J Pediatr Nephrol* **7**, 91–93.
84. Tejani A, Butt KMH, Khawar MR, Phadke K, Adamson O, Hong J, Fusi M, Trachtman H (1986). Cyclosporine experience in renal transplantation in children. *Kidney Int* **30**, 35–43.
85. Brodehl J, Bokenkamp A, Hoyer PF, Offner G (1992). Long-term results of cyclosporin A therapy in children. *J Am Soc Nephrol* **2**, S246–S254.
86. Popovich RP, Moncrief JW, Decherd JF, Bomar JB, Pyle WK (1976). The definition of a novel portable, wearable equilibrium peritoneal dialysis technique. *Am Soc Artificial Internal Organs* **5**, 64A.
87. Diaz-Buxo JA, Walker PJ, Farmer CD, Chandler JT, Holt KL (1981). Continuous cyclic peritoneal dialysis (CCPD). *Kidney Int* **19**, 145A.
88. Oreopoulos DG, Katirtzoglou A, Arbus G, Cordy PE (1979). Dialysis and transplantation in young children. *Br Med J* **1**, 1628–1629.
89. Kohout EC (1981). Continuous ambulatory peritoneal dialysis: a preliminary pediatric experience. *Am J Diseases Children* **135**, 270–271.
90. Broyer M, Donckerwolcke RA, Brunner FP, Brynger H, Challah S, Gretz N, Jacobs C, Kramer P, Selwood NH, Wing AJ (1983). Combined report on regular dialysis and transplantation of children in Europe, XII, 1982. *Proc Eur Dialysis Transplant Assoc* **20**, 79–108.
91. Eschbach JW, Egrie JC, Downing MR, Browne JK, Adamson JW (1987). Correction of the anemia of end-stage renal disease with recombinant human erythropoietin: results of a combined Phase I and II clinical trial. *New Engl J Med* **316**, 73–78.

92. Eschbach JW, Kelly MR, Haley NR, Abels RI, Adamson JW (1989). Treatment of the anemia of progressive renal failure with recombinant human erythropoietin. *New Engl J Med* **321**, 158–163.

93. Kreusser W, Weinkauff R, Mehls O, Ritz E (1982). Effect of parathyroid hormone, calcitonin and growth hormone on cAMP content of growth cartilage in experimental uremia. *Eur J Clin Invest* **12**, 337–343.

94. Mehls O, Ritz E (1983). Skeletal growth in experimental uremia. *Kidney Int* (Suppl.) **15**, S53–S62.

95. Koch VH, Lippe BM, Nelson PA, Boechat MI, Sherman BM, Fine RN (1988). Accelerated growth after recombinant human growth hormone treatment of children with chronic renal failure. *J Pediatr* **115**, 365–371.

96. Tonshoff B, Mehls O, Schauer A, Heinrich U, Blum W, Ranke M (1989). Improvement of uremic growth failure by recombinant human growth hormone. *Kidney Int* (Suppl.) **36**, S201–S204.

97. Rotundo A, Nevins TE, Lipton M, Lockman LA, Mauer SM, Michael AF (1982). Progressive encephalopathy in children with chronic renal insufficiency in infancy. *Kidney Int* **21**, 486–491.

98. Miller LC, Bock GH, Lum CT, Najarian JS, Mauer SM (1982). Transplantation of the adult kidney into the very small child: long-term outcome. *J Pediatr* **100**, 675–680.

99. Trompeter RS, Bewick M, Haycock GB, Chantler C (1983). Renal transplantation in very young children. *Lancet* **1**, 373–375.

100. Nevins TE (1987). Transplantation in infants less than 1 year of age. *Pediatr Nephrol* **1**, 154–156.

101. Najarian JS, Frey DJ, Matas AJ, Gillingham KJ, So SKS, Cook M, Chavers B, Mauer SM, Nevins TE (1990). Renal transplantation in infants. *Ann Surg* **212**, 353–367.

102. Briscoe DM, Kim MS, Lillehei C, Eraklis AJ, Levey RH, Harmon WE (1992). Outcome of renal transplantation in children less than two years of age. *Kidney Int* **42**, 657–662.

103. Galfre G, Milstein C (1981). Preparation of monoclonal antibodies: strategies and procedures. *Methods Enzymol* **73**, 3–46.

104. Cosimi AB, Colvin RB, Burton RC, Rubin RH, Goldstein G, Kung P, Hansen WP, Delmonico FL, Russell PS (1981). Use of monoclonal antibodies to T cell subsets for immunologic monitoring and treatment in recipients of renal allografts. *New Engl J Med* **305**, 308–314.
105. Ortho Multicenter Transplant Study Group (1985). A randomized clinical trial of OKT3 monoclonal antibody for acute rejection of cadaveric renal transplants. *New Engl J Med* **313**, 337–342.
106. Delmonico FL, Cosimi AB (1988). Monoclonal antibody treatment of human allograft recipients. *Surg Gynecol Obstetr* **166**, 89–98.
107. Benvenisty AI, Cohen D, Stegall MD, Hardy MA (1990). Improved results using OKT3 as induction immunosuppression in renal allograft recipients with delayed graft function. *Transplantation* **49**, 321–327.
108. Frey DJ, Matas AJ, Gillingham KJ, Canafax D, Payne WD, Dunn DC, Sutherland DER, Najarian JS (1991). MALG vs OKT3 following renal transplantation: a randomized prospective trial. *Transpl Proc* **23**, 1048–1049.
109. Goldstein G, Barnes L, Hirsch RL (1987). OKT3 monoclonal antibody reversal of renal and hepatic rejection in pediatric patients. *J Pediatr* **111**, 1046–1050.
110. Leone MR, Alexander SR, Barry JM, Henell K, Funnell MB, Goldstein G, Norman DJ (1987). OKT3 monoclonal antibody in pediatric kidney recipients with recurrent and resistant allograft rejection. *J Pediatr* **111**, 45–50.
111. Alexander SR, Arbus GS, Butt KMH, Conley S, Fine RN, Greifer I, Gruskin AB, Harmon WE, McEnery PT, Nevins TE, Nogueira N, Salvatierra O, Jr, Tejani A (1990). The 1989 report of the North American pediatric renal transplant cooperative study. *Pediatr Nephrol* **4**, 542–553.
112. Jensen CWB, Jordan ML, Schneck FX, Shapiro R, Tzakis A, Hakala TR, Starzl TE (1991). Pediatric renal transplantation under FK 506 immunosuppression. *Transpl Proc* **23**, 3075–3077.

113. Shapiro R, Jordan M, Fung J, McCauley J, Johnston J, Iwaki Y, Tzakis A, Hakala T, Todo S, Starzl TE (1991). Kidney transplantation under FK 506 immunosuppression. *Transpl Proc* **23**, 920–923.

114. Starzl TE, Fung J, Jordan M, Shapiro R, Tzakis A, McCauley J, Johnston J, Iwaki Y, Jain A, Alessiani M, Todo S (1990). Kidney transplantation under FK 506. *JAMA* **264**, 63–67.

115. Schneck FX, Jordan ML, Jensen CWB, Shapiro R, Tzakis A, Scantlebury VP, Ellis D, Gilboa N, Simmons RL, Hakala TR, Starzl TE (1992). Pediatric renal transplantation under FK 506 immunosuppression. *J Urol* **147**, 1585–1587.

116. Tzakis A G, Reyes J, Todo S, Nour B, Shapiro R, Jordan M, McCauley J, Armitage J, Fung JJ, Starzl TE (1993). Two-year experience with FK 506 in pediatric patients. *Transpl Proc* **25**, 619–621.

117. Benfield MR, Stablein DM, Tejani A (1999). Trends in immunosuppressive therapy: a report of the North American pediatric renal transplant cooperative study (NAPRTCS). *Pediatr Transpl* **3**, 27–32.

118. Ransom JT (1995). Mechanism of action of mycophenolate mofetil. *Ther Drug Monit* **17**, 681–684.

119. European Mycophenolate Mofetil Cooperative Study Group (1995). Placebo-controlled study of mycophenolate mofetil combined with cyclosporine and corticosteroids for prevention of acute rejection. *Lancet* **345**, 1321–1325

120. US Renal Transplant Mycophenolate Mofetil Study Group (1995). Mycophenolate mofetil for prevention of acute rejection in primary cadaveric renal allograft recipients. *Transplantation* **60**, 225–232.

121. Ettenger R, Cohen A, Nast C, Mouton L, Marik J, Gales B (1997). Mycophenolate mofetil as maintenance immunosuppression in pediatric renal transplantation. *Transpl Proc* **29**, 340–341.

122. Weber LT, Shipkova M, Lamersdorf T, Niedmann PD, Wiesel M, Mandelbaum A, Zimmerhackl LB, Schutz E, Mehls O, Oellerich

M, Armstrong VW, Tonshoff B (1998). Pharmacokinetics of mycophenolic acid (MPA) and determinants of MPA free fraction in pediatric and adult renal transplant recipients. German study group on mycophenolate mofetil therapy in pediatric renal transplant recipients. *J Am Soc Nephrol* **9**, 1511–1520.

123. Antoniadis A, Papachristou F, Gakis D, Takoudas D, Sotiriou I (1998). Comparison between mycophenolate mofetil and azathioprine based immunosuppression in pediatric renal transplantation from living related donors. *Transpl Proc* **30**, 4085–4086.
124. Wahlberg J, Wilczek HE, Fauchald P, Nordal KP, Heaf JG, Olgaard K, Hansen JM, Lokkegaard H, Mueller EA, Kovarik JM (1995). Consistent absorption of cyclosporine from a microemulsion formulation assessed in stable renal transplant recipients over a one-year study period. *Transplantation* **60**, 648–652.
125. Soulillou JP, Cantarovich D, Le Manff B, Giral M, Hourmant M, Hirn M, Jacques Y (1990). Randomized controlled trial of a monoclonal antibody against the interleukin 2 receptor (33B3.1) as compared with rabbit antithymocyte globulin for prophylaxis against rejection of renal allografts. *New Engl J Med* **322**, 1175–1182.
126. Le Mauff B, Gascan H, Olive D, Moreau J-F, Mawas C, Soulillou J-P, Jacques Y (1987). Parameters of interaction of a novel monoclonal antibody (33B3.1) with the human IL2-receptors: interrelationship between 33B3.1, anti-tac and IL2 binding sites. *Hum Immunol* **19**, 53–68.
127. Soulillou JP, Peyronnet P, Le Mauff B, Hourmant M, Olive D, Mawas C, Delaage M, Hirn M, Jacques Y (1987). Prevention of rejection of kidney transplants by monoclonal antibody directed against interleukin 2. *Lancet* **1**, 1339–1342.
128. Vincenti F, Kirkman R, Light S, Bumgardner G, Pescovitz M, Halloran P, Neylan J, Wilkinson A, Ekberg H, Gaston R, Backman L, Burdick J (1998). Interleukin-2 receptor blockade with daclizumab to prevent acute rejection in renal transplantation. Daclizumab Triple Therapy Study Group. *New Engl J Med* **338**, 161–165.

129. Ettenger RB (1998). New immunosuppressive agents in pediatric renal transplantation. *Transpl Proc* **30**, 1956–1958.
130. Ettenger R, Potter D, Knechtle S, Pescovitz M, Colombani P, Nevins T, Alexander S, Ramos E, Light S (1998). Humanized monoclonal anti-interleukin-2 receptor (IL-2R) antibody daclizumab (Zenapax) in pediatric renal transplantation. *J Am Soc Nephrol* **9**, 673A.
131. Nashan B, Moore R, Amlot P, Schmidt AG, Abeywickrama K, Soulillou JP (1997). Randomized trial of basiliximab versus placebo for control of acute cellular rejection in renal allograft recipients. CHIB 201 International Study Group. *Lancet* **350**, 1193–1198.
132. Kahan BD, Rajagopalan PR, Hall M (1999). Reduction of the occurrence of acute cellular rejection among renal allograft recipients treated with basiliximab, a chimeric anti-interleukin-2 receptor monoclonal antibody. US Simulect Renal Study Group. *Transplantation* **67**, 276–284.
133. LeFrancois N, Raffaele P, Martinenghi S, Martin X, Betuel H, Touraine JL, Pozza G, Dubernard JM (1990). Prophylactic polyclonal versus monoclonal antibodies in kidney and pancreas transplantation. *Transpl Proc* **22**, 632–633.
134. O'Donaghue DJ, Johnson RWG, Mallick NP, Gokal R, Ballardie FW, Bakran A, Pearson R, Scott P (1989). Rabbit anti-thymocyte globulin treatment of steroid resistant rejection in renal allograft recipients immunosuppressed with cyclosporine A. *Transpl Proc* **21**, 1736–1737.
135. Gaber AO, First MR, Tesi RJ, Gaston RS, Mendez R, Mulloy LL, Light JA, Gaber LW, Squiers E, Taylor RJ, Neylan JF, Steiner RW, Knechtle S, Norman DJ, Shihab F, Basadonna G, Brennan DC, Hodge EE, Kahan BD, Kahana L, Steinberg S, Woodle ES, Chan L, Ham JM, Stratta RJ, Wahlstrom E, Lamborn KR, Horn HR, Moran HB, Pouletty P, Schroeder TJ (1998). Results of the double-blind, randomized, multicenter, phase III clinical trial of thymoglobulin versus Atgam in the treatment of acute

graft rejection episodes after renal transplantation. *Transplantation* **66**, 29–37.

136. Broyer M, Gagnadoux MF, Guest G, Niaudet P (1987). Triple therapy including cyclosporine A versus conventional regimen — a randomized prospective study in pediatric kidney transplantation. *Transpl Proc* **19**, 3582–3585.
137. Bell L, Giardin C, Sharma A, Goodyer P, Mazer B (1997). Lymphocyte subsets during and after rabbit antithymocyte globulin induction in pediatric renal transplantation: sustained T cell depletion. *Transpl Proc* **29**, 6S–9S.
138. Sehgal SN (1995). Rapamune (Sirolimus, Rapamycin): an overview and mechanism of action. *Ther Drug Monit* **17**, 660–665.
139. Sehgal SN (1998). Rapamune (RAPA, rapamycin, sirolimus): mechanism of action immunosuppressive effect results from blockade of signal transduction and inhibition of cell cycle progression. *Clin Biochem* **31**, 335–340.
140. Kahan BD, Podbielski J, Napoli KL, Katz SM, Meier-Kriesche HU, van Buren CT (1998). Immunosuppressive effects and safety of a sirolimus/cyclosporine combination regimen for renal transplantation. *Transplantation* **66**, 1040–1046.
141. Groth GC, Backman L, Morales JM, Calne R, Kreis H, Lang P, Touraine JL, Claesson K, Campistol JM, Durand D, Wranner L, Brattstrom C, Charpentier B (1999). Sirolimus (rapamycin)-based therapy in human renal transplantation: similar efficacy and different toxicity with cyclosporin. Sirolimus European Renal Transplant Study Group. *Transplantation* **67**, 1036–1042.
142. Starzl TE, Kaupp HA, Jr, Brock DR, Lazarus RE, Johnson RV (1960). Reconstructive problems in canine liver homotransplantation with special reference to postoperative role of hepatic venous flow. *Surg Gynecol Obstet* **111**, 733–743.
143. Moore FD, Wheeler HB, Demissianos HV, Smith LL, Blankura O, Abel K, Greenberg JB, Bammin GJ (1960). Experimental whole organ transplantation of the liver and spleen. *Ann Surg* **152**, 374–387.

144. Starzl TE, Marchioro TL, von Kaulla KN, Hermann G, Brittain RS, Waddell WR (1963). Homotransplantation of the liver in humans. *Surg Gynecol Obstet* **117**, 659–676.
145. Starzl TE, Marchioro TL, Huntley RT, Rifkind D, Rowlands DT, Jr, Dickinson TC, Waddell WR (1964). Experimental and clinical homotransplantation of the liver. *Ann NY Acad Sci* **120**, 739–765.
146. Moore FD, Birtch AG, Dagher F, Keith F, Krisher JA, Order SE, Shucart WA, Dammin GJ, Couch NP (1964). Immunosuppressive and vascular insufficiency in liver transplantation. *Ann NY Acad Sci* **120**, 729–738.
147. Starzl TE, Groth CG, Brettschneider L, Penn I, Fulginiti VA, Moon JB, Blanchard H, Martin AJ, Jr, Porter K (1968). Orthotopic homotransplantation of the human liver. *Ann Surg* **168**, 392–415.
148. Eiseman B, Liem DS, Raffucci F (1965). Heterologous liver perfusion in treatment of hepatic failure. *Ann Surg* **162**, 329–345.
149. Abouna GM, Kirkley JR, Hull CJ, Ashcroft T, Kerr DNS (1969). Treatment of hepatic coma by extracorporeal pig-liver perfusion. *Lancet* **1**, 64–68.
150. Condon RE, Bombeck CT, Steigmann F (1970). Heterologous bovine liver perfusion therapy of acute hepatic failure. *Am J Surg* **119**, 147–154.
151. ACS/NIH Organ Transplant Registry (1972). Second Scientific Report. *JAMA* **221**, 1486–1491.
152. Wall WJ, Calne RY, Herbertson MB, Baker PG, Smith DP, Underwood J, Kostakis A, Williams R (1977). Simple hypothermic preservation for transporting human livers long distances for transplantation. A report of 12 cases. *Transplantation* **23**, 210–216.
153. Anonymous (1979). Present state of liver transplantation. *Br Med J* **1**, 1441–1442.
154. Calne RY, Williams R (1979). Liver transplantation. *Curr Prob Surg* **16**, 1–44.

155. Starzl TE, Iwatsuki S, van Thiel DH, Gartner JC, Zitelli BJ, Malatack JJ, Schade RR, Shaw BW, Jr, Hakala TR, Rosenthal JT, Porter KA (1982). Evolution of liver transplantation. *Hepatology* **2**, 614–636.
156. Gordon RD, Iwatsuki S, Esquivel CO, Tzakis A, Todo S, Starzl TE (1986). Liver transplantation across ABO blood groups. *Surgery* **100**, 342–348.
157. Bismuth H, Houssin D (1984). Reduced-sized orthotopic liver graft in hepatic transplantation in children. *Surgery* **95**, 367–370.
158. Emond JC, Whitington PF, Thistlethwaite JR, Alonso EM, Broelsch CE (1989). Reduced-size orthotopic liver transplantation: use in the management of children with chronic liver disease. *Hepatology* **10**, 867–872.
159. Starzl TE, Todo S, Fung J, Demetris AJ, Venkataramman R, Jain A (1989). FK 506 for liver, kidney, and pancreas transplantation. *Lance*t **2**, 1000–1004.
160. Fung JJ, Todo S, Tzakis A, Alessiani M, Abu-Elmagd K, Jain A, Bronster O, Martin M, Gordon R, Starzl TE (1991). Current status of FK 506 in liver transplantation. *Transpl Proc* **23**, 1902–1905.
161. Goss JA, Shackleton CR, McDiarmid SV, Maggard M, Swenson K, Seu P, Vargas J, Martin M, Ament M, Brill J, Harrison R, Busuttil RW (1998). Long-term results of pediatric liver transplantation. An analysis of 569 transplants. *Ann Surg* **228**, 411–420.
162. Otte JB, Yandza T, de Ville de Goyet J, Tan KC, Salizzoni M, de Hemptinne B (1988). Pediatric liver transplantation: report on 52 patients with a 2-year survival of 86%. *J Pediatr Surg* **23**, 250–253.
163. Otte J-B, de Ville de Goyet J, Reding R, Hausleithner V, Sokal E, Chardot C, Debande B (1994). Sequential treatment of biliary atresia with Kasai portoenterostomy and liver transplantation: a review. *Hepatology* **20**, 41S–43S.

164. Kocoshis SA, Tzakis A, Todo S, Reyes J, Nour B (1993). Pediatric liver transplantation. History, recent innovations, and outlook for the future. *Clin Pediatr* **32**, 386–392.
165. Otte JB, de Ville de Goyet J, Sokal E, Alberti D, Moulin D, de Hemptinne B, Veyckemans F, van Obbergh L, Carlier M, Clapuyt P, Claus D, Jamart J (1990). Size reduction of the donor liver is a safe way to alleviate the shortage of size-matched organs in pediatric liver transplantation. *Ann Surg* **211**, 146–157.
166. Emond JC, Whitington PF, Broelsch CE (1991). Overview of reduced-size liver transplantation. *Clin Transpl* **5**, 168–173.
167. Broelsch CE, Whitington PF, Emond JC, Heffron TG, Thistlethwaite JR, Stevens L, Piper J, Whitington SH, Lichtor JL (1991). Liver transplantation in children from living related donors. *Ann Surg* **214**, 428–439.
168. Lower RR, Shumway NE (1960). Studies on orthotopic homotransplantation of the canine heart. *Surg Forum* **11**, 18–19.
169. Hurley EJ, Dong E, Jr, Stofer RC, Shumway NE (1962). Isotopic replacement of the totally excised canine heart. *J Surg Res* **2**, 90–94.
170. Lower RR, Stofer RC, Shumway NE (1961). Homovital transplantation of the heart. *J Thorac Cardiovasc Surg* **41**, 196–204.
171. Lower RR, Done E, Jr, Shumway NE (1965). Long-term survival of cardiac homografts. *Surg* **58**, 110–119.
172. Hardy JD, Chavez CM, Kurrus FD, Neely WA, Eraslan S, Turner MD, Fabian LW, Labecki TD (1964). Heart transplantation in man: developmental studies and report of a case. *JAMA* **188**, 1132–1140.
173. Barnard CN (1967). The operation. A human cardiac transplant: an interim report of a successful operation performed at Groote Schuur Hospital, Cape Town. *S Afr Med J* **41**, 1271–1274.
174. Barnard CN (1968). What we have learned about heart transplants. *J Thorac Cardiovasc Surg* **56**, 457–468.
175. Kantrowitz A, Haller JD, Joos H, Cerruti MM, Carstensen HE (1968). Transplantation of the heart in an infant and an adult. *Am J Cardiol* **22**, 782–790.

176. Haller JD, Cerruti MM (1968). Heart transplantation in man: compilation of cases. January 1, 1964 to October 23, 1968. *Am J Cardiol* **22**, 840–843.
177. Cooley DA, Hallman GL, Bloodwell RD, Nora JJ, Leachman RD (1968). Human heart transplantation. Experience with twelve cases. *Am J Cardiol* **22**, 804–810.
178. Nora JJ, Cooley DA, Fernbach DJ, Rochelle DG, Milam JD, Montgomery JR, Leachman RD, Butler WT, Rossen RD, Bloodwell RD, Hallman GL, Trentin JJ (1969). Rejection of the transplanted human heart. Indexes of recognition and problems in prevention. *New Engl J Med* **280**, 1079–1086.
179. DeBakey ME, Diethrich EB, Glick G, Noon GP, Butler WT, Rossen RD, Liddicoat JE, Brooks DK (1969). Human cardiac transplantation: clinical experience. *J Thorac Cardiovasc Surg* **58**, 303–317.
180. Haller JD, Cerruti MM (1969). Heart transplantation in man: compilation of cases — II. January 23, 1964 to June 22, 1969. *Am J Cardiol* **24**, 554–563.
181. Messmer BJ, Nora JJ, Leachman RE, Cooley DA (1969). Survival-times after cardiac allografts. *Lancet* **1**, 954–956.
182. English TAH (1983). Is cardiac transplantation suitable for children? *Pediatr Cardiol* **4**, 57–58.
183. Baum D, Kaye MP, Miller WW (1993). Pediatric heart, heart-lung, and lung transplantation: historical pespective. *Prog Pediatr Cardiol* **2**(4), 1–3.
184. Griffith BP, Hardesty RL, Deeb GM, Starzl TE, Bahnson HT (1982). Cardiac transplantation with cyclosporine A and prednisone. *Ann Surg* **196**, 324–329.
185. Pennington DG, Sarafian J, Swartz M (1985). Heart transplantation in children. *Heart Transpl* **4**, 441–445.
186. Bailey LL, Nehlsen-Cannarella SL, Concepcion W, Jolley WB (1985). Baboon-to-human cardiac xenotransplantation in a neonate. *JAMA* **254**, 3321–3329.
187. Bailey LL (1985). Role of cardiac replacement in the neonate. *J Heart Transpl* **4**, 506–509.

188. Cooley DA, Frazier OH, van Buren CT, Bricker JT, Radovancevic B (1986). Cardiac transplantation in an 8-month-old female infant with subendocardial fibroelastosis. *JAMA* **256**, 1326–1329.
189. Cooley DA (1991). Session I: Pediatric heart transplantation in historical perspective. *J Heart Lung Transpl* **10**, 787–790.
190. Fricker FJ, Griffith BP, Hardesty RL, Trento A, Gold LM, Schmeltz K, Beerman LB, Fischer DR, Mathews RA, Neches WH, Park SC, Zuberbuhler JR, Lenox CC, Bahnson HT (1987). Experience with heart transplantation in children. *Pediatrics* **79**, 138–146.
191. Addonizio LJ, Rose EA (1987). Cardiac transplantation in children and adolescents. *J Pediatr* **111**, 1034–1038.
192. Starnes VA, Stinson EB, Oyer PE, Valantine H, Baldwin JC, Hunt SA, Shumway NE (1987). Cardiac transplantation in children and adolescents. *Circulation* **76**, V43–V47.
193. Kaye MP, Kriett JM (1991). Session VIII: Pediatric heart transplantations — The world experience. *J Heart Lung Transpl* **10**, 856–859.
194. Boucek MM, Kanakriyeh MS, Mathis CM, Trimm RF, III, Bailey LL and the Loma Linda University Pediatric Heart Transplant Group (1990). Cardiac transplantation in infancy: donors and recipients. *J Pediatr* **116**, 171–176.
195. Pahl E, Fricker FJ, Armitage J, Griffith BP, Taylor S, Uretsky BF, Beerman LB, Zuberbuhler JR (1990). Coronary arteriosclerosis in pediatric heart transplant survivors: limitation of long-term survival. *J Pediatr* **116**, 177–183.
196. Pahl E, Zales VR, Fricker FJ, Addonizio LJ (1994). Post-transplant coronary artery disease in children. A multicenter national survey. *Circulation* **90**, II56–II60.
197. Sigfusson G, Fricker FJ, Bernstein D, Addonizio LJ, Baum D, Hsu DT, Chin C, Miller SA, Boyle GJ, Miller J, Lawrence KS, Douglas JF, Griffith BP, Reitz BA, Michler RE, Rose EA, Webber SA (1997). Long-term survivors of pediatric heart transplantation: a multicenter report of sixty-eight children who have survived longer than five years. *J Pediatr* **130**, 862–871.

198. Embleton ND, O'Sullivan JJ, Hamilton JR, Dark JH, Summerfield GP (1997). High prevalence of anemia after cardiac transplantation in children. *Transplantation* **64**, 1590–1594.
199. Boyle GJ, Michaels MG, Webber SA, Knisely AS, Kurland G, Cipriani LA, Griffith BP, Fricker FJ (1997). Post-transplant lymphoproliferative disorders in pediatric thoracic organ recipients. *J Pediatr* **131**, 309–313.
200. Gajarski RJ, Kearney DL, Price JK, Denfield SW (1997). Update on pediatric heart transplantation. Long-term complications. *Tex Heart Inst J* **24**, 260–268.
201. Braunlin EA, Shumway SJ, Bolman RM, McDonald KM, Ring WS, Olivari MT, Nakhleh RE (1998). Usefulness of surveillance endomyocardial biopsy after pediatric cardiac transplantation. *Clin Transpl* **12**, 184–189.
202. Penson MG, Winter WE, Fricker FJ, Harker K, Kahler DA, Kubilis PS, Schowengerdt KO (1999). Tacrolimus-based triple-drug immunosuppression minimizes serum lipid elevations in pediatric cardiac transplant recipients. *J Heart Lung Transpl* **18**, 707–713.
203. Webber SA (1997). 15 years of pediatric heart transplantation at the University of Pittsburgh: lessons learned and future prospects. *Pediatr Transpl* **1**, 8–21.
204. Serrano-Ikkos E, Lask B, Whitehead B, Eisler I (1998). Incomplete adherence after pediatric heart and heart-lung transplantation. *J Heart Lung Transpl* **17**, 1177–1183.
205. Laine J, Jalanko H, Ronnholm K, Sairanen H, Salmela K, Leijala M, Holmberg C (1998). Paediatric kidney transplantation. *Ann Med* **30**, 45–57.
206. Meyers KE, Thomson PD, Weiland H (1996). Noncompliance in children and adolescents after renal transplantation. *Transplantation* **62**, 186–189.
207. Caplan AL (1994). Current ethical issues in organ procurement and transplantation. *JAMA* **272**, 1708–1709.

XII

THE HISTORY OF SKIN TRANSPLANTATION

Felix T Rapaport[†]
Jean Dausset

Introduction

Skin has played a relatively modest role in clinical transplantation, largely involving the resurfacing of denuded body areas following thermal injury. This tissue has, however, been of key importance in our understanding of allograft recognition and rejection. The immunological basis of allograft rejection was documented initially by Medawar[1,2] with skin grafts in rabbits. Similar studies[3–6] extended this experience to man, and provided the centerpiece for the definition of the HLA system of human histocompatibility. Much of the *in vivo* relevance of the serological and cellular data on HLA compatibility was a direct outcome of the extensive skin grafting studies performed in carefully HLA-typed volunteers. This effort has been recognized as the basis for the immunological "Laws of Transplantation" in man.[7]

Experimental Skin Transplantation

The skin grafts used for studies in human subjects were full-thickness specimens, excised from the donors under local anesthesia, with an 11 mm Castroviejo trephine. The grafts were placed on the anterior surface of the forearms of the recipients, on sites prepared by a similar trephine (Fig. 1). The superficial vessels on the graft surface and the rate and extent of blood flow through their lumina were examined daily with a dissecting microscope, through a mineral oil interface.[4] During the first few days, there was no significant difference in the appearance of autografts or allografts. All were pale on the first day, and showed some pink areas within the next 48 hours. By the third or fourth day, the grafts were uniformly pink (Fig. 2); this was followed by epidermal regeneration and a gradual merging of graft and host skin. In the case of autografts, the early dilated vessels seen shortly after transplantation disappeared, and were replaced by small vessels with active blood flow. These reached a maximum by the eighth day, and then gradually decreased in number in the autografts, where they were replaced by a vascular pattern which approximated the appearance of normal skin.[4] In contrast, skin allografts lost their healthy pink color by the fifth day, progressing to cyanosis and edema and surrounded by an acute inflammatory reaction. The events of rejection progressed rapidly, with appearance of small thrombi in the graft by the eighth or ninth day, rapidly followed by hemorrhages and escharification (Fig. 3). On stereomicroscopy, the allograft vessels became progressively more detailed. Slowing of blood flow occurred by the ninth day, rapidly followed by thrombosis and vascular disruption, with appearance of multiple subepidermal hemorrhages. The cessation of blood flow was considered the endpoint for rejection, and provided a precise and reproducible criterion for graft rejection. The diagnosis was confirmed by the subsequent cascade of hemorrhagic events and escharification.[4]

Rejection of a skin allograft produced in the recipient a state of generalized hypersensitivity to additional skin grafts from the same donor. If a waiting period of ten to 13 days was allowed after first-set graft rejection, a second graft from the same donor suffered an accelerated response termed the "second-set phenomenon".[6] The events of first graft rejection were telescoped into the first four of five days after transplantation, with rapid formation of a black hemorrhagic

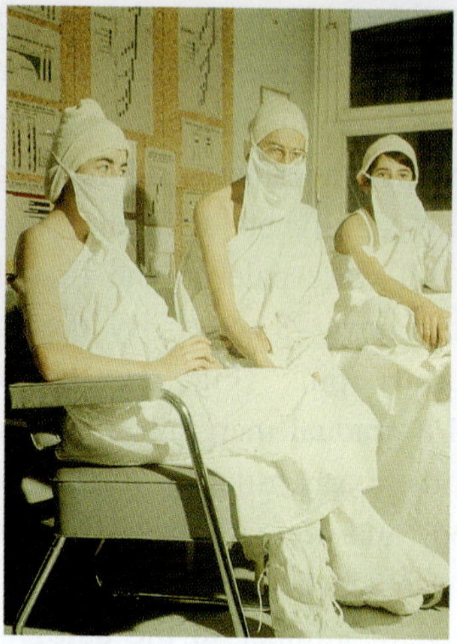

(a)

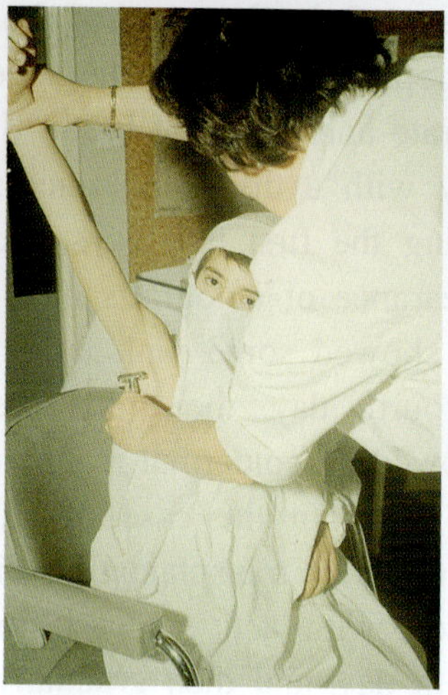

(b)

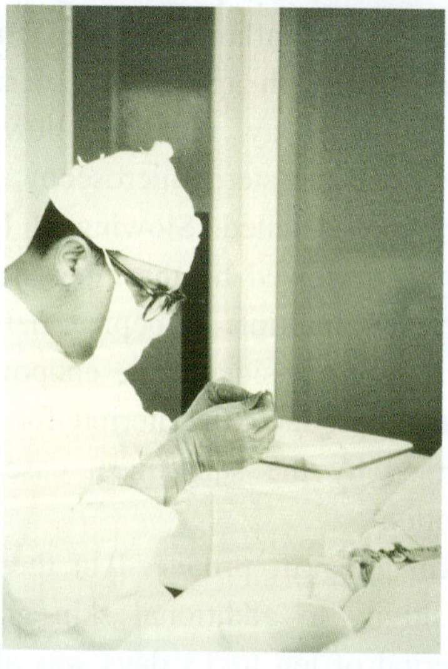

(c)

Fig. 1 Skin grafting procedures and preparation in Professor Dausset's anteroom and office.

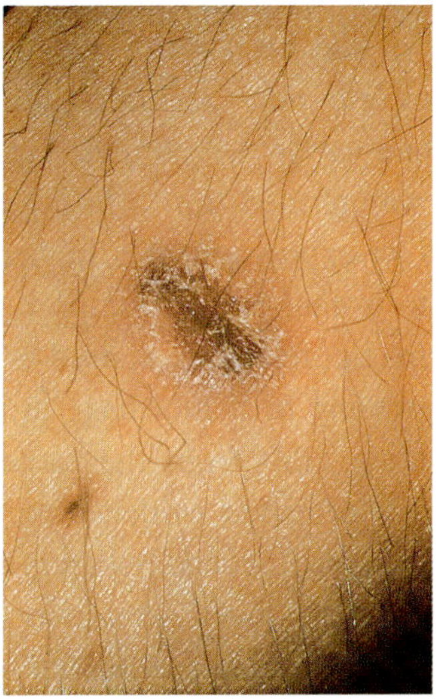

Fig. 2 Appearance of pink allografts on the 6th postoperative day.

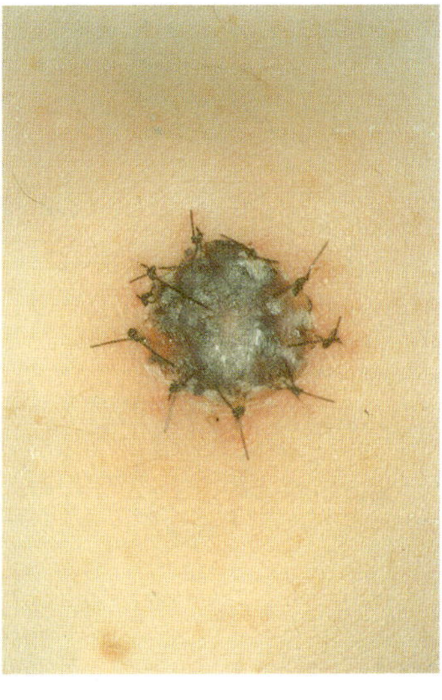

Fig. 3 Acute hemorrhagic skin allograft rejection site.

eschar.[5] If only one to five days were allowed after rejection of a first-set graft, re-challenge of the host with a second graft from the same donor resulted in complete non-vascularization of the graft, with a characteristic dead-white color which desiccated into a tan-colored eschar (Fig. 4). This was termed the "white graft" response (Fig. 5), and was interpreted as an expression of a higher level of sensitivity in the recipient, leading to instant or hyperacute rejection.[6]

Second-set hemorrhagic rejection caused a flare-up of erythema and edema, often progressing to hemorrhage at the site of rejection of the immediately preceding graft from the same donor. This reaction, termed the "recall flare"[5] provided evidence of the generalized donor-specific state of sensitization induced in recipients by first-set rejection, and suggested that the induced sensitivity was associated with a persistence at the first rejection site of donor antigen(s), which triggered a local reaction upon rejection of the second graft from the same source.[5] The white graft reaction did not elicit a recall flare.[6]

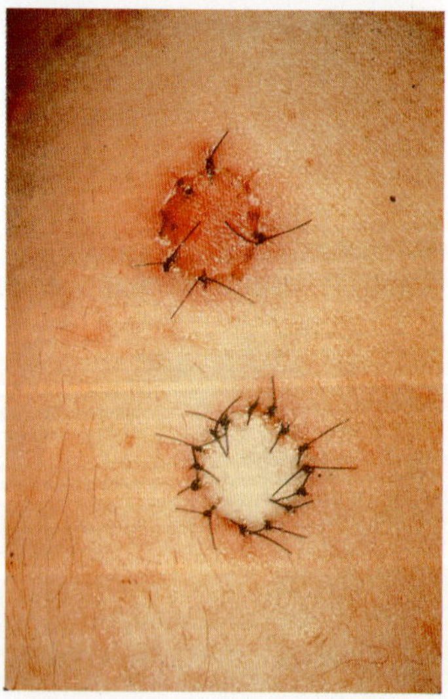

Fig. 4 Contrast between a white graft reaction and a normal pink graft from another donor.

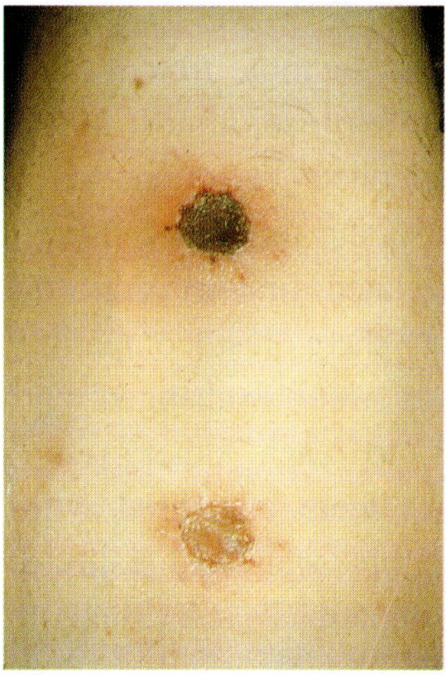

Fig. 5 Contrast between the hemorrhagic eschar of first-set rejection and the tan-colored eschar of the white graft rejection.

Development of a standard technique for the study of small experimental skin allografts in man,[4] and reliable criteria for diagnosis of graft survival time led to an investigation of the range of allograft survival in unrelated human subjects.

Sharing of Histocompatibility Antigens by Unrelated Individuals

Skin grafts performed in randomly selected normal unrelated individuals had a median graft survival of eight to 11 days, with a range of six to 21 days.[8, 9] The greater the number of recipients tested with skin grafts from any one given donor, the greater was the incidence of longer as well as shorter skin graft survivals in the recipients, suggesting that randomly selected, unrelated individuals might share varying concentrations of histocompatibility antigens.

Normal subjects were then sensitized with skin grafts from one donor, and the specificity of the induced sensitivity was tested by challenging the recipients with second-set grafts obtained from the same donor, and with four or more *first-set* grafts obtained from *other* unrelated donors. To our surprise, almost 40% of these "first-set" skin grafts underwent accelerated rejection, as if the recipients had aslo become sensitized to the tissues of these donors. This raised the possibility that the unrelated first-set graft donors involved in such "cross-reactions" might share histocompatibility antigens with the sensitizing donor. Grafts were exchanged between donors who had elicited cross-reactions to each other (i.e. possibly "compatible" subjects) and the results were compared with the survival of grafts exchanged between donors who had failed to induce such cross-sensitization (i.e. "incompatible" individuals). The grafts exchanged between "compatible" individuals regularly survived for a longer time than skin grafts exchanged between putatively incompatible subjects[10] (Fig. 6). This experiment was the first demonstration of the existence of shared antigens or tissue groups in unrelated human subjects,[7] and pointed to the potential usefulness of identifying optimally histocompatible unrelated subjects on a prospective basis prior to transplantation.[10] These developments provided the rationale for an assessment of the relevance of *in vitro* serological and cellular histocompatibility testing techniques to the *in vivo* data obtained by skin grafting studies.[7]

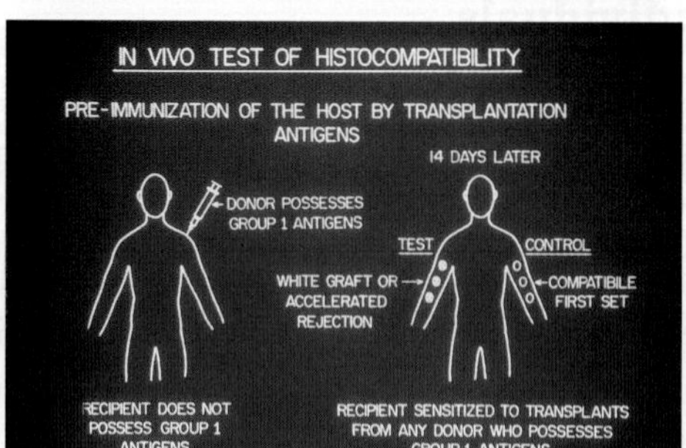

Fig. 6 *In vivo* test of histocompatibility of sharing transplantation antigens.

Human Histocompatibility Antigens

The observation of Dausset and Nenna[11] that serum from a polytransfused patient agglutinated leukocytes from another subject provided the first evidence that antibodies resulting from immunization with blood transfusions[12] or multiple pregnancies[13] could produce group-specific reactions in unrelated human subjects. The first serological reagent capable of recognizing a leukocyte antigen group was named Mac by Dausset in 1958,[14] and a number of other leukocyte antigen groups was reported shortly thereafter,[12, 13] heralding a search for a correlation between *in vivo* skin graft responses and the results of testing with these *in vitro* reagents. This goal provided the background for an extensive trans-Atlantic collaborative project between the authors, using Dausset's HLA-typed volunteer blood and leukocyte donors in Paris, and the skin grafting techniques developed in New York by Rappaport. Beginning in 1963, one of us (FTR) and his team commuted to Paris four times each year, spending three to five weeks at the Hospital Saint-Louis, and conducting veritable skin grafting marathons each time. During these episodes, Dausset's large office became an operating room, with strict enforcement of sterile precautions. The prospectively selected volunteers were skin grafted, and had to return daily to the hospital for dressing changes and study of their grafts, until the time of rejection. It is with enormous gratitude, that we can state today that all grafting, bleeding and sensitization sessions, over an 18-year span, were completed without a single untoward effect or complication in the donors and recipients. It is remarkable, in retrospect, to consider how many key advances in our understanding of HLA were a direct product of the coordinated use of experimental skin grafts *in vivo* and serologic-cellular studies *in vitro*, all made possible by this unique group of volunteers.

The skin grafting studies provided a major test of the validity of the *in vitro* HLA-serologic and cellular histocompatibility data.[7] The adventure began with grafts placed in recipients on the basis of relatively crude serologic criteria of compatibility; the endeavor continued in parallel with concurrent progress in the identification of new HLA specificities. For example, after definition of van Rood's 4A, 4B antigents,[13] two siblings homozygous for these antigens were grafted with skin grafts obtained from 4A and 4B donors. The result provided the first definitive evidence of a difference in the survival time of serologically HLA compatible versus incompatible grafts.[15] As additional HLA antigens were isolated, their activity in transplantation was then assessed through a pre-immunization protocol developed in the New York Universities laboratories.[16] Recipients were now immunized with leukocytes or leukocyte extracts from donors who were incompatible with the recipients for a given HLA antigen specificity. Two weeks later, each recipient was given skin grafts bearing the same HLA incompatibility and grafts from "compatible" donors who did not bear these antigen(s). Leukocyte group-incompatible skin grafts were regularly rejected in an accelerated or white graft fashion, while "compatible" grafts survived as first-set grafts.[16] A valuable by-product of these experiments was the production of new serological reagents in the recipients, which permitted a further systematic *in vitro* analysis of the HLA products.

The availability of families in whom donor-recipient compatibility could be assessed on the basis of simplified genetic relationships (sibling-to-sibling, parent-to-sibling grafts) led to a series of graft survival studies in 268 families of known HLA genotypes, in whom over 625 grafts were done under controlled conditions of known donor-recipient HLA haplotype identity or difference. The relative roles of individual products of the HLA A and B loci in the control of skin graft survival were thus demonstrated, with full statistical significance ($P < 0.0001$).[17–21] As shown in Table 1, skin grafts exchanged between genotypically HLA-identical siblings survived for 23 days, while grafts exchanged between HLA-different siblings were rejected within 13 days.[19]

The parallel evolution of *in vitro* cellular studies during this period provided new impetus to these investigation, culminating in the discovery by Bach and Amos of the HLA Class II antigens.[22] Here again, skin grafting showed a high correlation between graft survival in haploidentical donor-recipient pairs of individuals and the intensity of the *in vitro* MLC reactions between lymphocytes from these individuals.[23, 24]

As the volume of information on products of Class II HLA determinants grew, and serological reagents became available for the identification of these products, all volunteer families given skin grafts at Hospital Saint Louis underwent DR typing, and the behavior of skin grafts within these families, as well as a number of additional confirmatory skin grafting experiments was analyzed for DR compatibility. The results showed that compatibility for Class II antigens (HLA-DR compatibility) was consistently associated with prolonged skin graft survival, with an effect roughly similar to that exerted by Class I antigens. There were aslo significant additive effects in the capacity of HLA-A, B and/or Dr incompatibilities to produce decreases in graft survival, as summarized in Table 1.

Over the years, more than 1000 skin grafts have been performed in over 750 recipients, within the 238 families of fully HLA genotyped volunteers in Paris, as well as in patient volunteers on the Fracture Service at Bellevue Hospital in New York, and in medical and dental

Table 1 Skin Graft Survival Under Varying Conditions of Donor-Recipient HLA Incompatibility

Level of Incompatibility	HLA Haplotype Involved	Skin Graft Survival Time
3 Incompatibilities	A, B, DR	10 days
2 Incompatibilities	A, DR A, B B, DR	13 days
1 Incompatibility	A B DR	16 days
0 Incompatibility	HLA genetically different	20 days
	HLA identical siblings	25 days

students at New York University. The credit for the present state of the art in HLA and in clinical transplantation truly belongs to this extraordinary group of volunteers, drawn from all walks of life, in Paris and in New York. This is an enormous debt, which we acknowledge with our deepest gratitude, in the name of the many thousands of transplanted patients who have benefitted from the results of these labors.

References

1. Medawar PB (1994). *J Anat* **78**, 176.
2. Medawar PB (1945). *J Anat* **78**, 155.
3. Rapaport FT, Converse JM (1968). In Rapaport FT, Dausset J (eds.) *Human Transplantation.* New York: Grune & Stratton, p. 304.
4. Converse JM, Rapaport FT (1956). *Ann Surg* **143**, 306.
5. Rapaport FT, Converse JM (1957). *NY Acad Sci* **64**, 836.
6. Rapaport FT, Converse JM (1958). *Ann Surg* **147**, 273.
7. Dausset J (1990). In Terasaki P (ed.) *History of HLA: Ten Recollections*. Los Angeles, CA: University of California Press, p. 1.
8. Rapaport FT, Thomas L, Converse JM, *et al. Fed Proc* **10**, 36.
9. Rapaport FT, Lawrence HS, Thomas L, *et al. Ann NY Acad Sci* **99**, 564.
10. Rapaport, FT, Lawrence, HS, Thomas, L, *et al.* (1962). *J Clin Invest* **41**, 2166.
11. Dausset J, Nenna A (1952). *Compt Rendus Soc Biol (Paris)* **146**, 1539.
12. Payne R, Trapp M, Weigle J, *et al.* (1964). *Cold Spring Harbor Symp Quant Biol* **29**, 285.
13. van Rood JJ, van Leeuwen AJ (1963). *Clin Invest* **42**, 1382.
14. Dausset J (1958). *Acta Haematol (Basel)* **20**, 156.
15. Dausset J, Rapaport FT, Colombani J, *et al. Transplantation* **3**, 701.

16. Dausset J, Rapaport FT, Ivanyi P, *et al. Histocompatibility Testing.* Copenhagen: Munksgaard, p. 63.
17. Dausset J, Rapaport FT (1969). *Transpl Proc* **1**, 649.
18. Dausset J, Rapaport FT, Legrand L (1969). *Nouv Rev Franc D'Hematologie* **9**, 215.
19. Dausset J, Rapaport FT, Legrand L, *et al.* (1970). In Terasaki P (ed.) *Histocompatibility Testing*. Copenhagen: Munksgaard, p. 381.
20. Rapaport FT, Casson PR, Legrand L, *et al.* (1970). *Fed Proc* **29**, 507.
21. Legrand L, Dausset J, Rapaport FT (1971). *Transfusion* **11**, 233.
22. Bach FH, Amos DB (1967). *Science* **156**, 1506.
23. Hirschhorn K, Rapaport FT, Lawrence HS, *et al.* (1964). *Ann NY Acad Sci* **120**, 303.
24. Sasportes M, Lebrun A, Rapaport FT, *et al.* (1972). *Transpl Proc* **4**, 209.

XIII

THE HISTORY OF CORNEAL TRANSPLANTATION

Chad K Rostron

Evolution of Ophthalmic Surgery

Cataract surgery

One of the advantages that ophthalmologists enjoy is that they are able to directly view the pathological processes of the eye. Medical observers have been able to document the two most common causes of blindness, cataract and corneal scarring, since records began. Perhaps it was the chance observation that the spontaneous dislocation of a mature cataract from the visual axis gave some restoration of sight that prompted the technique of "couching", a form of cataract operation first described in Sanskrit manuscripts 2000 years ago. With the aid of a sharp instrument to penetrate the eye and dislocate the lens, navigational vision could be restored in an otherwise blind eye.

Corneal surgery

Effective treatment of corneal scarring was more elusive, since complete removal of a diseased cornea requires its replacement with similar transparent tissue if the integrity of the globe is to be preserved. Although the Greek physician Galen (AD 130–200) documented removal of corneal scars by superficial keratectomy, it was not until

1837 that the first successful transplantation of a cornea was recorded by an Irish physician, Bigger.[1] He had been held captive by some Bedouin in the Sahara, during which time he managed to restore the sight of a pet gazelle with a homograft from another animal. It was inevitable perhaps that the earliest corneal transplants attempted in humans were xenografts. Transient success was claimed by Kissam in the U.S., who performed a graft with porcine tissue in 1838. Further experimental work in Germany by von Hippel included a partially successful lamellar xenograft using full thickness rabbit cornea into a lamellar bed. He developed a clockwork trephine with which to cut both donor and recipient cornea.[2] The first successful full-thickness corneal allograft was performed in 1905 by a surgeon named Zirm working near Prague, who restored the sight of a 45-year-old man with bilateral corneal scarring from lime burns.[3] This early success was a spur to further experimentation, but it would not be for another 50 years that corneal transplantation became a reproducible procedure.

Early surgical experiences

The introduction of cocaine anesthesia in 1884 by ophthalmologist Karl Koller in Vienna, inspired by his colleague in neurology Sigmund Freud, was a potent stimulus to development of ophthalmic surgery. Safe and adequate anesthesia for corneal and other anterior segment procedures could now readily be achieved, and cataract extraction through a 180° corneal incision became a standard procedure. Iridectomy and other glaucoma drainage procedures were also to become commonplace within a few years. The chief technical problem with early corneal grafting was that of fastening the transplanted tissue in place. Although sharp trephines and scissors could be manufactured to excise the diseased central cornea and to prepare the donor tissue, sutures at that time were not suitable for such fine work, particularly when watertight wound closure was required. The tissue was held in place by "stay" sutures across the surface of the eye, which was kept closed with padding while the patient rested in bed for many days until wound closure was established. In the absence of an understanding of corneal physiology, many early corneal grafts failed because of corneal endothelial damage. This occurred either during preparation of the donor tissue, or secondary to leaking wounds with a flat anterior chamber as the lens/iris diaphragm pressed forward onto the endothelium. Between 1913 and 1955, some 3500 keratoplasties were carried out by Filatov in Russia with a success rate of around 60%.[4] In France, progress was made particularly in the application of lamellar keratoplasty by Paufique.[5]

Indications for Grafting

The main causes of corneal failure worldwide are trachoma, vitamin A deficiency, herpes simplex and other types of infectious keratitis.[6] These diseases destroy the optical function of the cornea by scarring and opacification, and by stromal melting and thinning that cause surface topographic irregularity. Although the effect of an external keratitis on vision may be profound, on corneal endothelial function it is often minimal. When grafting is carried out for these conditions it is necessary to replace only the diseased superficial stromal layers to restore normal corneal clarity and optical function. By avoiding penetration of the globe, lamellar keratoplasty gives freedom from many of the complications of a penetrating graft.

However conditions that do cause corneal endothelial depletion and failure, such as Fuchs' corneal endothelial dystrophy, and iatrogenic endothelial failure following intraocular surgery, do require endothelial replacement. For such cases a penetrating keratoplasty with transplantation of a viable donor endothelial cell layer has historically been the only effective treatment. In the absence of a clear differentiation of the types of corneal pathology being treated, penetrating keratoplasty became the standard treatment for all types of corneal disease.

Corneal physiology

Post-war technological advances during the 1950s brought a surge of developments in all areas of medicine. Careful studies of corneal physiology were made by David Maurice that demonstrated the essential role of the corneal endothelial pump function in maintaining corneal transparency. He observed that the cornea of a whole donor globe held in a refrigerator at 4°C becomes oedematous and hazy as aqueous diffuses amongst the stromal fibres and epithelial cells. As the globe is brought back to room temperature the endothelial pump function recommences, and transparency is restored as the fluid is pulled out of the cornea — the so-called "temperature reversal effect". Maurice was also a pioneer in specular microscopy, a technique which allowed visualisation of the corneal endothelium both *in vitro* and *in vivo*[7] (Fig. 1). Armed at last with a clear understanding of normal corneal function, clinicians could take rational steps to enhance corneal transplantation technique, and regular success with penetrating keratoplasty became a real possibility.

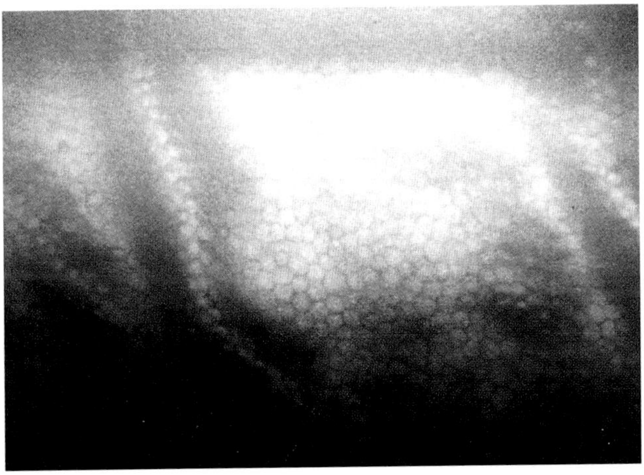

Fig. 1 Specular micrograph of corneal endothelium following penetrating keratoplasty showing a somewhat depleted cell density of 1500 cells/mm^2.

Pharmacological Advances

Pharmacological advances at that time brought the benefits of both antibiotics and steroids. Although iatrogenic infection associated with corneal transplantation is rare, it is a devastating complication because organisms inoculated directly into an eye produce an endophthalmitis which will often destroy the eye before the infection can be brought under control. However, the most important factor in enhancing the success of penetrating keratoplasty has been the introduction of steroids. In the eye it is possible to achieve high local tissue concentrations of topically applied drugs, so that there is virtually no risk of the significant systemic side-effects that often occur with the systemic treatment given following solid organ transplant surgery. In the majority of corneal grafts, topical steroids alone are sufficient to control postoperative inflammation and to act as prophylaxis against, or if necessary as treatment for, immune rejection. Nevertheless, in high dose or with prolonged use, steroids can take their toll on the eye. Their chief local side-effects are cataract formation, and steroid-induced glaucoma. Both, of course, lead to progressive visual degradation and introduce additional complications when assessing the quality of outcome of the transplantation procedure.

Advances in Instrumentation

Many small incremental advances in instrumentation and surgical technique made a substantial impact on graft outcome. Although surgeons had used magnifying loupes from the outset of corneal transplantation, the introduction of the operating microscope gave a new dimension to the accuracy of the technique. Improved microsurgical instrumentation reduced tissue damage during surgery. The introduction of nylon monofilament as a suture material, enabling perfect corneal wound closure with negligible tissue reaction, was a major step forward. Corneal wound healing is slow, and a suture may need to be left *in situ* for as long as 12 to 18 months before adequate wound strength is attained if the healing process has been inhibited by steroid treatment

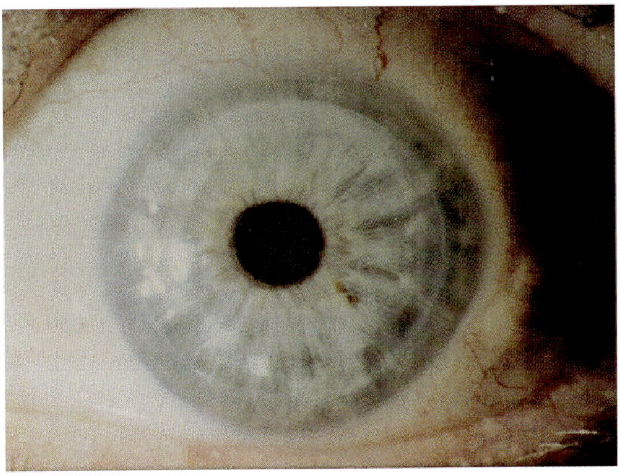

Fig. 2 A penetrating corneal graft with multiple interrupted 11/0 polyester sutures. These have the advantage over nylon of being non-biodegradeable, and can give long term support to the wound.

(Fig. 2). Another advance has been the introduction of the viscoelastic substance sodium hyaluronate, which provides ophthalmic surgeons with a method of handling delicate eye tissues with virtually no trauma during intraocular surgery.[8]

Contact Lenses

The development of contact lenses has also had beneficial effects, both direct and indirect, on keratoplasty. They are now the mainstay in the management of the corneal ectasias such as keratoconus, and are also frequently invaluable in correcting post-keratoplasty optical defects. The earliest contact lenses were of glass, with generally extremely limited wearing times. The introduction of moulded scleral lenses made from polymethylmethacrylate (PMMA) in the 1930s gave greatly enhanced acceptability, and the subsequent development of the corneal contact lens by Tuohy further extended their use. In the 1970s the "gas-permeable" polymers replaced the impermeable PMMA, but even today the large scleral haptic lens designs are still sometimes required in advanced corneal ectasia.

Eye Banking

Alongside the changes in surgical technique came the introduction of eye banking and developments in tissue handling and storage. Historically, donor globes were harvested up to 24 hours post-mortem and the whole eye was then kept in a moist chamber at 4°C for a further 24 to 48 hours, before the corneal button for transplantation was excised for immediate transplantation. In 1974, McCarey and Kaufman in the U.S. demonstrated that by excising the whole cornea from the globe and placing it in a tissue culture medium at 4°C, that the endothelium could remain viable for several days.[9] Their "MK medium" which contained TC199, Earle's salts, HEPES buffer and gentamicin remained the standard corneal preservation medium for some 15 years. This was subsequently superseded by other commercial preparations, such as KSol and Optisol, containing osmotic agents to limit corneal tissue swelling, and offering extended preservation times of a week or ten days.[10]

Techniques of Preservation

In the quest for longer term storage, much work has explored the possibility of cryopreservation of the cornea. The first successful graft using cryopreserved tissue was reported by Eastcott *et al.* in 1954.[11] In the 1960s, Casey in England and Kaufman in the U.S. published short series of clinical cases using cryopreserved tissue, with what were reasonable outcomes for the era. However, no further human studies have been published since then, despite extensive laboratory work on the subject. A different approach to extending corneal tissue preservation times is that of organ culture at 34°–37°C. Pioneering studies were carried out by Doughman in Minnesota,[12] and the system further developed by Pels[13] in Holland and Easty in England. By placing corneas into organ culture, tissue can be preserved for four to six weeks, and the risk of transmission of infective endophthalmitis is probably reduced in comparison to 4°C storage (Fig. 3). However the organ culture system is labour intensive, and whilst it has become the favored

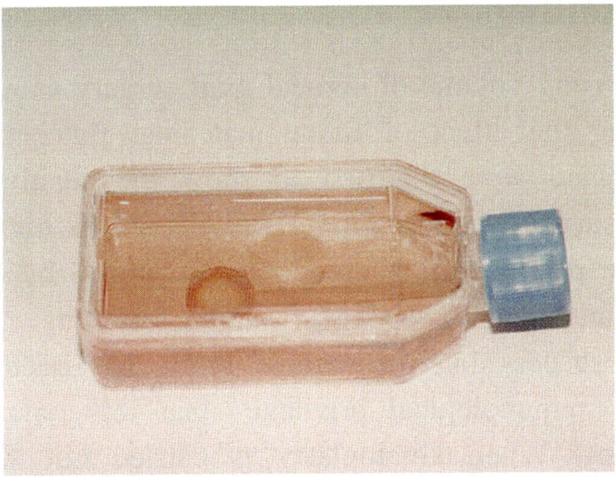

Fig. 3 A corneo-scleral segment held in organ culture medium at 34°C can be preserved for several weeks, and the corneal endothelium can be examined through the floor of the tissue culture flask.

preservation method for European eye banks, 4°C storage is still used by most eye banks in the U.S.

Eye Bank Associations

The massive increase in cataract surgery during the latter half of the 20th century brought with it a tide of complications. Surgical damage to the corneal endothelium can lead to corneal failure manifest as aphakic and pseudophakic bullous keratopathy, and these became the prime indications for keratoplasty in some institutions. The development of eye banks and eye bank organizations such as the American and European Eye Bank Associations has done much to improve the standards and quantity of tissue provided for keratoplasty, and to encourage positive public attitudes towards tissue and organ donation. Eye banks in the U.S. and most European countries currently maintain supplies of corneal tissue at a level commensurate with demand. Globally however, most corneal blindness is found in countries without the resources for transplantation services.

Tissue Matching

With the increasing volume of corneal graft surgery came the possibility of studying the effect of tissue matching. Batchelor and Casey *et al.* in England were the first to show the positive benefits on corneal graft survival of matching HLA Class I in high risk cases.[14] Although these results were subsequently confirmed by other workers, the results of matching HLA Class II were conflicting.

The high success rate of corneal grafts has been attributed to the apparent "immunological privilege" that this type of transplant seems to enjoy. When keratoplasty is carried out without tissue-matching into an uninflamed eye with an avascular cornea (as for instance in keratoconus), a two-year graft survival with topical steroid treatment can only be >90%. The situation changes dramatically when transplantation is carried out into a recipient where there has been previous inflammation and vascularization, and in this scenario graft rejection rates can be >50% even with systemic and topical immunosuppression.[15]

Our understanding of the mechanisms of graft rejection has grown considerably in the past 15 years. The immunological privilege of the cornea can be attributed to a constellation of a wide variety of factors including the avascularity of the cornea which diminishes access of immune cells to the anterior segment, the absence of lymphatic vessels in the cornea, the paucity of Langerhans or other antigen presenting cells, and the low level of expression of MHC antigens in the cornea. In addition, there are active mechanisms by which the immune response is controlled, some of which are unique to the eye. These include factors that inhibit T-cell and complement activation, and others which can induce apoptosis of stimulated T-cells. Presentation of intraocular antigen may also lead to a tolerogenic form of immunity termed anterior chamber-associated immune deviation (ACAID).[16] It has been shown in experimental penetrating keratoplasty that the development of sensitization to donor antigens is virtually universal. The fact that such sensitization does not always lead to rejection indicates the importance of the tolerogenic mechanisms including ACAID.

Sensitization of the recipient in low risk keratoplasty is usually by the "indirect pathway" in which donor antigens are processed by recipient antigen presenting cells and presented in the context of host MHC to T-cells generating self-restricted or indirect alloreactive T-cells. This explains why minor histocompatibility antigens are a more significant barrier to graft success than MHC antigens in this scenario.

In the high risk setting however, the immune privilege of the cornea is eroded and host sensitization may occur through both direct and indirect pathways. Expression of donor MHC is upregulated and becomes more important in transplant rejection. The failure of the American Collaborative Corneal Transplantation Study to show any benefit from MHC matching may have been because the intensive steroid treatment used in the protocol suppressed MHC expression to the extent that it was no longer a significant factor in graft survival.[17]

Refractive Errors After Keratoplasty

The restoration of corneal transparency, although a wonderful achievement, is not synonymous with the restoration of vision, as many clinicians have become painfully aware. Keratoplasty may often leave considerable refractive error in the eye. The refractive properties of the cornea, and the possibility of manipulation of the kerato-refractive status of the eye, has been the subject of research for almost as long as the history of keratoplasty itself. In the 1940s, Jose Barraquer in Colombia devised the operation of keratomileusis to flatten the cornea as a means of correcting myopia, and the operation of keratophakia to steepen the cornea (using donor tissue) as a means of correcting hypermetropia. Because of their complexity, these techniques were never widely adopted. Refractive keratoplasty finally became a reality for corneal surgeons with the introduction of epikeratophakia.[18] In this procedure, a pre-lathed onlay graft is transplanted onto the de-epithelialised surface of a normally functioning cornea. Multicenter trials revealed that for the correction of high hypermetropia, commonly found in aphakia, this procedure was effective,[19, 20] but that for the correction of myopia it was less satisfactory, due to inaccurate optical outcomes. The procedure was also modified to correct corneal irregularity in early keratoconus by onlaying a graft lathed such that it had no optical power.[21] The multicenter trials of epikeratophakia used freeze-dried corneal tissue. The grafts were transplanted onto corneas which had normally functioning endothelium, and were shown to become covered with host epithelium and repopulated with host keratocytes. No rejection of the transplanted stromal matrix was seen, and animal studies suggested that although dead donor keratocytes were transplanted, that they did not sensitize the recipient to any donor antigens.[22] Although epikeratoplasty remains an effective treatment for early keratoconus, epikeratophakia has now been supplanted by improved intraocular lens implantation techniques.

Correction of post-keratoplasty refractive errors

In penetrating keratoplasty, achieving a satisfactory optical outcome remains a challenge. Clinicians continue to be frustrated in achieving the desired optical outcome by primary intention. Part of the problem is the slowness and weakness of the corneal wound healing process. Although pre- and postoperative adjustment of suture tension can improve short-term optical performance, the situation can only really be judged finally when all sutures have been removed, and often there will be considerable residual astigmatism. The development of corneal topographic analysis systems has done much to help our understanding and management of post-graft ametropia[23] (Fig. 4). Historically, adjustment of residual astigmatism has been carried out through a variety of approaches, including wound relaxing incisions, tension sutures, and astigmatic keratotomy.[24] In the past few years, the introduction of excimer Laser *In Situ* Keratomileusis (LASIK) has at last given surgeons the ability to correct residual spherical and astigmatic refractive defects and to restore the normal optical function of the eye.[25, 26] Nevertheless, this is secondary surgery, which carries the risk of precipitating wound dehiscence or graft rejection, and so cannot be considered ideal.

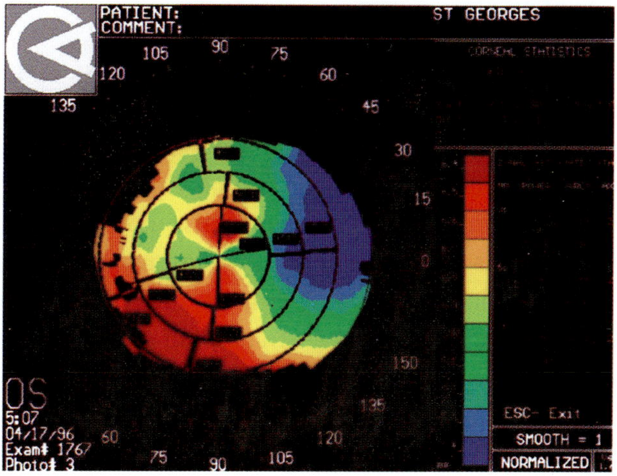

Fig. 4 Corneal topography following keratoplasty for keratoconus showing persistent inferior steepening and irregular astigmatism, although within the optical zone the astigmatism is regular enough to allow good spectacle corrected acuity.

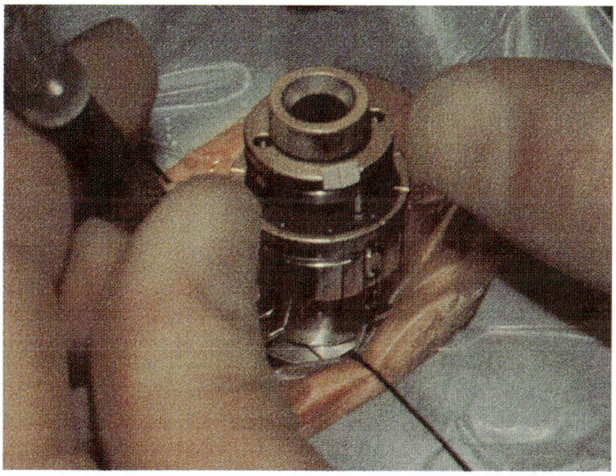

Fig. 5 The guided trephine system is stabilised on the eye by a suction ring, and allows precise vertical trephination of the recipient cornea. The donor is cut in a similar manner whilst the tissue is held in an artificial anterior chamber.

Guided trephine system

A hi-tech approach to penetrating keratoplasty is exemplified by Krumeich's guided trephine system. This allows extremely accurate trephination of the host cornea with the trephine system attached and stabilized on the globe by a vacuum suction ring[27] (Fig. 5). The donor cornea is trephined in a similar manner whilst mounted on an artificial anterior chamber. As a final step to try and ensure sphericity of the grafted tissue, a stainless steel ring can be sutured in the wound between graft and host. The added cost and complexity of this approach underlines the problem of achieving not only graft clarity but restoration of effective visual function.

New Lamellar Techniques

Studies of graft survival following penetrating keratoplasty have shown an increased risk of rejection with larger grafts, possibly due to increased proximity of the grafted tissue to the vascular limbus, and to increased donor tissue antigenic load. To enhance graft survival, surgeons try to keep graft size small. However, small grafts create bigger optical problems due to the proximity of the graft wound to the visual axis. This can particularly be a problem when the recipient cornea has an eccentrically situated ectasia, as is commonly found in keratoconus. A new approach to this problem is now offered as surgeons use advances in microsurgical technique to explore the full potential of lamellar keratoplasty. Resection of the recipient stroma can now be made right down to the level of Descemet's membrane, and today the procedure is referred to as deep lamellar keratoplasty. The cleavage of Descemet's membrane from the posterior stroma is achieved by air dissection,[28] hydro-dissection,[29] or visco-dissection[30] (Fig. 6). Deep lamellar keratoplasty is suitable for all corneal pathologies in which endothelial function is intact, e.g. keratoconus, corneal scarring, etc. Endothelial rejection is not an issue since the recipient endothelium is not replaced, and large grafts can be used, ensuring complete excision of ectactic or diseased tissue.

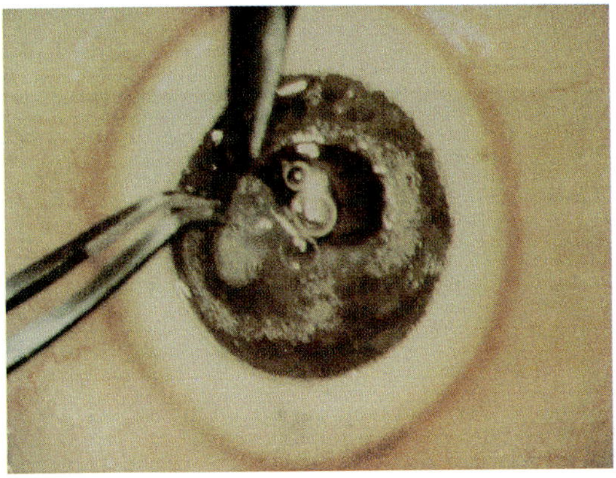

Fig. 6 Deep lamellar keratoplasty: the cornea has been injected with air rendering it opaque (white). The bulk of the central stroma has been resected (a few residual fine bubbles seen). The central stroma is being excised with scissors whilst Descemet's membrane is held clear by viscoelastic. Larger bubbles are seen centrally in the anterior chamber beneath Descemet's membrane.

Epithelial Transplantation

The fine optical function of the cornea is dependent not only on the stromal contour but on the health of the corneal epithelium and its overlying tear film. When there is corneal epithelial failure, the surface of the cornea becomes conjunctivalised, vascularised, and irregular. In 1977, Thoft described conjunctival transplantation for ocular surface disease, and in the 1980s devised an operation that he named "keratoepithelioplasty".[31] In this procedure, thin lenticules of superficial donor stroma with their overlying corneal epithelium were transplanted to restore normal epithelial cover on the recipient cornea. With the subsequent development of an understanding that the corneal epithelium originates from stem cells located at the limbus, a more rational categorization of the many variants of corneal epithelial and conjunctival epithelial transplantation procedures can be made. Allografts of corneal epithelial stem cells require major systemic immunosuppression and thus have a more problematical risk/benefit ratio

(A)

Fig. 7(a) A microkeratome shown mounted on a suction ring. The suction ring is made adherent to the eye before commencing the cut.

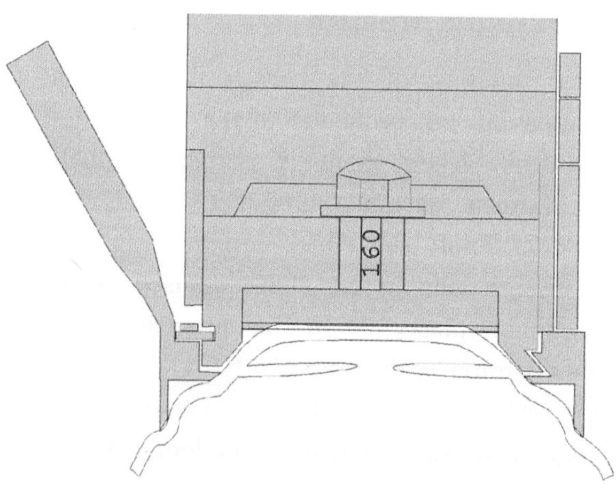

(B)

Fig. 7(b) Cross-sectional diagram showing how the microkeratome cuts a lamellar section through the cornea by flattening the central corneal area.

for the treatment of a non-life threatening condition. A recent report from Italy suggests that laboratory cultured epithelial stem cells may in the future be the ideal treatment for stem cell failure.[32] Their technique was based on earlier laboratory work by Lindberg *et al.*[33]

Endothelial Transplantation

The desirability of replacing the corneal endothelium only, when there is endothelial cell failure, has been apparent for some years. The obstacles to this have been the difficulty of getting human corneal endothelial cells to replicate in the laboratory, and the technical challenge of delivering endothelium to the diseased eye. Two recently devised methods of corneal endothelial sheet transplantation have been described. The first, described by Jones and Culbertson in the U.S., uses a microkeratome to create a large flap in the recipient cornea, and the exposed posterior stroma is then trephined. The donor tissue is also resected with a microkeratome, enabling a 250 μ thick button of posterior stroma/Descemet's/endothelium to be prepared and transplanted. The large flap minimizes postoperative astigmatism[34] [Fig. 7(a) and (b)]. The second technique, from Melles in Holland, requires a deep intrastromal pocket to be formed from a limbal wound, and the posterior stroma is then cut with a low profile intrastromal trephine. The posterior lamellar button is extracted from the wound, and the prepared donor lamella introduced through the same route. This technique gives minimal disruption to the external corneal contour.[35] It is likely that in the future, penetrating keratoplasty will be progressively supplanted by such techniques whereby only the diseased part of the cornea is replaced.

Conclusion

The first hundred years in the history of corneal transplantation have seen huge steps in our understanding of corneal physiology and pathology, the miracle of successful transplantation, and the mysteries of graft rejection. Further scientific advance will no doubt enable even greater success in what we can achieve. The ten million people blind from corneal disease globally remain a huge unanswered challenge.[6]

References

1. Bigger SLL (1837). An inquiry into the possibility of transplanting the cornea with a view of relieving blindness. *Dublin J Med Sci* **11**, 408–417.
2. Von Hippel A (1888). Eine neue Methode der Hornhaut Transplantation. *Graefes Arch Ophthal* **34**, 108–130.
3. Zirm E (1906). Eine erfolgreiche totale Keratoplastik. *Archiv Ophthalmol* **64**, 580–593.
4. Filatov VP (1935). Transplantation of the cornea. *Arch Ophthalmol* **13**, 321–347.
5. Paufique L, Sourdille GP, Offret G (1948). *Les Graffes de la Cornee.* Paris: Masson.
6. Smith GTH, Taylor HR (1991). Epidemiology of corneal blindness in developing countries. *Refract Corneal Surg* **7**, 436–439.
7. Maurice DM (1974). A scanning slit optical microscopy. *Invest Ophthalmol Vis Sci* **13**, 1033–1037.
8. Holmberg AS, Philipson BT (1984). Sodium hyaluronate in cataract surgery. I. Report on the use of Healon in two different types of intra-capsular surgery. *Ophthalmology* **91**, 45–52.
9. McCarey BE, Kaufman HE (1974). Improved corneal storage. *Invest Ophthalmol* **13**, 165–173.
10. Kaufman HE, Beuerman RW, Steinemann TL, Thompson HW, Varnell ED (1991). Optisol corneal storage medium. *Arch Ophthalmol* **109**, 864–868.
11. Eastcott HHG, Cross AG, Leigh AG, North DP (1954). Preservation of corneal grafts by freezing. *Lancet* **1**, 237.
12. Doughman DJ (1980). Prolonged donor preservation in organ culture: long term clinical evaluation. *Trans Am Ophthalmol Soc* **78**, 567–628.

13. Pels L, Schuchard Y (1993). Organ culture in the Netherlands. In Brightbill FS (ed.) *Corneal Surgery: Theory, Technique, and Tissue*. St. Louis: Mosby, pp. 622–632.
14. Batchelor JR, Casey TA, Werb A, Gibbs DC, Prasad SS, Lloyd DF, James A (1976). HLA matching and corneal grafting *Lancet* **1**, 551–554.
15. Mader TH, Stulting RD (1991). The high-risk penetrating keratoplasty. *Ophthalmol Clin North Am* **4**, 411–426
16. Streilein JW (1995). Immunological non-responsiveness and acquisition of tolerance in relation to immune privilege in the eye. *Eye* **9**, 236–240.
17. The Collaborative Transplantation Studies Research Group (1992). The collaborative corneal transplantation studies (CCTS): effectiveness of histocompatibility matching in high-risk corneal transplantation. *Arch Ophthalmol* **110**, 1392–1403.
18. Kaufman HE (1980). The correction of aphakia. XXXVI Edward Jackson Memorial Lecture. *Am J Ophthalmol* **89**, 1–10.
19. McDonald MB, Kaufman HE, Aquavella JV, *et al.* (1987). The nationwide study of epikeratophakia for aphakia in adults. *Am J Ophthalmol* **103**(3 Pt 2), 358–365.
20. Morgan KS, McDonald MB, Hiles DA, *et al.* (1987). The nationwide study of epikeratophakia for aphakia in children. *Am J Ophthalmol* **103**(3 Pt 2), 366–374.
21. McDonald MB, Kaufman HE, Durrie DS, *et al.* (1986) Epikeratophakia for the correction of keratoconus: the nationwide study. *Arch Ophthalmol* **104**: 1294
22. Moore MB, Gebhardt BM, Verity SM, McDonald MB (1987). Fate of lyophilized xenogeneic corneal lenticules in intra-stromal implantation and epikeratophakia. *Invest Ophthalmol Vis Sci* **28**, 555–559.
23. Klyce SD (1984). Computer-assisted corneal topography: high resolution graphical presentation and analysis of keratoscopy. *Invest Ophthalmol Vis Sci* **25**, 1426–1435.

24. Troutman RC (1983). Corneal wedge resections and relaxing incisions for post-keratoplasty astigmatism. *Int Ophthalmol Clin* **23**, 161–168.
25. Pallikaris IG, Papatzanaki ME, Stathi EZ, *et al.* (1990). Laser *in situ* keratomileusis. *Lasers Surg Med* **10**, 463–468.
26. Guell JL, Gris O, de Muller A, Corcostegui B (1999). LASIK for the correction of residual refractive errors from previous surgical procedures. *Ophthalmic Surg Lasers* **30**, 341–349.
27. Ruhswurm I, Scholz U, Pfleger T, Zehetmayer M, Hanselmayer G, Skorpik C (1999). Three year clinical outcome after penetrating keratoplasty for keratoconus with the Guided Trephine System. *Am J Ophthalmol* **127**, 666–673.
28. Archila E (1985). Deep lamellar keratoplasty dissection of host tissue with intrastromal air injection. *Cornea* **3**, 217–218.
29. Sugita J, Kondo J (1994). Deep lamellar keratoplasty with complete removal of pathological stroma for vision improvement. *Br J Ophthalmol* **81**, 184–188.
30. Morris E, Kirwan JF, Sujatha S, Rostron CK (1998). Corneal endothelial specular microscopy following deep lamellar keratoplasty with lyophilised tissue. *Eye* **12**, 619–622.
31. Thoft RA (1984). Keratoepithelioplasty. *Am J Ophthalmol* **97**, 1–6.
32. Pellegrini G, Traverso CE, Franzi AT, *et al.* (1997). Long-term restoration of damaged corneal surfaces with autologous cultivated corneal epithelium. *Lancet* **349**, 990–993.
33. Lindberg KL, Brown ME, Chaves HV, *et al.* (1993). *In vitro* propagation of human ocular surface epithelial cells for transplantation. *Invest Ophthalmol Vis Sci* **34**, 2672–2679.
34. Jones D, Culbertson W (1998). Endothelial lamellar keratoplasty (ELK). *Invest Ophthalmol Vis Sci* **39**, S76.
35. Melles GR, Lander F, Beekhuis WH, Remeijer L, Binder PS (1999). Posterior lamellar keratoplasty for a case of pseudophakic bullous keratopathy. *Am J Ophthalmol* **127**(3), 340–341.

XIV

THE HISTORY OF BONE MARROW TRANSPLANTATION

Shaun R McCann

"All is flux, nothing is stationary": Heracleitus (513 BC)

History

I am never sure if knowledge of history serves to reinforce old ideas and prejudices or leads to an open mind. Hopefully a knowledge of preceding events will help to explain how we have arrived at our present state of knowledge and perhaps will provide a stimulus for further exploration.

Since the earliest of days, many civilizations have had a special regard for blood. Bad blood was deemed to be a cause of disease and the removal of blood, phlebotomy, has been practiced since the days of Hippocrates. The Romans believed that great strength could be derived from blood and drank the blood of slain gladiators. Although many commentators believed in close links between blood and life, it was only in the last 100 years that a link between blood and bone marrow was discovered. However even before the links between bone marrow and blood were discovered the marrow was often referred to as the "essence". There are many ancient references to the cooking and eating of marrow presumably because of its perceived nutritious qualities. During the 17th century, philosophers/scientists such as Robert Boyle took a great interest in the possibility of blood transfusion.

Boyle suggested a series of experiments which were recorded by Samuel Pepys, one of the original Fellows of the Royal Society (14 November, 1666):

> "At Gresham College tonight...there was a pretty experiment of blood of one dog let out (till it dies) into the body of another on one side while his own ran out the other side. The first died upon the place and the other very well and likely to do well. This did give occasion to many pretty wishes of the blood of a Quaker to be let into an Archbishop and such like but as Dr Crome says may, if he takes be of a mighty use to man's health, for the amendment of bad blood by borrowing from a better body".

The Physiology of Blood and Marrow

In 1886, Ernst Neumann was credited with the recognition that marrow is the seat of blood formation and more importantly that blood formation is a continuous process, occurring during postnatal life. The opening paragraph of his 1868 note reads: "The present work intends to demonstrate the physiologic importance of the bone marrow and that it is an important organ for blood formation which has not been recognised. It operates continually in a *de novo* formation of red blood cells". As with many radical claims in medicine his views were vehemently opposed.

Almost at the same time, Giulio Bizzozero, a 22-year-old graduate from Pavia, claimed that non-nucleated red cells were formed from nucleated cells in the bone marrow and that white cells were also derived from the bone marrow. Many discussions and arguments took place subsequently, mostly based on morphology and cytochemistry.

Ironically, research conducted preceding and during World War II led to the realisation that ionizing radiation could be lethal to animals and that death was due to bleeding. These observations opened up a whole new area of research that was to culminate in one of the most spectacular advances in medicine in the last 50 years. By the early 1950s, a number of experiments on small laboratory animals conclusively demonstrated that an animal, exposed to a lethal dose of total body irradiation, could be protected from fatal haemorrhage by shielding of the spleen (a haemopoietic organ in rodents). Thus, the development of the Atomic Bomb and the knowledge of its devastating effects served as a stimulus to our understanding of the physiology of bone marrow and blood formation and led to the development of bone marrow transplantation as a new and radical modality of therapy for leukaemia and bone marrow failure syndromes.

Jacobsen, in his seminal experiments in 1949 not only demonstrated that mice could be protected from the lethal effects of ionising irradiation by shielding the spleen with lead, but that this effect was much greater than when any other areas were similarly protected. As is common in medicine, it is very difficult to have a new idea and this is illustrated by the neglected experiments of Fabrious-Muller in the early 1920s. He demonstrated that guinea pigs developed severe thrombocytopenia following irradiation and that this effect could be largely abrogated by shielding the animal's legs from the irradiation.

In 1951, Jacobsen extended the earlier observations by showing that intraperitoneal injections of isologous or autologous spleen cells provided similar protection from the lethal effects of irradiation. A debate subsequently took place as to origin of the protective effect of infused spleen cells. One group favoured the idea that a "humoral" substance was responsible for stimulating the growth of haemopoietic tissue and the other believed that the cellular infusion was responsible for the regrowth of the bone marrow. A number of ingenious experiments were devised to solve the conflict, but it was in 1956 that Ford *et al.* provided the final proof of the cellular origin of the protective effect of marrow or spleen cell infusions.[1] By using a chromosomal marker, they were able to clearly demonstrate that the transplanted cells were of donor origin and later showed a gradual increase in the numbers of donor cells over a time period. Animals which were "rescued" from the lethal effects of ionising irradiation by these

Fig. 1 Chimera.

infusions of donor cells were known as "radiation chimeras". In Greek mytho-logy, the Chimera was one of the five progeny of Typhon and Echidna and has traditionally been composed of the head of a lion, the body of a goat and the tail of a snake (Fig. 1). Chimeras had been described in 1945 by Owen when he demonstrated the occurrence of two different blood types in cattle which shared the same placental blood supply. A similar occurrence was documented in humans in 1953.

Further Developments

From the earliest experiments, it became clear that transplantation of bone marrow or other sources of haemopoietic stem cells was intrinsically different from transplantation of solid organs. The main problem facing those trying to perfect solid organ transplantation is graft rejection due to immunological differences between donor and host. It has become clear that life-long immunosuppression of the recipient is necessary for continued survival of the donor graft. The major differences between transplantation of solid organs and haemopoietic stem cells is that haemopoietic stem cell grafts contain large numbers of immunologically competent cells which recognise immunologic differences between the graft and host. This leads to two potential problems, namely the possibility of graft rejection, as in solid organ transplantation, and, far more importantly, the possibility of the immunologically competent cells from the graft "attacking" host tissues and establishing a new phenomenon which we call Graft versus Host Disease (GvHD). The unraveling of the HLA system and the definition of class I and II, so-called major histocompatibility (MHC) antigens, helped to develop transplant strategies that would minimise the possibilities of graft rejection and the occurrence of GvHD. MHC antigens were shown to be encoded on chromosome 6 and are composed of a series of highly polymorphic cell surface glycoproteins. The genes that encode HLA are tightly linked and are inherited in families as "haplotypes" with low recombination frequencies. For a given individual, there is a 25% chance that he or she will inherit the same haplotype as a sibling. So-called minor histocompatibility antigens are also important in the genesis of GvHD. These antigens are not genetically linked, and therefore there is a distinct probability of differences between siblings who share the same MHC haplotype. The possibility of differences in minor histocompatibility antigens between non-related individuals is even greater.

Bone Marrow Structure and Function

Other important discoveries in the story of bone marrow transplantation were the demonstration of the "pluripotent stem cell" by Till and McCulloch in 1961[2] and the work of Metcalf,[3] and others which began to clarify the growth factors and cell-cell interactions which occur in the marrow and govern the differentiation and proliferation of haematopoietic tissue. Not only did these experiments provide insight into the normal physiology of human bone marrow but also led to the development of recombinant growth factors for use in the treatment of haematological and other malignancies, the support of patients following bone marrow transplantation and the mobilisation and collection of human haemopoietic progenitors from peripheral blood.

Other Animal Models

Small laboratory animals such as mice, rats and guinea pigs are not the ideal models from which to build a human bone marrow transplant programme and the development of a dog model by E. Donal Thomas, Rainer Storb and colleagues in Seattle really paved the way for the clinical transplant programmes that are in worldwide use today. The dog proved to be a particularly useful animal. They are large enough to obtain serial blood and marrow samples from, the major histocompatibility complex has been elucidated, spontaneous haemopoietic malignancies arise similar to those occurring in man and they live for a long time thus allowing studies of the long-term effects of haemopoietic transplantation to be undertaken.

Early Attempts at BMT

The first published attempt at human bone marrow transplantation was by the doyen and Nobel Prize winner E. Donal Thomas in 1957.[4] Although this attempt was unsuccessful, it did demonstrate that large amounts of marrow could be safely infused intravenously. In 1959, the same author reported on the attempt to treat a twin with acute lymphoblastic leukaemia with total body irradiation (TBI) and an infusion of marrow from her identical sister (syngeneic).[5] Although the leukaemia recurred after a short interval, the report was able to show that an infusion of marrow could protect against the lethal effects of TBI. A critical observation by the authors was that TBI alone was unlikely to cure leukaemia and that an "allogeneic" effect or "immunological" effect mediated by marrow from a non-identical source might have an anti-leukaemic effect. Once again this was an extension of the ideas of

Barnes and colleagues who, in 1956, suggested that a graft versus leukaemia effect might be necessary to cure leukaemia in a mouse model.[6] In 1959, Georges Mathe and colleagues reported the first attempt to use bone marrow transplantation for the treatment of a non-malignant disorder.[7] In 1957, a highly publicised accident occurred at a nuclear reactor in the Nuclear Centre in Vinca, Belgrade. Six laboratory workers were accidentally exposed to heavy radiation doses from neutrons and gamma rays. A group of physicians led by Georges Mathe offered bone marrow transplantation as a form of treatment. One patient was treated initially with spleen and liver cells from a five-month-old fetus. Subsequently, he was given 8.5×10^9 allogeneic bone marrow cells. He died from bleeding. Four of the other workers received the same or higher doses of nucleated marrow cells one month after their exposure to the radiation. Two patients eventually recovered normal marrow function although it is not clear what contribution, if any, the transplanted marrow made to their recovery.

Again in 1959, J.J. McGovern and colleagues described the first attempt at autologous "transplantation".[8] (Although the authors did not use this terminology but correctly referred to "the infusion of stored autologous marrow", the term autologous transplantation, surely an oxymoron, has gradually become accepted in the medical literature.) In spite of these early attempts, a report by Mort Bortin in 1970 is somewhat depressing.[9] Of the 203 cases of bone marrow transplants reviewed, 154 had been reported between 1959 and 1962. One hundred and fifty-two of the patients had died by the time the report was written and of the 73 cases of bone marrow transplantation for aplastic anaemia, there was no evidence that marrow transplantation had cured any of them. As a result of these early failures, very few transplants were carried out between 1962 and 1968.

Two of the major factors contributing to early failures were the lack of knowledge of the human histocompatibility complex (MHC) and the relatively primitive armamentarium available to supply support care to the patient in the post-transplant period. The development of serological and cellular methods for evaluating compatibility between donor and host, the discovery of new and potent antibiotics, the revolutionary changes in transfusion medicine and blood banking (particularly the developments in platelet procurement, storage and availability) together with what might seem to be a rather simple development, long-term venous access via a right atrial catheter by Hickman,[10] all played very important roles in the successful subsequent achievements of modern bone marrow transplantation.

The New Era

In 1968, the team of Robert Good carried out a bone marrow transplant, in Minneapolis, based on results of histocompatibility tests in a child with severe immunological deficiency.[11] Of course, because of the profound immunodeficiency, the patient did not require the "conditioning therapy" which would have been necessary for a patient with leukaemia. Similar reports appeared from the Netherlands in 1969[12] and from E.D. Thomas and colleagues from Seattle in 1970.[13] In 1975, the major publications from the Seattle team in the New England of Medicine which showed, unequivocally, the efficacy of bone marrow transplantation in significant numbers of patients with leukaemia set the stage for a worldwide explosion in the application of bone marow transplantation.[14] The demonstration that Busulfan could be substituted for TBI by George Santos in Baltimore was a critical discovery as it allowed many teams to develop transplant strategies without having access to irradiation facilities. Teams in Europe and the U.S. quickly developed and happily, perhaps because of the complexity of the procedure and the fact that new clinical observations were continually being made, the cooperation between transplant teams around the world was maintained. The founding of the European Group for Bone Marrow Transplantation (EBMT) in 1973 by Bruno Speck and a small group of like-minded physicians and, much later, the American Society for Blood and Marrow Transplantation (AMBTR) ensured that there was excellent cooperation between teams and facilitated the conduct of many clinical trials which paved the way for important developments. One of the most important developments was the establishment of an international "registry" based in Milwaukee and known as the International Bone Marrow Transplant Registry (IBMTR). Together with the EBMT, large amounts of data on all aspects of bone marrow were collected. These data proved to be invaluable when planning novel techniques ranging from donor selection criteria to preservation of frozen gonadal tissue. In 1997, the IBMTR, under the direction of Mary Horowitz, collected data from 6017 allotransplants from 292 teams in 44 countries, truely a remarkable achievement (Fig. 2).

The major clinical problems were, and to a large extent still are, finding a histocompatible sibling donor, prevention of graft rejection, support care with blood products, antibiotics, antivirals and antifungals, prevention and treatment of graft versus host disease and

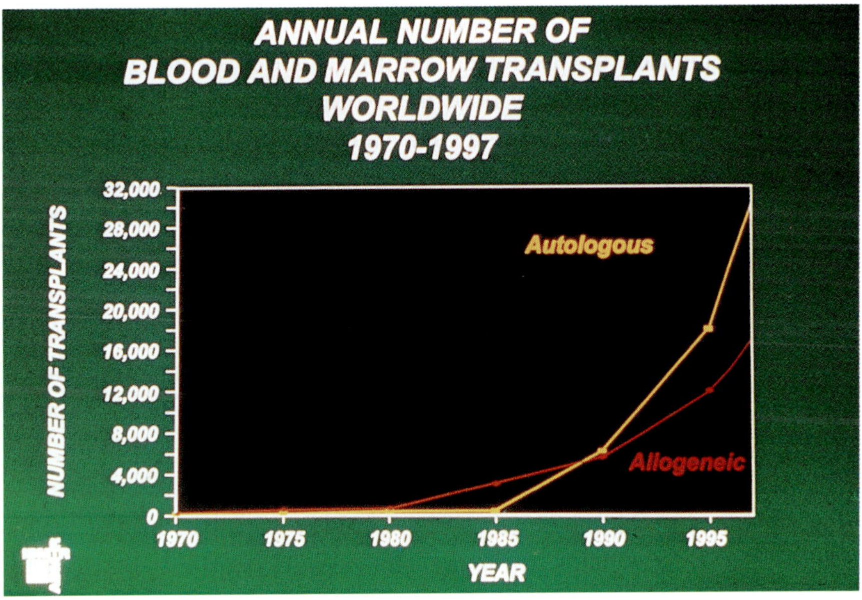

Fig. 2 Annual number of blood and marrow transplants worldwide, 1970–1977 (reproduced with kind permission of M. Horowitz, IBMTR).

veno-occlusive disease of the liver, and the long-term management of gonadal failure and sterility in adults and growth retardation in children and of course, worst of all, leukaemia relapse.

In most countries, the chance of finding a histocompatible sibling donor is about 25%, thus the majority of patients will not have a suitable donor. There are exceptions, however, for example in Ireland, where because of large family size, a histocompatible donor can be found for 75% of adult patients with chronic myeloid leukaemia.[15] In order to increase the possibility of finding a compatible donor, registries of volunteer donors were established. The first such registry of unrelated volunteer donors who underwent "tissue typing" was established by Shirley Nolan in England in 1973, the mother of Anthony Nolan who died from Wiscott-Aldrich Syndrome after failure to find a compatible bone marrow donor. Following this, Jeffrey McCulloch on behalf of a consortium, including the American Association of Blood Banks, the American Red Cross and the Council of Community Blood Centers, established the National Marrow Donor Program (NMDP) in Minneapolis in 1986 and in 1988 the World Marrow Donor Association (WMDA) was established by Jan van Rood and colleagues.

Better techniques for tissue typing including the application of "molecular techniques" has raised the hope of many patients who previously would have been denied a transplant because of the unavailability of a histocompatible donor. Of course, all the technology in the world would be useless if individuals were not altruistic enough to offer to donate marrow to a person they would never meet, often living in a foreign country. There are over 5.5 million volunteer donors between the BMDW and the NMDP, and since 1987, the IBMTR has collected data on over 11,000 transplants from unrelated volunteers (Fig. 3). Many problems remain to be solved (e.g. patient management) as there are undoubtedly far more clinical problems associated with unrelated transplants, including logistics of marrow transport from one country to another, ethical problems, medicolegal issues and expense. However, this international effort is a remarkable achievement and

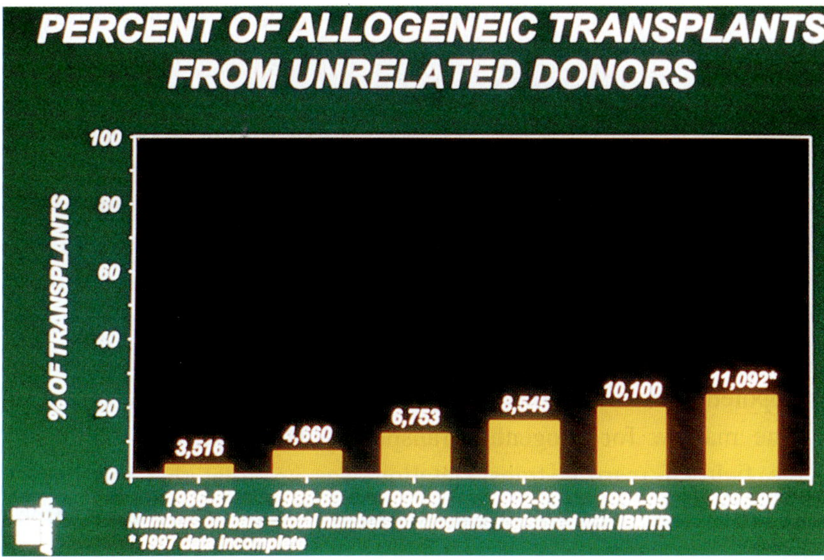

Fig. 3 Percent of allogeneic transplants from unrelated donors, 1986–1997 (reproduced with kind permission of M. Horowitz IBMTR).

stands as a testimony to the generosity of spirit of patients, their families, bone marrow donors and the medical and scientific community.

A detailed description of the history of the other major developments alluded to earlier is beyond the scope of this chapter. The development of cyclosporine for the prevention of GvHD, the manufacture of acyclovir as the first effective antiviral agent early in the 1980s, and the ability of transfusion services to provide "unlimited" supplies of blood products stand out as major milestones.

New Developments

To obtain adequate amounts of bone marrow for transplantation, approximately one litre of marrow is removed from the iliac crest of the donor. This is a relatively cumbersome procedure and usually requires general anaesthesia. Other sources of haemopoietic stem cells were sought. As early as 1962, Goodman demonstrated the presence of stem cells in mouse blood.[16] This observation was largely ignored however, until the late 1970s when Richman and colleagues found granulocyte progenitors in the peripheral blood of patients recovering from chemotherapy.[17] These observations were augmented by the development of haemopoietic growth factors which were shown to "mobilise" very large numbers of haemopoietic stem cells into the peripheral blood. These stem cells were sufficient to rescue patients from myeloablative therapy in the setting of an autograft. Further developments showed that large numbers of stem cells could be "mobilised" into the blood of normal volunteers following the administration of recombinant haemopoietic growth factors. These stem cells have been used in place of bone marrow for allogeneic transplants. The impact of stem cells on the field of marrow transplantation was so great that the EBMT changed it's name in 1993 to the European Group for Blood and Bone Marrow Transplantation. A large number of questions remain to be answered and again, many clinical trials are being conducted internationally to evaluate the efficacy of peripheral blood compared with bone marrow as a source of haemopietic stem cells.

Another source of haemopoietic stem cells is the placental or umbilical cord blood. A conversation between Edward A. Boyle, Hal E. Broxmeyer and Judith Bard in 1982 resulted in the suggestion that human umbilical cord blood might be used as an alternative source of stem cells if a histocompatible sibling was unavailable.[18] Numerous experiments were carried out in mice to test the hypothesis, and the first human umbilical cord transplant was carried out by Eliane Gluckman and her team in Paris in 1988.[19] The cord blood was obtained from a histocompatible sibling in North Carolina, cryopreserved in Indiana and flown to Paris where the patient underwent a successful umbilical cord transplant. Since then, a number of "cord banks" have been established in the U.S. and Europe, and umbilical cord blood has been used successfully in unrelated transplants for leukaemia and other bone marrow diseases.[20] Because of the relatively small volumes of cord blood obtained, the majority of transplants to date have been carried out in children. As with any source of unrelated stem cells, a number of questions are still being debated including ethical issues, logistics, expense and of course efficacy compared with more traditional sources of human haemopoietic stem cells. It is difficult to predict whether umbilical cord blood will have a large impact, until methods of "*in vitro* expansion" are perfected to allow umbilical blood to be used for adult transplantation, and until the financial problems of long-term cryopreservation are solved.

So You Think You Have a New Idea?

The first recorded bone marrow transplant was carried out in Ireland in the 8th century A.D. The event is clearly described in the Tain Bo Cualgne as the longest and most heroic tale of the Ulster Cycle, a series of legends which tell of ancient Ireland. It tells of a foray made by Queen Medb from Connaught into the territory of Ulster to carry off the bull, Donn Cualgne, in a daring cattle raid. Much death and destruction took place and a famous warrior Cethern Mac Fintain was critically wounded.

> "Fingin Faithliaig asked Cu Chulainn for a marrow mash to cure and heal Cethern Mac Fintain. Cu Chulainn proceeded to the encampment of the men of Ireland and brought from there all he found/of their herds/and flocks/ and droves. And made them a mash/flesh/and bones/and hide all together. And Ceithern Mac Fintain was placed in the marrow mash for the space of three days and three nights and he began to soak up the marrow mash which was about him. And the marrow entered into his wounds and gashes and sores and many stabs. Then after three days and three nights he arose from the marrow mash. And thus he arose with the boards of his chariot pressed to his belly to prevent his entrails from falling out (Fig 4).

Whether he developed graft versus host disease or not is not recorded!

Fig. 4 The first bone marrow transplant. Cethern Mac Fintain "soaking up the marrow mash".

Fig. 5 Dr E. Donnall Thomas and King Gustaf of Sweden. Dr Thomas and Dr Joseph Murray (not shown in this picture) shared the Nobel Prize for Physiology or Medicine in 1990, recognising the field of organ transplantation. Dr. Thomas was honoured for his endeavours in experimental and clinical bone marrow transplantation. (reproduced with kind permission of Blackwell Science, Inc. *Hematopoietic Cell Transplantation*, 2nd Ed. (eds.) E. D. Thomas, K. G. Blume, S. J. Formar).

Conclusion

In conclusion, the story of Bone Marrow Transplantation is one of a combination of dedication, perseverence and hard work on the part of physicians, trust on the part of patients and their families, and careful experimental observations on the part of scientists and doctors. It represents a truely major human victory and illustrates very clearly the possibilities of human endeavour and the results of international cooperation. I am priviledged proud to be part of this "world marrow transplant group" and am priviledged to have worked for Dr E.D. Thomas and his colleagues (Fig. 5).

Acknowledgments

I wish to thank James Cogan for Figures 1 to 4.

References

1. Ford CE, *et al.* (1956). *Nature* **177**, 452–454.
2. Till JE, McCulloch EA (1961). *Radiat Res* **14**, 213–222.
3. Metcalf D (1984). In Metcalf D (ed.) *Hematopoietic Colony Stimulating Factors*. New York: Elsevier, pp. 27–54.
4. Thomas ED, *et al.* (1957). *N Engl Med.* **257**, 491–496.
5. Thomas ED, *et al.* (1959). *J Clin Invest* **38**, 1709–1716.
6. Barnes DWH, *et al.* (1956). *Br Med J* **2**, 626–627.
7. Mathe G, *et al.* (1959). *Rev Franc Etudes Clin Biol* **IV**, 226–238.
8. McGovern JJ, *et al.* (1959). *N Engl Med* **260**, 675–683.
9. Bortin M (1970). *Transplantation* **9**, 571–578.
10. Hickman RO, *et al.* (1979). *Surg Gynecol Obstet* **148**, 871–875.
11. Gatti RA, *et al.* (1968). *Lancet* **2**, 1366–1369.
12. de Koning J, *et al.* (1969). *Lancet* **1**(7608), 1223–1227.
13. Thomas ED, *et al.* (1971). *Trans Assoc Am Physicians* **84**, 248–261.
14. Thomas ED, *et al.* (1975). *N Engl J Med* **292**, 832–843.
15. O'Riordan JR, *et al.* (1992). *Bone Marrow Transpl* **9**, 27–30.

16. Goodman JW, Hodgson GS (1962). *Blood* **19**, 507–515.
17. Richman C, *et al.* (1976). *Blood* **47**, 1031–1039.
18. Broxmeyer HE (1998). In Broxmeyer HE (ed.) *Cellular Characteristics of Cord Blood and Cord Blood Transplantation*, AABB Press, pp. 1–9.
19. Gluckman E, *et al.* (1989). *N Engl J Med* **321**, 174–178.
20. Gluckman E, *et al.* (1997). *N Engl J Med* **337**, 373–381.

XV

THE HISTORY OF ARM TRANSPLANTATION

Nadey S Hakim

Earl Owen, world-renowned pioneer of microsurgery and a good friend of mine, first voiced the idea of a hand transplant more than 30 years ago in a speech at Edinburgh University, but it was not until the mid-1990s that he decided it was both technically and immunologically possible. The only reported previous hand transplant attempt had taken place in Ecuador in 1964 and had failed due to initial absence of any anti-rejection drug use, allowing immediate severe rejection. Earl trained and practised in London for many years (and was awarded a Hunterian Professorship in 1968) and hoped to perform the world's first hand transplant here.

With this in mind, and with encouraging results of successful skin, nerve and joint human transplants, he approached me in 1998 and I contacted the Royal College of Surgeons and the U.K.'s Transplant Service Authority to ask for permission for such an operation to go ahead. However, British authorities were unconvinced and had doubts about the ethics of the procedure for a single limb. They believed that putting a healthy patient on potentially dangerous immunosuppressants could not be justified at this stage, despite the long term successes with experimental animal composite tissue transplants with a cocktail of recently proven powerful drugs.

Earl was keen to form a team of experts in preparation for the procedure, so I suggested we contact Max Dubernard in Lyon, Head of Transplantation and Urology at the city's Edouard Herriot Hospital, where Earl had been a visiting professor for 20 years. Max had carried out France's first pancreas transplant in 1976 and was not only one of France's top surgeons, but also a prominent local politician who ran his own hospital department. He was excited by Earl's suggestion and agreed that the proposed forearm transplant could go ahead in his department in Lyons. Earl quietly began assembling a skilled international team composed of transplant, orthopedic and hand microsurgeons, anaesthetists, a psychiatrist and a psychologist specialising in body image disturbances.

The hand surgeons knew they could attach a hand and the transplant surgeons knew about controlling the rejection of donated internal organs. However, what we could not predict was how the body would react to receiving a hand — an organ made up of many tissue types: nerve, artery, tendon, muscle, bone and the tissue that rejects most violently of course, the skin.

The results of experiments on primates using earlier immunosuppressives had been abysmal. Most died secondary to complications from the immunosuppressant drugs. Primates may be the closest animal model to man but following transplantation of any organ they require more immunosuppression than man. These experiments were conducted in the mid-1980s, with just the one drug, but since then we have been using combination therapy, where much lower doses of several drugs are used, with improved results.

There was still the ethical objection to putting someone on potentially life-threatening immunosuppressants for a "quality-of-life" operation.[1] However, one may argue that such operations are already routine — kidney and pancreas transplants are done so that people do not have to endure a lifetime of dialysis or insulin injections and are not in themselves life-saving operations. Other members of the "international team" preferred our first case to be a double amputee, where the benefits would potentially be wider for the patient, but Earl conservatively argued that the area of skin on two forearms and hands may be too much for the immunosuppressive cocktail to protect.

If this were the case and the transplant failed because of it, it would set back limb transplantation many years. We agreed to go ahead with a single hand transplant knowing that skin is the most difficult body tissue to immunosuppress.

When internal organs are given for transplantation, patients are of course very grateful for the gift. However, when the donor has provided his hands, a patient would need to be able to cope psychologically with the constant reminder that "their" hands have come from a dead person and been used to carry out the day-to-day activities of that other person's life. The recipient would need to be carefully chosen and have certain characteristics to sustain his enthusiasm through the early postoperative months. Whereas organ transplants need to work as soon as transplanted, yet cannot be seen, a hand transplant cannot function for some considerable time (as nerve regeneration slowly recovers), yet it is immediately and permanently seen.

A gentleman called Clint Hallam was well known to Earl and was one of three people he shortlisted for a hand transplant. Hallam, determined for many years to have the operation, had contacted Earl and confirmed his willingness to travel for the surgery anywhere and at any time. Now he underwent another series of psychological and physical checks and was commenced on a course of forearm strengthening exercises. Hallam was considered a suitable patient by the consultant psychiatrist, with a "strong constitution and solid resolve". Augmented consent forms and legal contracts were drawn up in conjunction with Australian and French lawyers. These contracts cleared the surgeons of responsibility for any risks and complications involved in the operation. Mr Hallam agreed to adhere to a regular schedule of postoperative physiotherapy and a strict regimen of immunosuppressive medication. He agreed to attend regular follow-up appointments, blood drug level tests and biopsies, and to undergo tests to monitor nerve regeneration and hand function.

The surgical team was drawn from around the world and therefore its members needed to agree on a time to convene and then wait for a donor to become available. The date was set for the beginning of

September 1998. However, after a week there was still no suitable donor whose hand would match Hallam's for size, shape, colour and hair distribution and he became restless and agitated. With the proposal to reconvene again in November, the team began to disperse. The Australians were about to board a plane to Sydney, Max Dubernard was on a train to Paris and Marco Lanzetta was driving back to Milan when news came through that a suitable donor had been found by the excellent Etablissment Francais des Greffes organisation, which had placed their resources on the alert for such a donor.

A 41-year-old Frenchman had been in a motorbike accident, suffered a fractured skull and died from a blood clot on the brain. His next-of-kin was traced and permission for his organs to be removed was given, the family feeling that at least some good might come of an unhappy occurrence.

Hallam was told to prepare for surgery the following morning. The Australians were stopped from boarding their flight, and Max went from Paris to harvest the forearm. In London, I got a call at midnight and, after performing two early morning kidney transplants, I flew to France later that day, joining Marco, who had reached Milan and turned back to drive again over the Alps to Lyon to join the others.

The actual operation went very smoothly, a tribute to the whole team, which had rehearsed the procedure many times. This team included the very experienced theatre staffs from both the transplant and the orthopaedic surgery and the anaesthetic departments, whose heads of divisions, as well as their most experienced staff members, all volunteered readily to be available for the procedure "at any time while the team was in waiting".

The operation commenced mid-morning and ended before midnight on 23rd September, 1998. The preparation of the donor and recipient ends was both intriguing and obsessive, taking more time than was expected from the previous practice sessions in the Lyon university's anatomy school. The orthopaedic team joined the forearm bones 10 cm from the wrist with a plate and six screws for each bone. Several muscles and tendons were joined before the Zeiss motorised operating microscope was focused and the ulnar and radial arteries were joined with microsutures. A large vein was then joined and the time came for the release of the tourniquet! This was the most anxious part of the procedure and Earl describes the moment he gave the order as defining for him a new concept of the word "stress". The white limb gradually became pink and the team relaxed as the anaesthetists checked all parameters as the team took a break.

The operation proceeded after that at a more leisurely pace as the median and ulnar nerves were joined some 21 cm from the wrist under the microscope as accurately as possible. This was interesting as the donor nerves were thinner and coloured a pale yellow, greatly contrasting with the recipient's fatter thicker white fascicles. After this, the tendons, some muscles and finally the skin were repaired. Skin grafts were taken from the right thigh to both allow for postoperative swelling and to provide observable islands of recipient skin within the donor skin surrounds. A loose dressing support was applied and the patient attended the recovery ward. A brachial block had been administered before he awoke so that the patient could not immediately test out our joins by using his forearm muscles, which could jeopardise the joins. The block was continued for the first week.

The surgery was technically successful, and so was the adherence to the immunosuppressive regime which had started preoperatively with intravenous anti-lymphocyte serum and first doses of the immunosuppressive cocktail of drugs.[2,3]

"My hand's back," said Hallam on emerging from the anaesthetic. "It's almost like I'd lost an old friend years ago, and suddenly it's back."

Congratulations arrived from Jacques Chirac, the French President. Dan Rather, CBS News presenter, stated: "It's the kind of quantum leap in modern medicine that, when it happens, is stunning. Transplant surgery that is simply staggering because it holds so much promise". In the first three months following the operation, Clint Hallam was under close supervision and adhered to his schedules of observations, exercises and drugs. The team was surprised by the evidence of the rapid return of nerve regeneration, and the fast rate of nail growth in the donor hand. Clint Hallam left the hospital area on 1st January 1999 assuring us that he would be continuing the very satisfactory progress of his transplant and was expected to return to Australia for continuation of the protocol.

It was now the team's task to perform the first successful double arm transplant — another world first! In the last few months of 1998, we awaited a suitable donor for the potential double hand transplant patient — a man who had lost his arms in an explosion. Each time a suitable donor was found, permission was not granted by the next-of-kin. Technically, the transplant surgeons did not need permission, because in France, one is presumed to have consented to organ donation on one's death unless otherwise stated, but we felt it was necessary to make an exception and have the relatives' cooperation for these unusual cases.

The eminent hand and transplant surgeons flew into Lyon from around the world three times since June 1999 but finally, 12 days into 2000, the call came to say a donor had been found. The team members were alerted immediately. I had to abandon a lecture tour in California, Earl flew out of Sydney, and four Italian specialists drove from Milan to join Max Dubernard and our augmented French colleagues.

The father of the donor involved in the world's first double hand transplant was paralysed down one side of his body. He understood what it was like to not have the use of a limb, and he agreed to the operation when many other parents had refused. His 19-year-old son had fallen from a bridge. He was on a life-support machine for several days and his family hoped for a recovery, but sadly this did not occur.

The recipient of the new hands, Denis Chatelier, was a 33-year-old house painter and father of two, whose hands were blown off when a home-made model rocket he was showing to his nephews exploded prematurely. He longed to live a normal life with his family again. When he saw the television report of the first single hand transplant in 1998, Chatelier had contacted Max Dubernard.

Chatelier described himself as a *battant*, a fighter, and a *croyant*, a believer. He was fit and in good shape for the 17-hour operation and up to two years of physiotherapy. Max agreed to go ahead: "He is strong-willed and tenacious", he told colleagues.

In the early hours of Thursday, January 13th, along with his kidneys, liver and heart, the young donor's hands and forearms were removed. Prosthetic hands, supplied by our team, were fitted to provide a normal appearance for his funeral. The donor limbs were perfused with preservation fluid, placed in dry plastic bags and carried in a cool box.

Earl had led the first transplant. This time Max was to be in charge. We began at 6 am. Fifty surgeons, specialists and nurses worked on Chatelier, who lay with arms outstretched on the hand surgery operating mini-tables as if crucified; a separate team of four surgeons for each donor hand and each host arm. He had been sedated and then anaesthetised. The procedures followed the protocol of the Hallam operation. Some of the details are now described.

Tourniquets were applied to his arms and incisions were made on each to expose the deeper structures. The nerves, arteries, veins and muscles were dissected and tagged and the forearm bones transversely cut with an electric saw so that the grafted limbs would be the same length. At the same time, in an adjoining theatre, the donor hands were similarly dissected and tagged. About 10 cm of the donor arm would also be used. First the two forearm bones, the radius and ulna, were joined, using small titanium plates with holes for 4.5 mm stainless steel screws to go through; then the arteries and cephalic veins were connected and the first arterial clamps were removed. The tourniquets were loosened and the hands' cool white turned pink. The surgeons left them undisturbed for 20 minutes covered with warm, wet sponges. Next were the muscles, then the tendons, interwoven with the muscles where possible, and finally, the skin. It took until 11pm.

Although confident of attaching the hands, we had been wary of any unforeseen complications. Patients can lose a lot of blood when two wound sites are opened, and if either of our first two transplant patients had died, it would have meant calling a halt to any further such transplants. However, at midnight in each case we could relax and watch as the patient's arms were bandaged following the successful operation.

In the days that followed, Chatelier's progress was good. A cocktail of four immunosuppressants and steroids was given, the same mix used in the first transplant 15 months before, and one that had proved very effective. The patient was told not to try to move his fingers. That would come later, after extensive passive physiotherapy and then supported at first with thin rubber bands attached to provide gentle resistance and then to be followed by intensive more active physiotherapy over the next three months.

At the time of writing, a total of 20 new arms have been transplanted in seven countries, and 19 are reported to be doing well. I am still trying to organise Britain's first single transplant. In Austria and Italy, other operations are being planned. The international experience has included the following centres: Lyon (France), Louisville (USA), Guangzhou (China), Insbruck (Austria), Monza (Italy), Guangxi (China), Harbin (China), Brussels (Belgium) and Malaysia. An International Registry on Hand and Composite Tissue Transplantation has been initiated at the recent Hand Transplantation and Composite Allograft IV International Symposium which took place in Varenna Italy in September 2002. Several topics were discussed at this meeting including the ethical, medicolegal, psychological and immunological implications of composite tissue allografts.

Unfortunately, the final chapter in the story of the world's first successful hand transplant patient, Mr Hallam, has ended less than satisfactorily. Over the months following his operation, it became clear that Mr Hallam had been less than honest with the transplant team. He had originally told the team that his hand had been lost in an industrial accident; however it later emerged that it had been severed during an accident in prison in New Zealand, where Mr Hallam was serving a sentence following convictions for fraud. The psychiatrist later admitted that Mr Hallam had actually not been a good choice for the transplant. Mr Hallam proved to be unreliable. He irregularly disappeared and travelled around the world and could not be contacted for weeks on end. He did not turn up for check-ups and he did not comply with his anti-rejection drug therapy, taking the medication as and when he felt like it, therefore causing illness and gradual rejection.

At the transplant's peak function, Mr Hallam could use the hand to grab objects, use a knife and fork, write with a pen and even play basketball with his children. Following his non-compliance with treatment, however, he began to lose some of the function and the hand became inflamed from chronic rejection. In November 2000, Mr Hallam asked for the hand to be amputated, but then changed his mind and pledged that he would try once again and would take the necessary medication. He was given a supply of drugs, but when these ran out he did not obtain more. On 31st January 2001, more than two years after the transplant surgery, Mr Hallam contacted me once more. One look at his hand convinced me that there was no alternative but to amputate, as the chronic rejection was severe but, surprisingly, there were no symptoms of a systemic rejection reaction. The hand was painful, suggesting, I believe, successful nerve regeneration.

It was with sadness that I removed the hand. We had given Mr Hallam the chance of having the new hand he had repeatedly requested, and the surgery was successful. He destroyed his chances by failing to take his medication. I was just relieved that there had been no serious complications and that Mr Hallam was still in good health. He is now to be fitted with a prosthesis and he admits that he is fully responsible for the rejection. Pathologists are studying the amputated hand to determine the rates of rejection of the various types of tissue.

However, Mr Chatelier is doing well and is adhering to the anti-rejection therapy. The hands have no signs of rejection, are functioning particularly well with good feeling and he is in good health. His town in France's west coast has given him full support and he is already making woodwork model seagulls for sale!

In the last few years we have taken significant steps in the transplantation of hands and much has been learned. Although the necessary expertise and the immunosuppressant drugs have been available for some time, hand transplantation has not been attempted until recently. The hand is still a special symbol, representing personal identity in a way that other tissues do not. The hand represents a sort of halfway house in terms of personal identity, between internal organs and brain tissue. Opponents of brain tissue transplants reject the argument that they are not simply about restoring normal function to the recipient: they may also alter the recipient's identity in a profoundly problematic way — so that the person who gave consent to receiving the tissue is no longer the same person after the transplant. Similarly, we need to consider the wider function of the hand in relation to identity, as an instrument of physical intimacy, of contact with others, of consummate skills in artists and musicians, of agency itself (as witness the use of "hand" to represent agency in such phrases as "the hand of Fate", "by his own hand", "the hand of God"). It is given in marriage, washed to escape blame, expresses our feelings, and carries out our commands. The intimacy function of the hand might even, more speculatively, be thought in some to give rise to rights in those with whom the donor and recipient of the transplanted hand have been intimate.

To the hand, man owes his survival and achievements.

I have written this chapter on behalf of the original international transplant team of which I was part, and which included:

Professor Earl Owen (Sydney)
Professor Jean Michel Dubernard (Lyon)
Professor G Herzberg (Lyon)
Professor Xavier Martin (Lyon)
Drs Hari Kapila (Sydney), Marco Lanzetta (Milan and Sydney) and Dr M. Dawahra (Lyon)

References

1. Dickenson D, Hakim NS (1999). Ethical issues in limb transplants. *Postgrad Med J* **75**, 513–515.
2. Dubernard JM, Owen E, Herzberg G, Hakim NS, *et al.* (1999). Hand transplant: six-month report. *Lancet* **353**, 1315–1320.
3. Dubernard JM, Owen E, Lefrancois N, Hakim NS, *et al.* (2000). First human hand transplantation. *Transpl Int* **13**(Suppl 1), S521–S524.

XVI

THE HISTORY OF XENOTRANSPLANTATION

Robert A Holzknecht
Jeffrey L Platt

The Early Years of Transplantation

The modern era of transplantation arguably began in the early years of the 20th century when experimental surgeons such as Ehrlich, Ullmann, Guthrie and Carrel developed vascular anastomosis (the surgical connection of cut ends of blood vessels) that enabled the repair of traumatic wounds which would otherwise have led to severe disability or death. For example, the severing of major arteries feeding the extremities would have required amputation in order to avoid gangrene, whereas, the repair of arteries could greatly hinder that outcome. Vascular anastomosis would also enable surgeons to penetrate deeper into body cavities. However, the major implication of vascular anastomosis perceived in the early decades of the 20th century was the possibility that the technique might be the critical advance needed to allow surgeons to replace ailing organs with healthy ones, i.e. organ transplantation. In fact, it was early but ultimately unsuccessful experiments in organ transplantation that provided the basis for Alexis Carrel winning the Nobel Prize in medicine and physiology in 1912.

In the early years of the 20th century and, indeed, until the 1960s, when the development of vascular anastomosis made it possible to transplant organs between different individuals, it brought to the fore the question of where those organs might come from. With the advent of brain death laws, there were serious ethical concerns that precluded most opportunities to obtain human organs for clinical transplantation. Those concerns reflected the knowledge that even after the heart ceases beating and a person has been declared dead, cells in various organs remain alive. It was seen as questionable whether "living organs" could be ethically harvested from a deceased person. In this context, early efforts to apply transplantation for the treatment of human disease were conducted by using organs of animals rather than humans. In 1906, Jaboulay attempted two separate heterotopic transplants of pig and goat kidneys into human patients.[1] In both cases, the xenografts functioned only briefly, if at all, yet the era of clinical transplantation and xenotransplantation had begun.

If the early experiments in clinical transplantation failed in reversing the course of kidney disease, they succeeded in stimulating further investigation into this new field. Experiments in various laboratories between 1906 and 1912 demonstrated that vascular connections could be accomplished between individuals of the same species and individuals of disparate species, but in no case could enduring organ function be achieved. The failure of organ transplant function for more than minutes in the case of xenotransplants, and weeks in the case of allotransplants, discouraged further efforts at transplantation with a few notable exceptions. Among these exceptions, in 1912, Dr Leslie Hammond attempted the transplantation of a cadaveric kidney from an accident victim into a patient dying of renal tuberculosis.[2] In 1908, Carrel demonstrated that kidney autografts, in contrast to allografts or xenografts, could function enduringly, thus demonstrating that organ transplantation was technically and biologically feasible which suggested that genetic factors might impede success.[3]

The "Middle" Years of Transplantation

Discouraged by the failure of clinical and experimental allografts and xenografts, few attempts at clinical organ transplantation were made from 1910 to 1960. This period, however, was not fallow as investigators began asking and finding answers to the question of why organ transplants fail. One possibility favored by early investigators, and considered a serious factor well into the 1950s, was the existence of fundamental biological incompatibilities between different individuals that stood in the way of successful transplantation. Championed by Leo Loeb,[4,5] this theory seemed to explain why xenografts functioned poorly compared to allografts. Although experiments by Peter Gorer in tumor transplant models suggested the importance of immunology in the failure of transplants,[6,7] Charles Guthrie "speculated many years before that the cause of the failure of allografts might be related to the immune system".[8]

Immunological basis for the failure of transplants was not given full credence for several reasons. Immune-mediated transplant failure was dismissed by some because experiments by Loeb in the 1920s demonstrated that sequential transplants into one recipient from different individuals would fail at the same rate; if the immune system had caused the failure, it was predicted that a second transplant would fail more quickly. Another reason for dismissing immunological causes of organ failure came from experiments testing the potential role of blood group A and B antigens in organ failure. Incompatibility for blood group A and B could be shown to underlie the destruction of transfused erythrocytes, yet the same incompatibilities could not be linked to the failure of skin transplants.[9] The dismissing of alloimmunity as a cause of transplant failure in the middle years of the 20th century is excusable as it was yet to be discovered that other more formidable tissue antigens (HLA as the human example of the class of major histocompatibility antigens in all mammals) are the major target of immune responses in tissues between individuals of the same species and that tissue transplants receive a blood supply by the ingrowth of recipient blood vessels and, thus, incompatibility blood group antigens might not constitute an important cause of tissue transplant failure.

In the early 1940s, however, Peter Medawar linked the failure of skin allografts to immune responses.[10,11] Medawar demonstrated that transplants between individuals of the same species might survive for two weeks, whereas second transplants from the same transplant donors would fail within a few days. This very rapid failure of second transplants demonstrated that immunological recall or memory is characteristic in immunological reactions. While this result suggested why transplants might fail, they did not provide the means to overcome transplant failure. Thus, subsequent efforts to transplant kidneys between disparate individuals without immune manipulation failed, whereas kidneys transplanted between identical human twins survived.[12] The era of clinical transplantation had to await the development of safe and effective immunosuppression.

The Era of Clinical Transplantation

The era of clinical transplantation began, arguably, with the discovery of 6-mercaptopurine, and a related drug, azathioprine, and the findings that these drugs could suppress immune responses.[13, 14] Once 6-mercaptopurine and azathioprine were demonstrated to impair immune responses in animals, immunosuppression without whole body irradiation was immediately initiated with human subjects receiving kidney transplants.[15] These transplants succeeded and the era of clinical organ transplantation had begun. As in the early years of this century, the early years of the era of clinical transplantation were marked with the question of where organs would be obtained. The absence of "brain death" laws prior to 1965 severely limited the number of human organs available for transplantation until cadaver organs became available in greater numbers. As a result, surgeons such as Reemtsma carried out a series of kidney transplants from chimpanzees into human patients undergoing renal failure in the hope that these xenotransplants would succeed. However, it was noted that these kidney transplants seemed to suffer more frequent and severe rejection episodes than human kidneys, where a typical organ functioned for months, the longest lasting nine months.[16, 17] Of greater importance, perhaps, than this partial success in graft survival was the finding that these organs were relatively normal when the patients died of other intercurrent causes such as infection and stroke. It seemed, thus, that xenotransplantation might be possible. Reemtsma's experiments prompted others to attempt xenografts using baboons and monkeys, the supply of which is more plentiful than that of chimpanzees. Thomas Starzl carried out a series of kidney transplants from baboons into patients with renal failure, however, these transplants suffered more severely from immune rejection. The reason why xenografts suffered so severely compared to allografts from immunological rejection was suggested by the work of Najarian and Perper.[18, 19] Organs transplanted between widely disparate species, it seemed, were subject to immediate failure by a process known as hyperacute rejection, whereas organs transplanted between closely related species were not susceptible to hyperacute rejection. Calne referred to the transplants between disparate species susceptible to hyperacute rejection as being "discordant" and the transplants between closely related species, not susceptible to hyperacute rejection, as being "concordant".[20] Hyperacute rejection was thought, by some, to be caused by antibodies present in the recipient directed against the donor. This is based on the observations of Kissmeyer-Nielsen, who demonstrated in human patients that hyperacute rejection of human transplants reflected the occasional existence of these antibodies.[21] This type of allograft hyperacute rejection was typically caused by previous sensitization to antigens from blood transfusions prior to organ transplantation. One additional effort at xenotransplantation was undertaken by Starzl and colleagues who transplanted baboon livers into two patients with hepatitis who were not considered suitable recipients for human liver transplants. With the use of more modern and effective immunosuppressive agents, these transplants functioned for several months, but the patients ultimately died.[22]

While information regarding the immunological reactions that stood as a barrier to xenotransplantation accumulated, albeit slowly, in the 1960s and the 1970s, serious efforts at xenotransplantation were no longer undertaken. The advent of brain death laws provided a more plentiful source of human organs and, in any case, there was little success in overcoming the severe immunological reactions that

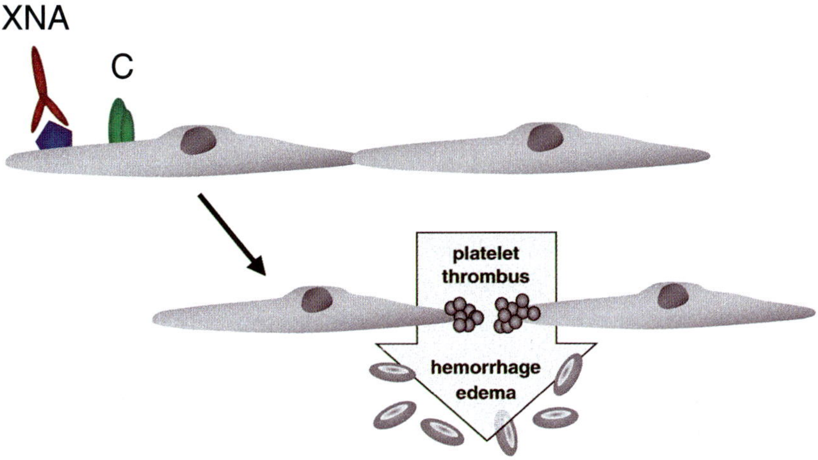

Fig. 1 Pathogenesis of xenoreactive hyperacute rejection. The binding of xenoreactive antibodies (XNA) to graft endothelium with concurrent activation of complement (C) cause, within minutes, the loss of heparan sulfate and the formation of intercellular gaps. These changes lead to hemorrhage and edema with binding and activation of platelets to form platelet thrombi.

destroyed xenografts. Notable among scientific accomplishments, however, were the observations of Moberg and Giles that depletion of anti-donor antibodies might prolong the survival of xenografts,[23, 24] and of Gewurz that inhibition of complement could delay the destruction of xenografts.[25] These observations established a model shown in Fig. 1 for the mechanism of xenograft rejection which anti-donor antibodies bind to the xenograft and activate the complement system, a series of 20 proteins that ordinarily attach to and destroy infectious microbes.

The "Modern" Era of Xenotransplantation

With the exception of a small number of clinical experiments mentioned below, xenotransplantation is not a regular part of medical practice and probably will remain so for the foreseeable future.

However, a number of factors have led to a recrudescence of interest for xenotransplantation and renewed hopes that success in the field may yet be possible. This section is entitled the "modern" era with full appreciation for the hazards when using that term.

One question that should be addressed is which species among various possibilities might best serve as a source of organs for xenotransplantation. Early studies by Reemtsma, Starzl and others focused on non-human primates as these animals are close to humans genetically, and it might be surmised that the immune reaction against those organs would be lower in magnitude than the immune reaction against more distant species. Despite this logic, the selection of species shifted in the 1990s from non-human primates to the pig. The most important reason for focusing on the pig was that pigs are potentially available in large numbers with short gestation times. In contrast to this, many species of non-human primates are available in only small numbers and some not at all because the species are threatened with extinction. However, the use of pigs instead of primates or humans as a source of organs raises some daunting barriers to successful transplantation. These barriers, as presently conceived, include the immune response of the recipient against the graft, physiologic incompatibilities between the recipient and the graft, and the possibility of transferring infectious organisms from the graft to the recipient.

Renewed interest in xenotransplantation has stemmed in large part from the success of human-to-human transplantation which is now the preferred treatment for severe failure of the heart, kidney, liver, lungs and may soon become the preferred treatment of diabetes caused by failure of the endocrine pancreas. The success of allotransplantation has increased the demand for human organs to the extent that as few as 5% of the organs needed in the United States are available each year, and the percentage in many countries, especially countries in Asia, may be far lower. The dramatically increased need for organs for transplantation has coincided with increasing knowledge regarding the immunological hurdles to xenotransplantation and the development of incisive strategies for adjusting those hurdles. As the scope of this essay does not permit a full description of the biological hurdles of xenotransplantation, only the briefest summary of progress will be presented here and the reader referred to other sources for more complete information.[26, 27]

The biological hurdles to xenotransplantation as they are currently understood are summarized in Fig. 2. Hyperacute rejection remains the first hurdle to the transplanting of organs from pigs into humans. Hyperacute rejection, in this setting, is now known to be mediated by antibodies specific for Galα1-3Gal, a sugar expressed on lower mammals such as pigs but not on humans.[28] Species that do not synthesize Galα1-3Gal have antibodies much like anti-blood group antibodies directed against that sugar. The formal demonstration that these antibodies cause hyperacute rejection can be credited to Ye, who showed that infusion of sugars recognized by some of these antibodies impaired hyperacute rejection.[29] This observation was confirmed by Sachs and Lin, who demonstrated that depletion of antibodies directed

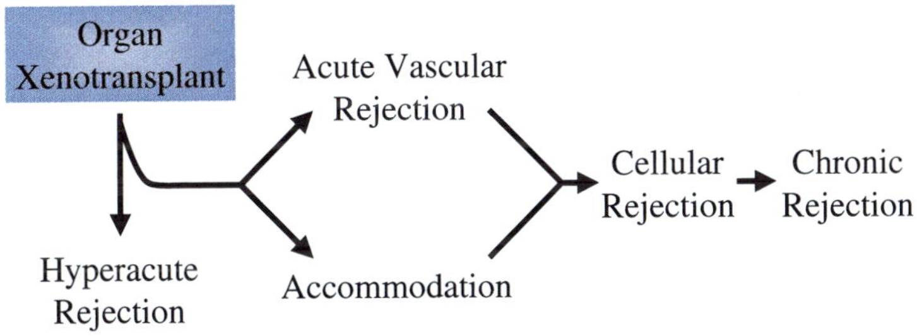

Fig. 2 The biological responses to xenotransplantation. A porcine organ transplanted into a human would be subject to one or more of the immunological reactions shown here. Transplantation into an untreated recipient would give rise to immediate "hyperacute" rejection, depicted in Fig. 1. If hyperacute rejection is bypassed, for example by the inhibition of complement, the graft becomes subject to acute vascular rejection. If anti-donor antibodies are depleted from the recipient the graft may achieve accommodation, a condition in which acute vascular rejection does not occur despite the return of anti-donor antibodies in the circulation of the recipient. If acute vascular rejection is bypassed, the graft may undergo cellular mediated rejection or chronic rejection.

against Galα1-3Gal, with columns that left the complement system intact, would also prevent hyperacute rejection.[30, 31]

Another highlight of the modern era of xenotransplantation has been the discovery that complement-regulatory proteins, which ordinarily protect cells against injury by homologous complement,[32] may function aberrantly in the xenotransplant and thus contribute to the susceptibility of the transplant to complement-mediated injury.[33, 34] The solution to this problem proposed by Delmasso *et al.*[35] was the genetic engineering of pigs to express human complement regulatory proteins. Genetic regulation of the genetic engineering of pigs and testing of this idea was first reported in 1995,[36] sparking heightened interest in xenotransplantation. Other highlights of the past decade in this field have included the discovery that other forms of rejection, notably acute vascular rejection, which may prove more daunting than hyperacute rejection as a barrier to xenotransplantation (Fig. 2).[37] As for physiologic incompatibilities, the demonstration that heterotopic cardiac transplants and kidney transplants can function for greater than two months[38] suggest that there might not be insurmountable biological hurdles to xenotransplantation other than the immune response of the recipient against the graft.

The modern era has also been characterized by the surfacing of more ominous concerns. In 1999, Robert Weiss reported that an endogenous retrovirus of pigs (PERV), which had raised little interest over the past several decades, could infect human cells in culture.[39] This finding crystallized concerns that xenotransplantation could serve as a source of infection for the transplant recipient and raised questions whether that concern might be extended to others in the population who contacted the xenograft recipient. Although recent studies have failed to show evidence that PERV can infect humans,[40] it remains unclear at this juncture whether the general concerns raised by Weiss and others regarding infectious disease will prove to be warranted.

The modern era of xenotransplantation has also seen further clinical trials with arguably better results. In the early 1990s, several groups used pig livers as *ex vivo* devices for the treatment of fulminant hepatic failure by perfusing the organ with a patient's blood. It was hoped that the patient's condition might improve to such a point as to allow the transplantation of a human liver. "Xenoperfusion," as such, had been attempted in the past but with uncertain outcome and was still limited by the availability of human livers for eventual transplantation. With this procedure, several patients were kept alive until human livers became available for transplant. Thus, the procedure could be viewed as a successful use of animal organs, albeit not as a transplant.

In various trials, pig tissues (a tissue is a cluster of cells in anatomical specific array as opposed to an intact organ, which is a group of tissues) were used as clinical xenografts. Pig skin was used as a covering for the treatment of severe burns; porcine pancreatic islets were used for the treatment of diabetes;[41] and porcine substantia nigra for the treatment of Parkinson's disease.[42] The results in the use of pig skin have not been reported. Porcine islets functioned briefly or not at all in the treatment of patients with diabetes.[41] Porcine substantia nigra was found to survive for many months in patients with Parkinson's disease,[43] and the patients possibly exhibited some improvement.[44] While none of the recent xenotransplants have proven an unqualified success, they have contributed to progress in the field for several reasons. First, the more recent xenotransplants provided an opportunity to test the immune response of the human against the pig. This response was shown to involve the heightened production of antibodies against Galα1-3Gal, but in the case of tissue xenografts, the response did not lead to hyperacute rejection. In fact, it would seem from experimental and preliminary clinical observations that hyperacute rejection does not afflict cell and tissue xenografts, and as a result, these types of xenografts may be implanted far more easily and with more likely success than organ xenografts. The clinical trials of 1990s also presented the opportunity to evaluate the hazard of infectious disease. Up to this point, no evidence has been forthcoming that PERV or other unusual and unexpected organisms are transmitted by pig xenografts in humans.[45] Further research and monitoring are needed in this area to determine the true threat posed by zoonotic agents in xenotransplantation.

Past, Present and Future of Xenotransplantation

It might seem odd at this juncture to present a "history" of xenotransplantation, as only a few experiments in humans have been conducted and as animal organs are not, at present, being transplanted in human patients anywhere in the world. While it is impossible to predict by any means if xenotransplantation, particularly organ xenotransplantation, will ultimately succeed, it is possible to gain some important lessons from the past and present in this field. It remains to be seen if this is truly the "modern era" or just another middle chapter in the history of xenotransplantation.

References

1. Jaboulay M (1906). De reins au pli du coude par soutures arterielles et veineuses. *Lyon Med* **107**, 575–577.
2. Anonymous (1911). Dr. Hammond gives patient new kidney. In *The New York Times,* New York.
3. Carrel A (1908). Transplantation in mass of the kidneys. *J Exp Med* **10**, 98–140.
4. Loeb L (1930). Transplantation and individuality. *Physiol Rev* **10**, 547–616.
5. Loeb L (1937). The biological basis of individuality. *Science* **86**, 1–5.
6. Gorer PA (1938). The antigenic basis of tumour transplantation. *J Pathol Bacteriol* **47**, 231–252.
7. Gorer PA (1950). Studies in antibody response of mice to tumour inoculation. *Br J Cancer* **4**, 372–379.
8. Guthrie CC (1912). Blood-vessel surgery and its applications. New York: Longmans, Green & Co.

9. Platt JL (1995). Xenotransplantation. In Wood K (ed.) *Handbook of Transplant Immunology.* Bucks: Medical & Scientific Publications, pp. 259–299.
10. Gibson T, Medawar PB (1943). The fate of skin homografts in man. *J Anat* **77**, 299–310.
11. Medawar PB (1944). A behaviour and fate of skin autografts and skin homografts in rabbits. *J Anat* **78**, 176–199.
12. Owen R (1945). Immunogenetic consequences of vascular anastomoses between bovine twins. *Science* **102**, 400–401.
13. Calne RY (1960). The rejection of renal homografts: inhibition in dogs by 6-mercaptopurine. *Lancet* **1**, 417–418.
14. Schwartz R, Stack J, Dameshek W (1958). Effect of 6-mercaptopurine on antibody production. *Proc Soc Exp Biol Med* **99**, 164–167.
15. Merrill JP, Murray JE, Takacs FJ, Hager EB, Wilson RE, Dammin GJ (1963). Successful transplantation of kidney from a human cadaver. *J Am Med Assoc* **185**, 347–353.
16. Reemtsma K (1969). Renal heterotransplantation from nonhuman primates to man. *Ann NY Acad Sci* **162**, 412–418.
17. Reemtsma K, McCracken BH, Schlegel JU, Pearl M (1964). Heterotransplantation of the kidney: two clinical experiences. *Science* **143**, 700–702.
18. Perper RJ, Najarian JS (1966). Experimental renal heterotransplantation. I. In widely divergent species. *Transplantation* **4**, 377–388.
19. Perper RJ, Najarian JS (1966). Experimental renal heterotransplantation. II. Closely related species. *Transplantation* **4**, 700–711.
20. Calne RY (1970). Organ transplantation between widely disparate species. *Transpl Proc* **2**, 550–556.
21. Kissmeyer-Nielsen F, Olsen S, Petersen VP, Fjeldborg O (1966). Hyperacute rejection of kidney allografts, associated with pre-existing humoral antibodies against donor cells. *Lancet* **ii**, 662–665.
22. Starzl TE, Fung J, Tzakis A, Todo S, Demetris AJ, Marino IR, Doyle H, Zeevi A, Warty V, Michaels M, Kusne S, Rudert WA, Trucco M (1993). Baboon-to-human liver transplantation. *Lancet* **341**, 65–71.

23. Giles GR, Boehmig HJ, Lilly J, Amemiya H, Takagi H, Coburg AJ, Hathaway WE, Wilson CB, Dixon FJ, Starzl TE (1970). Mechanism and modification of rejection of heterografts between divergent species. *Transpl Proc* **2**, 522–537.
24. Moberg AW, Shons AR, Gewurz H, Mozes M, Najarian JS (1971). Prolongation of renal xenografts by the simultaneous sequestration of preformed antibody, inhibition of complement, coagulation and antibody synthesis. *Transpl Proc* **3**, 539–541.
25. Gewurz H, Clark DS, Finstad J, Kelly WD, Varco RL, Good RA, Gabrielsen AE (1966). Role of the complement system in graft rejections in experimental animals and man. *Ann NY Acad Sci* **129**, 673–713.
26. Platt JL (1996). The immunological barriers to xenotransplantation. *Crit Rev Immunol* **16**, 331–358.
27. Platt JL (1998). New directions for organ transplantation. *Nature* **392** (Suppl), 11–17.
28. Galili U, Clark MR, Shohet SB, Buehler J, Macher BA (1987). Evolutionary relationship between the natural anti-Gal antibody and the Galα1-3Gal epitope in primates. *Proc Nat Acad Sci USA,* **84**, 1369–1373.
29. Ye Y, Neethling FA, Niekrasz M, Koren E, Richards SV, Martin M, Kosanke S, Oriol R, Cooper DKC (1994). Evidence that intravenously administered α-Galactosyl carbohydrates reduce baboon serum cytotoxicity to pig kidney cells (PK15) and transplanted pig hearts. *Transplantation* **58**, 330–337.
30. Kozlowski T, Shimizu A, Lambrigts D, Yamada K, Fuchimoto Y, Glaser R, Monroy R, Xu Y, Awwad M, Colvin RB, Cosimi AB, Robson SC, Fishman J, Spitzer TR, Cooper DK, Sachs DH (1999). Porcine kidney and heart transplantation in baboons undergoing a tolerance induction regimen and antibody adsorption. *Transplantation* **67**, 18–30.
31. Lin SS, Weidner BC, Byrne GW, Diamond LE, Lawson JH, Hoopes CW, Daniels LJ, Daggett CW, Parker W, Harland RC, Davis RD, Bollinger RR, Logan JS, Platt JL (1998). The role of

antibodies in acute vascular rejection of pig-to-baboon cardiac transplants. *J Clin Invest* **101**, 1745–1756.

32. Atkinson JP, Oglesby TJ, White D, Adams EA, Liszewski MK (1991). Separation of self from non-self in the complement system: a role for membrane cofactor protein and decay accelerating factor. *Clin Exp Immunol* **86** (Supp. 1), 27–30.
33. Miyagawa S, Hirose H, Shirakura R, Naka Y, Nakata S, Kawashima Y, Seya T, Matsumoto M, Uenaka A, Kitamura H (1988). The mechanism of discordant xenograft rejection. *Transplantation* **46**, 825–830.
34. Platt JL, Vercellotti GM, Dalmasso AP, Matas AJ, Bolman RM, Najarian JS, Bach FH (1990). Transplantation of discordant xenografts: a review of progress. *Immunol Today,* **11**, 450–456.
35. Dalmasso AP, Vercellotti GM, Fischel RJ, Bolman RM, Bach FH, Platt JL (1992). Mechanism of complement activation in the hyperacute rejection of porcine organs transplanted into primate recipients. *Am J Pathol* **140**, 1157–1166.
36. McCurry KR, Kooyman DL, Alvarado CG, Cotterell AH, Martin MJ, Logan JS, Platt JL (1995). Human complement regulatory proteins protect swine-to-primate cardiac xenografts from humoral injury. *Nature Med* **1**, 423–427.
37. Platt JL, Lin SS, McGregor CGA (1998). Acute vascular rejection. *Xenotransplantation* **5**, 169–175.
38. Vial CM, Ostlie DJ, Bhatti FN, Cozzi E, Goddard M, Chavez GP, Wallwork J, White DJ, Dunning JJ (2000). Life supporting function for over one month of a transgenic porcine heart in a baboon. *J Heart Lung Transpl* **19**, 224–229.
39. Patience C, Takeuchi Y, Weiss RA (1997). Infection of human cells by an endogenous retrovirus of pigs. *Nature Med* **3**, 282–286.
40. Paradis K, Langford G, Long Z, Heneine W, Sandstrom P, Switzer WM, Chapman LE, Lockey C, Onions D (1999). Search for cross-species transmission of porcine endogenous retrovirus in patients treated with living pig tissue. *Science* **285**, 1236–1241.

41. Groth CG, Korsgren O, Tibell A, Tollemar J, Moller E, Bolinder J, Ostman J, Reinholt FP, Hellerstrom C, Andersson A (1994). Transplantation of porcine fetal pancreas to diabetic patients. *Lancet* **344**, 1402–1404.
42. Isacson O, Deacon TW, Pakzaban P, Galpern WR, Dinsmore J, Burns LH (1995). Transplanted xenogeneic neural cells in neurodegenerative disease models exhibit remarkable axonal target specificity and distinct growth patterns of glial and axonal fibers. *Nature Med* **1**, 1189–1194.
43. Deacon T, Schumacher J, Dinsmore J, Thomas C, Palmer P, Kott S, Edge A, Penney D, Kassissieh S, Dempsey P, Isacson O (1997). Histological evidence of fetal pig neural cell survival after transplantation into a patient with Parkinson's disease. *Nature Med* **3**, 350–353.
44. Schumacher JM, Ellias SA, Palmer EP, Kott HS, Dinsmore J, Dempsey PK, Fischman AJ, Thomas C, Feldman RG, Kassissieh S, Raineri R, Manhart C, Penney D, Fink JS, Isacson O (2000). Transplantation of embryonic porcine mesencephalic tissue in patients with PD. *Neurology* **54**, 1042–1050.
45. Patience C, Patton GS, Takeuchi Y, Weiss RA, McClure MO, Rydberg L, Breimer ME (1998). No evidence of pig DNA or retroviral infection in patients with short-term extracorporeal connection to pig kidneys. *Lancet* **352**, 699–701.

XVII

THE HISTORY OF IMMUNOSUPPRESSIVE DRUGS

Melissa Kamps

Introduction

The current achievements of organ transplantation would certainly not have been possible without the development of compounds that could safely suppress the immune system. Prior to their arrival, organ transplants had only been truly successful in identical twins. Many centers wanted to succeed in the impossible; to graft and maintain an organ from one non-identical human to another. Initial attempts to transplant organs in non-identical twins had consistently met with failure. The history of immunosuppression for human organ transplantation is rooted in these initial failures. Without the efforts of those who tirelessly worked to understand the complexities of the immune system and solve what was believed insolvable, the immunosuppressive drugs discussed below would not be with us today. The history of each drug's development was slightly different, however most stories combined a lot of hard work with a little bit of luck. The following paragraphs describe this story of research and innovation, leading to the compounds used in immunosuppressive therapy today.

Azathioprine and Prednisone

Early transplant groups began postulating that perhaps drugs that decrease white cell function could be effective immunosuppressants.

In 1963, Goodwin and associates[1] at UCLA reported their results of six cases of attempted kidney transplants between humans. While all the non-identical transplants eventually failed, they were able to show the positive effects steroids and compounds that reduced the white cell count had on reversing rejection. Steroids had been shown in earlier research[2] to inhibit protein synthesis. Their actions as anti-inflammatory compounds inhibited the inflammatory response of the antigen-antibody reaction during rejection. The new compounds that could reduce the white cell count and thus reduce the number of circulating lymphocytes could potentially delay the rejection response.

This information corroborated with the results of Calne[3] and his research into thiopurines. In 1961, Calne studied the actions of 6-mercaptopurine (MP), thioguanine, and two imidazole derivatives, BW 57-322 and BW 57-323, in canines. While thioguanine and BW 57-323 proved too toxic and ineffective in preventing rejection, the results for 6-MP and BW 57-322 proved much more interesting. 6-MP had appeared to stop the rejection response, but its effects were variable and inconsistent, and its bone marrow toxicity was significant. The effects of BW 57-322 were also variable, but appeared to be less toxic to bone marrow. One dog was still alive with no signs of rejection after four months of therapy. BW 57-322, now known to all as azathioprine, was able to interfere with purine synthesis, especially in rapidly dividing cells, and thus inhibit nucleic acid synthesis. This inhibition prevented the proliferation of lymphocytes, and prevented the rejection response.

With this information, research groups began to try azathioprine in humans, and in 1963, Murray and associates[4] published their experiences in 13 patients (ten cadavers, three living donors) treated solely with drugs as the means to prevent rejection. Azathioprine was used as the primary immunosuppressive agent, with actinomycin C and prednisone used for treatment of rejection. Azaserine, an alkylating agent, was given for the first three weeks in most patients. Four of ten cadaveric recipients had function for more than 20 days, and all three living donor kidneys successfully functioned. Actinomycin C and prednisone bursts were able to treat rejection episodes, and several patients that developed infections were able to receive treatment without stopping immunosuppression. Azathioprine was approved for the prophylaxis of rejection in 1969, and many studies later its place in transplantation medicine had been secured. The development and performance of azathioprine as an immunosuppressant gave researchers the first real hope that kidney transplantation could provide a true alternative to irreversible kidney failure.

Antilymphocyte Globulin

When researchers studied the rejection response, they noticed a pattern whereby patients that had made it through the first three months after transplant without rejection could tolerate reductions in immunosuppression.[5] Current theories proposed by researchers surmised whether other protective mechanisms had begun working or if a state of tolerance had developed over time. Researchers were also looking at other means of depleting the lymphocyte count at the time of transplant. With this depletion, there would be few if any activated lymphocytes left to continue the rejection process, and subsequently the rejection process would cease. These ideas led to the development of anti-human lymphocyte antibodies, animal proteins that when injected into humans, would deplete the number of circulating lymphocytes in the body. They could then be used as "induction agents," inducing an immunosuppressed state in the patient that would render him or her less likely to reject a transplant.

Several research groups in the 1960s began to develop therapeutic antibodies against human lymphocytes. There were a variety of compounds developed, using different animals to create the antibodies, and differing production methods. Starzl and associates[6] were one of the first to develop antibody formulations. They were able to reduce azathioprine and prednisone doses while improving rejection rates in their series. Najarian[7] and associates, using Minnesota anti-lymphocyte globulin (MALG) demonstrated reduced rejection rates as compared to standard therapy with azathioprine and prednisone controls. MALG became widely used in the United States at various centers interested in induction therapy. Research centers in Europe, using the rabbit and horse, also tested various antibodies against lymphocytes. Much of the early research with these antibodies was completed using small numbers of patients. Other centers could not demonstrate a benefit with these compounds; thus some centers doubted the benefits of antibody induction.[8] Larger studies involving antilymphocyte globulin were performed, but controversy continued regarding its use and place in immunosuppressive protocols. Many centers continued to use different products, however in the late 1960s into the 1970s, no single product was completely dominant. Atgam, a horse antilymphocyte antibody, was approved in the early 1980s in the U.S., and several products had been approved in Europe. Recently, one of the European products, Thymoglobulin (a rabbit antithymocyte antibody) was approved in the U.S. These products continue to be used in different centers while research continues in this area.

Cyclosporine

As kidney transplantation advanced into the 1970s, azathioprine and prednisone had become mainstays of immunosuppressive protocols around the globe. Graft survival was routinely in the mid 70% range for recipients of living-donor transplants. Other advances in transplantation medicine, i.e. histocompatability testing, had also contributed to the long term success rates seen in these patients, but few doubted the importance of immunosuppression. However, for more complicated kidney patients, and those requiring heart or liver transplants, transplantation was still not a viable option. Azathioprine and prednisone, at tolerable doses, could not provide adequate immunosuppression for these patients. Researchers, armed with new information on the workings of the immune system, wanted to find a compound that could selectively inhibit the rejection response, while preserving other functions of the immune system. In Basle, Switzerland, a discovery was made in the early 1970s, which would change the face of transplantation science.[9]

Soil samples taken from Wisconsin, U.S., and Hardanger Vidda, Norway, were being analyzed to see if they contained compounds with antifungal or antimicrobial properties. Dr Jean Borel, a scientist with Sandoz Pharmaceuticals, was screening these samples for biologic activity. The fungus *Tolypocladium inflatum Gams*, found in the soil sample from Norway, synthesized a group of compounds called cyclosporins.[9] Two of these cyclosporins, designated 24-556, were found by Dr Borel to have immunosuppressive properties. The active metabolite was finally purified and termed cyclosporin A (CSA). This was a difficult drug to formulate, as CSA is a very water insoluble compound. Animal studies were initially inconclusive, luckily Dr Borel was able to perfect the appropriate diluent; one that was both lipophilic yet would allow absorption of the drug.

When studied in animal models, the immunosuppressive properties of CSA were not linked with cytostatic activity. This was important, as previous compounds used in transplantation all interfered with cytostatic activity. In these animal studies, CSA inhibited the formation of plaque-forming cells, delayed the development of graft versus host disease and skin graft rejection in mice, and depressed haemagglutinen production.[10] These effects were seen without significantly lowering granulocyte counts. This drug could potentially affect one aspect of the immune system without significantly affecting others. CSA demonstrated a mechanism of action that had not yet been seen in transplantation. The significance of this finding was not lost on Dr Borel.

Based on these initial findings and other animal trials, CSA was piloted in human cadaver renal transplant patients. A formulation of the drug in olive oil was found to be both palatable to humans and allowed for systemic absorption.[9, 11] In humans, the initial doses used were found to be toxic to the kidneys and the liver, revealing the dose-limiting toxicity of CSA. Rejection rates, however, decreased in these kidney transplant patients, and CSA continued to be studied by multiple researchers until its approval in the early 1980s. Its history since then speaks for itself. Transplantation success rates for living related and cadaver kidney transplants would reach the 90–95% level, and become the benchmark for future transplantation successes. Rejection rates at one year would be decreased to 30–40%. The use of combination therapy with CSA, azathioprine and prednisone has enabled transplant centers to tailor immunosuppressive regimens, using lower doses of these agents to avoid toxicities seen at higher doses. The development of CSA was instrumental in making heart and liver transplantation an achievable reality, and it opened the doors for further research into transplant immunology.[12]

Tacrolimus and Sirolimus

Immunosuppressive therapy in the 1980s and into the early 1990s was dominated by CSA, in combination with azathioprine and prednisone, and possibly with antibody induction therapy.

The origins of CSA got some researchers thinking that there may be other compounds in soil fungus with antirejection properties, perhaps having less toxicity than CSA. Kino and associates[13] with Fujisawa Pharmaceutical Company began to scourer the countryside of Japan, looking for a compound that had as much immunological promise as CSA did. Their diligence was rewarded when a soil sample, taken from Tsukuba, Japan, contained a strain of fungus that produced a compound with immunosuppressive activity.[14] The fungus was called *Streptomyces tsukubaensis*, and the compound isolated from this fungus was named FK-506. Its structure was determined to be a macrolide compound, differing extensively from the polypeptide structure of CSA. The immunologic effects, however, were remarkably very similar.

In initial animal studies, tacrolimus was shown to have a similar mechanism of action as CSA.[15] Both compounds were found to inhibit the function of T-lymphocytes, suppressing interleukin-2 (IL-2) production and expression of IL-2 receptors, with FK-506 approximately 100 times as potent as CSA. Both drugs bind with a binding protein in the lymphocyte and inhibit the function of calcineurin, thus both are known as calcineurin inhibitors.[16] FK-506 was also shown to have the same effects in graft versus host disease and skin grafts as CSA.[13] Like CSA, FK-506 has very poor solubility in water, but a solid dispersion formulation was discovered that allowed for a solid oral capsule.

In the early 1990s, FK-506 was studied as both rescue therapy and as primary immunosuppression for liver transplant recipients. Most of the initial studies in this drug were done at the University of Pittsburgh Medical Center. Dr Fung and his group[17] reported the results of a comparative randomized trial with FK-506 versus CSA in liver transplant patients in 1991. While the numbers were small, the results were very promising for FK-506, and further multi-center trials commenced. The investigators were also interested in the drug's side effect profile. Early reports suggested that FK-506 would not display the same nephrotoxicity that CSA had shown.[18] Unfortunately, further research confirmed that the similarities of these two drugs carried into the adverse event profiles.[19] FK-506 exhibited the same type of nephrotoxicity that was reported for CSA, with nephrotoxicity as the rate-limiting step.

FK-506, now known as tacrolimus, was approved in 1993 for the prevention of rejection, and is now used extensively in liver, pancreas, heart, heart/lung and kidney transplantation. It has provided an alternative when CSA could either not be tolerated or was not effective, and at many centers it is the primary immunosuppressive agent used.

Tacrolimus' origins from a streptomyces fungus increased interest in another compound, rapamycin. Rapamycin was isolated from the strain *Streptomyces hydroscopicus*, from a soil sample taken from Rapa Nui, the Easter Island.[20] It has a similar structure as FK-506, but when scientists studied its immunosuppressive activity, very large differences between rapamycin and the calcineurin inhibitors became evident. Rapamycin was found to affect both T- and B-lymphocyte proliferation. It binds to the FK binding protein, and this resulting complex inhibits the progression of T-cell proliferation, affecting both calcium-dependent and calcium-independent reactions.[20, 21] The toxicity profile of this medication is different as well, as renal dysfunction is not the dose-limiting toxicity for rapamycin. Rather, hyperlipidemia, thrombocytopenia and leukopenia have been the main obstacles to rapamycin therapy.[21]

Rapamycin, renamed sirolimus, has been studied in kidney transplantation as a rescue agent and for primary prevention of rejection. It was approved in the fall of 1999 for the primary prevention of rejection in kidney transplantation. Its place in the immunologic arsenal of organ transplantation is yet to be determined.

Mycophenolate Mofetil

Mycophenolate mofetil (MMF) has for many years been a drug in search of an indication. The active moiety of MMF, mycophenolic acid (MPA), was actually originally isolated from a penicillium culture back in 1896 by Gosio.[22] Its structure was determined in 1952, and first interest in this compound surfaced in 1972, when animal trials expressed antiviral and antitumor activity.[23] MPA was then studied in some small studies in humans with advanced malignancies, however results in humans were poor. The size of the doses needed, the poor bioavailability of MPA, and the GI toxicity exhibited at high doses, may have prohibited any meaningful human data.

MPA went back onto the shelf and stayed there until 1976, when researchers reported some success with MPA in patients with resistant psoriasis. Lynch and associates[24] initiated a placebo-controlled study with MPA in 375 patients with psoriasis. The initial report of 37 patients showed promising results in terms of reduction in the psoriasis, however the side-effect profile was still a problem. GI toxicity (diarrhea, nausea, bloating, blood in the stool) was reported, along with an increase in herpes infections, fatigue, and an increase in carcinomas. Due to the toxicity of MPA, the drug still remained investigational, and went back on the shelf.

The immunosuppressive properties of MPA had not been lost on the immunology community. Doctor's Eugui and Allison[25] were looking for a compound that inhibited the activity of inosine monophosphate dehydrogenase (IMPD), an enzyme that was responsible for lymphocyte proliferation and *de novo* synthesis of guanosine diphosphate (GDP). MPA was found to express the properties of IMPD inhibition that the investigators were looking for, and *in vitro* studies were set up to further discern the activity of MPA on the immune system. One of the problems with MPA that had been discovered in previous trials was its poor bioavailability. A colleague working with Dr Allison was able to formulate a synthetic derivative, using a morpholinoethyl ester. The resulting compound, now known as mycophenolate mofetil (MMF) is rapidly hydrolyzed to MPA *in vivo*. With a better bioavailable formulation now available, initial trials using MMF in animals and humans commenced.

MMF, its research name RS-61443, was studied extensively by Dr Sollinger and colleagues[26] as a maintenance immunosuppressive drug and as a potential treatment for rejection. The first animal trials in dogs studied MMF alone or combination with CSA and steroids for prevention of rejection. Those dogs receiving MMF in combination with CSA and prednisone had significantly fewer rejections and better overall survival. GI toxicity was still a problem, as seen previously, but did appear to be dose-related. With promising results, MMF moved on to Phase I human trials, and was also used as a rescue agent for those patients whose rejections had been resistant to conventional therapy.

Sollinger and his colleagues[27] began to perform Phase I safety and efficacy studies with MMF, as well as using the drug for resistant rejection episodes. The safety studies showed that MMF was devoid of adverse events to the kidney and liver, and was generally well tolerated. MMF was also able to salvage some of the kidney transplants with refractory rejection, leading to the possibility that the mechanism of MMF may also include the capability of reversing or at least stopping the rejection process.

With these initial promising studies, MMF went on to the large multicenter trials, and was approved in 1995 for the prevention of rejection in kidney transplantation. As MMF, the more specific purine inhibitor came onto the market, many believe that azathioprine's glory days as the maintenance immunosuppressive agent has past. Azathioprine, the drug that had been touted as one of *the* most important developments in transplantation, has been largely supplanted by MMF in most centers. Rejection rates of 10% or less are now being reported with MMF-based immunosuppressive protocols.

Agents Making History Today

The field of transplantation medicine continues to change. Research into the mechanism of immunosuppressive medications is leading us to newer and even more powerful agents. The explosion of monoclonal and DNA technology now allow researchers to fashion potential drugs that previously existed only in their imagination. The research into IL-2 inhibition by CSA and tacrolimus has lead to the study of the IL-2 receptor itself. A new class of monoclonal antibodies, known as IL-2 receptor antagonists, has been approved and is being used in transplantation. Other receptors and proteins are being analyzed for their role in the rejection pathway, and more importantly, how this role can be modified to stop rejection in its tracks. Other agents coming in the future may completely change how we view and administer immunosuppression to our patients. The complete history of immunosuppression in organ transplantation is still being written.

References

1. Goodwin WE, Kaufman JJ, Mims MM, *et al.* (1963). Human renal transplantation. I. Clinical experiences with six cases of renal homotransplantation. *J Urol* **89**(1), 13–24.
2. Popowniak KL, Nakamoto S (1971). Immunosuppressive therapy in renal transplantation. *Surg Clin North Am* **51**(5), 1191–1204.
3. Calne RY (1961). Inhibition of the rejection of renal homografts in dogs by purine analogues, *Transpl Bull* **28**, 445–461.

4. Murray JE, John Merrill JP, Harrison JH, *et al.* (1963). Prolonged survival of human-kidney homografts by immunosuppressive drug therapy. *N Engl J Med* **268**(24), 1315–1323.
5. Murray JE, Wilson RE, Tilney NL, *et al.* (1968). Five years' experience in renal transplantation with immunosuppressive drugs, *Ann Surg* **168**(3), 416–434.
6. Starzl TE, Marchioro TL, Hutchinson DE, *et al.* (1967). The clinical use of antilymphocyte globulin in renal homotransplantation. *Transplantation* **5**(4) (Suppl.), 1100–1105.
7. Najarian JS, FK Merkel Moore GE, *et al.* (1969). Clinical use of antilymphoblasts. *Transpl Proc* **1**(1), 460–462.
8. Lucas ZJ, Palmer JM, Payne R, *et al.* (1970). Renal allotransplantation in humans I. Systemic immunosuppressive therapy. *Arch Surg* **100**(2), 113–125.
9. Borel JF (1982). The history of cyclosporin A and its significance. In White D (ed.) *Cyclosporine A* Amsterdam: Elsevier Biomedical Press, Chap. 2, pp. 4–17.
10. Borel JF (1976). Biological effects of cyclosporin A: a new antilymphocytic agent. *Agents and Actions* **6**(4), 468–475.
11. Calne RY, White D, Thiru S, McMaster P, *et al.* (1978). Cyclosporin A in patients receiving renal allografts from cadaver donors. *Lancet* **2**(8104–5) 1323–1327.
12. Borel JF (1983). Cyclosporine: historical perspectives. *Transpl Proc* **15**(4) (Suppl. 1), 2219–2229.
13. Kino T, Goto T (1993). Discovery of FK-506 and update. *Ann N Y Acad Sci* **685**, 13–21.
14. Kino T, Hatanaka H, Hashimoto M, *et al.* (1987). FK-506, a novel immunosuppressant isolated from a streptomyces: I. fermentation, isolation, and physico-chemical and biological characteristics. *J Antibiotics* **40**(9), 1249–1255.
15. Ochiai T, Nakajima K, Sakamoto K, *et al.* (1989). Comparative studies on the immunosuppressive activity of FK506, 15-deoxyspergualin, and cyclosporine. *Transpl Proc* **21**(1), 629–832.

16. Kino T, Hatanaka H, Miyata S, *et al.* (1987). FK-506, A novel immunosuppressant isolated from a streptomyces. II. Immunosuppressive effect of FK-506 *in vitro*. *J Antibiotics* **40**(9), 1256–1265.
17. Fung J, Abu-Elmagd K, Jain A, *et al.* (1991). A randomized trial of primary liver transplantation under immunosuppression with FK506 vs cyclosporine. *Transp Proc* **23**(6), 2977–2983.
18. Starzl TE, Fung J, Jordan M, *et al.* (1990). Kidney transplantation under FK506. *JAMA* **264**(1), 63–67.
19. Shapiro R, Jordan ML, Scantlebury VP, *et al.* (1995). A prospective randomized trial of FK506-based immunosuppression after renal transplantation. *Transplantation* **59**(4), 485–490.
20. Kahan BD, Chang JY, Sehgal SN (1991). Preclinical evaluation of a new potent immunosuppressive agent, rapamycin. *Transplantation* **52**(2), 185–191.
21. Anonymous (1999). *Rapamune (Sirolimus) Oral Solution Product Information*. Wyeth-Ayerst Pharmaceuticals Inc.
22. Knudtzon S, Nissen NI. Clinical trial with mycophenolic acid (NSC-129185): a new antitumor agent. *Cancer Chemother Rep Part 1* **56**(2), 221–227.
23. Sweeney MJ, Gerzon K, Harris PN, *et al.* (1972). Experimental antitumor activity and preclinical toxicology of mycophenolic acid. *Cancer Res* **32**, 1795–1802.
24. Lynch WS, Roenigk HH (1977). Mycophenolic acid for psoriasis. *Arch Dermatol* **113**, 1203–1208.
25. Eugui EM, Allison AC (1993). Immunosuppressive activity of mycophenolate mofetil. *Ann NY Acad Sci* **685**, 309–329.
26. Platz KP, Sollinger HW, Hullett DH, *et al.* (1991) RS-61443 — a new potent immunosuppressive agent. *Transplantation* **51**(1), 27–31.
27. Sollinger HW, Deierhoi MH, Belzer FO, *et al.* (1992). RS-61443 — a phase I clinical trial and pilot resscue study. *Transpl* **53**(2), 428–432.

XVIII

THE HISTORY OF INFECTIOUS DISEASES IN TRANSPLANTATION

Brendan Madden

Introduction

The first bona fide report in the medical literature of organ transplantation came from Vienna at the turn of the 20th century when Emerich Ullmann transplanted a dog's kidney from its normal position to the neck. This auto-transplant worked well but when Ullmann subsequently transplanted a kidney from one dog to another, its function was short-lived. Ullmann was later to successfully perform a xenograft by transplanting a dog's kidney to the neck of a goat. Alexis Carrel took Ullman's work a significant step further when he devised successful surgical techniques for suturing small blood vessels. In 1905 Carrel, working with Charles Guthrie, successfully performed a dog kidney auto-transplant, although the recipient subsequently died when the kidney failed. Research in transplantation was static until 1923 when Carlos Williamson concluded that some biological phenomenon, which had not yet been identified, was responsible for graft failure. He was the first to describe the characteristics of allograft rejection under the microscope. This work was taken further in the 1950s by Emile Holman, when he concluded that rejection of grafts was due to a special antibody directed against that tissue. Around the same time, MacFarlane Burnet in Australia and Peter Medawar in England, discovered the means by which a newborn animal could be permanently induced to tolerate a foreign protein. In 1954, a team at the Peter Bent Brigham Hospital in Boston, successfully transplanted a human kidney from an identical twin to his brother, who was dying of renal failure. Successful transplantation between non-related persons followed with increasing knowledge of the make-up and nature of the human immune system, together with the development of immunosuppressive therapies and immunomodulation techniques. As a result, many patients have successfully received allograft skin, liver, heart, lung, bone marrow, cornea, pancreas and small bowel transplantation. On-going and exciting research into xenografting is expected to broaden the frontiers of transplantation even further. However, the success of transplantation relies not only on careful pre-operative assessment and operative skills but also on detailed post-operative medical care. In this regard, the diagnosis and treatment of infectious conditions is of great importance. This chapter aims to review the history of our knowledge of infection and to outline important advances in this field, particularly in the area of transplant medicine.[1]

The Development of Our Knowledge of Infection

Our current knowledge of infection[2, 3] has evolved over thousands of years. Ancient Egyptians described diseases of the eye and diseases resulting from water and food contamination and they attributed importance to the study of sputum, urine, faeces and other body fluids. In the first century BC, Varro had said that swampy land was dangerous because "certain minute animals, invisible to the eye, breed there, and borne of the air reach the inside of the body by way of the mouth and cause disease".

In the Bible, plagues and epidemics were frequently mentioned and specific attention was given to leprosy, which was not only feared but led to isolation of the victim. This practice was carried on throughout the middle ages when shunning lepers and fleeing from areas of pestilence and segregating the severely ill represented an awareness that diseases could be transmitted. In the 16th century,

Fracastoro suggested that there were "seeds" in the environment which could multiply in the body and produce disease and it was suggested by his contemporary, Giralamo Cardano, that these "seeds of disease" were in fact living creatures. In 17th century Rome, a perceptive Jesuit called Athanasius Kircher documented that when vinegar and sour milk were examined under a microscope "worms" could be visualized and that, furthermore, small animalcula could be demonstrated in the blood of people who died of plague. It was not until the 19th century, however, that bacteria were established as important causes of disease and it became clear that spread of these agents between humans was possible.

A number of significant advances were necessary before infection could be understood and diagnosed. One of these was the development of the thermometer. It was well known since the days of Hippocrates that body temperature was an important adjunct to physical examination. In the middle ages, the qualities of the four humours were hot, cold, dry and moist, and fever was therefore considered an important observation, although human temperature could not be quantified at that stage. The first thermometer was probably constructed by Galileo in 1592. It was a crude device without a measurement scale, which was influenced by atmospheric pressure. In 1665, Christiaan Huygens suggested that the thermometer should be an instrument of temperature measurement with a scale between the freezing point of water (0° centigrade) and the temperature of boiling water (100° centigrade). In 1717, Gabriel Daniel Fahrenheit devised a scale which set the temperature of a mixture of ice and ammonium chloride as the lower fixed point and used smaller degree increments than in the centigrade scale. He found mercury more useful than water because its expansion and contraction are more rapid. Thermometry was first widely introduced into clinical practice by Hermann Boerhaave in Holland. In 1742, Anders Celsius re-introduced the centigrade standard into clinical practice and Karl August Wunderlich studied thousands of patients and documented that fever was a symptom of disease. The instrument was made narrower and more practical by Aitkin in 1852 and Thomas

Clifford Allbutt designed the size and shape of thermometers used today in 1870.

The microscope was one of the most important inventions in the development of medical science. In ancient times, the use of a ground lens as a magnifying glass was well known and eyeglasses had been made in the middle ages. The combination of more than one lens to increase power was described by Zacharias Janssen and his son, but these devices were crude and had maximal magnification of approximately ten-fold. Francisco Stelluti published the first scientific treatise using the microscope to describe the structure of the bee in 1625. In 1655, Pierre Borel referred to worm-like creatures he saw in the blood of patients with fevers, although what he was describing remains unknown to this day. Antony van Leeuwenhoek produced microscopes with a magnification power in excess of 270-fold. He was the first to recognise the blood corpuscles and also made a thorough study of spermatozoa and noted the striated appearance of skeletal muscle. Marcello Malpighi contributed widely in the fields of botany and biology and was the first to confirm by microscopic examination of the lungs the capillaries which Harvey postulated.

The discovery of vaccination was another significant advance in our understanding of infection. In the 18th century, bubonic plague, typhus, malaria, diphtheria and, particularly, smallpox were frequent causes of mortality. The latter was endemic and frequently assumed epidemic proportions in cramped towns where sanitation was poor. It had been known in Asia that the technique of variolation was important in conferring protection against smallpox. This technique consisted of pricking serum from the sore of a person with smallpox into another's skin. Resistance then followed a mild case of the illness. In 1798, Edward Jenner (1749–1823) published his famous paper on cowpox, demonstrating the benefits of inoculating humans with the fluid from the sores of vaccinia, a disease of cattle. Vaccination had none of the potential dangers associated with inoculation of material from smallpox itself.

In the 19th century Matthias Schleiden and Theodore Schwann demonstrated that living creatures were composed of living cells which were capable of independent activity. Rudolph Virchow was able to demonstrate that cells arise only from pre-existing cells. Virchow did, however, make too little of the role of microorganisms in producing disease and did not give his support to Semmelweis in his battle for asepsis. Agostino Bassi of Lodi was able to link a disease in silkworm with a fungal parasite (*botrytis paradoxa*) in the late 18th century. He postulated that many contagious diseases were also due to live organisms which remained to be discovered. Jacob Henle supported this view and theorised that the parasites which caused the contagious illnesses were probably plants. Bacteria have been so classified until very recently.

In the 18th century, Charles White of England and Joseph Clarke and Robert Collins of Ireland, had significantly reduced the occurrence of postpartum infection (puerperal fever) in their hospitals by strict regimes of personal and environmental cleanliness, limited vaginal examination during labour and cleansing of beds and linen. Their observations were by and large unheeded, and when Oliver Wendell-Holmes in 1843 suggested that puerperal fever was due to infection carried to the new mother by obstetricians from other infected persons, this too was largely ignored. One notable exception, however, was Ignaz Semmelweis (1818–1865) who assembled the facts and audited events on the obstetric wards of the Allgemeines Krankenhaus in Vienna. He wished to prove that postpartum infection was contagious. His observations showed that the annual mortality rate on one of the wards where medical students were trained exceeded 10% and that it reached almost 20% during some months, largely as a consequence of puerperal fever. When he compared this to the ward where midwives received instruction, the death rate never exceeded 3%. He then went on to discover that the doctors and students usually came to the ward to examine patients directly from the autopsy room. The midwives and their teachers on the second ward however, received clinical tuition without attending postmortems. When one of his colleagues

(Kolletschka) died following infection from a scalpel wound sustained during an autopsy on a lady with puerperal fever, Semmelweis noted that Kolletschka's organs at postmortem demonstrated the same features which were seen in patients who died of puerperal fever. Semmelweis's action was to ensure that physicians and students under his charge scrubbed their hands with soap and water and soaked them in chlorinated lime solution before entering the clinic or ward and to repeat this after each examination. Over the next few months, the obstetrical mortality rate declined from 20% to 1.2%. Unfortunately, rather than continuing on Semmelweis's regime throughout the hospital, the head of the department condemned Semmelweis and his publications were ridiculed. Even before the germ theory had arrived, Semmelweis, in his book *Die Aetiologie Der Begriff Und Die Prophylaxis Des Kindbettfiebers* in 1861, constructed a policy for asepsis. Sadly, Semmelweis was broken by the attacks on his work and died of septicaemia in 1865.

Joseph Lister (1827–1912) observed that broken bones, over which the skin was intact, usually healed without complication. However, compound fractures were commonly associated with infection. Although there was important work by Simpson, showing that pus came in some way from capillaries, and by Cohnheim subsequently proving that white blood cells migrated through the walls of tiny blood vessels into areas of inflammation, these authors had not yet shown that pus was stimulated as a response to agents which could be passed from one person to another. Lister also observed that severe infections could complicate operations such as amputations. He believed that this occurred because of wound contamination by circulating invisible particles in the air which he termed "disease dust". In 1860, he reviewed the work of Louis Pasteur and appreciated the connection between his own observations on wounds and the microscopic bacteria involved in the fermentation process. In 1867, Lister published a report on antisepsis during surgery in the *Lancet* of his experiences with 11 patients where he sprayed carbolic acid over each patient during the operation to kill any bacteria before they could grow in the wound.

Unfortunately, as with Semmelweis, Lister's reports were greeted with indifference or hostility. His views were subsequently, albeit subconsciously at first, absorbed into standard surgical practice. William Halsted later introduced rubber gloves into his surgical practice to protect the hands of his operating room nurse (a lady he later married) and one of his students suggested that the surgeons should use them too, since they could be sterilised. Wearing of surgical masks subsequently followed.

Lister was influenced by the work of Louis Pasteur (1822–1895). Pasteur firmly established the germ theory of disease and explained the effectiveness of asepsis and antisepsis of Semmelweis and Lister. Indeed, it was this work which laid the basis for preventative measures of the future. In 1854, he stated "Dans le champ de l'observation le hasard ne favorise que l'esprit preparé" (in the field of observation, chance favours the prepared mind). Pasteur was trained as a chemist and had major research interests in stereochemistry, proving that there were two types of tartaric acid crystals. He discovered that bacteria acted on each type of crystal differently and this observation led him to focus on the germ theory of disease. He proved that microscopic live creatures were responsible for fermentation and that some of these organisms grew in the presence of oxygen (e.g. aerobic), while others were able to live independently of oxygen (e.g. anaerobic). He also demonstrated that heating wine for a few moments at about 60°C (108°F) destroyed the organisms which produced spoilage — this process was later called "pasteurization". His work on diseases of silkworms saved the silkworm industry in France and throughout the world. When he suffered a stroke, he continued to work with great enthusiasm and he focussed his mind on the problem of anthrax (splenic fever) a fatal disease of sheep. He isolated the anthrax bacillus, confirming Koch's earlier work, but was unable to see a way of treating or preventing this disease. He was asked to investigate the cause of chicken cholera and, on return from a vacation, found that cultures of the chicken cholera organism, prepared before he left, were harmless when injected into healthy fowls but subsequent injections of fresh, virulent cultures into the same hens failed to produce the disease. Pasteur then treated cultures of the anthrax bacillus in various ways until he found that microbes grown in a particular temperature range became harmless without losing their capability of producing resistance in injected animals. He tested the validity of his discovery at a public demonstration organised by the Melun Agricultural Society in 1881. To an assembled audience, virulent cultures of the anthrax bacillus were injected into normal sheep and into an equal number of sheep previously inoculated with attenuated harmless cultures. In the next few days, all of the unprotected sheep died but all of the previously injected sheep remained well. The principle of immunity was therefore established. Although Jenner had obtained the same protection from smallpox by producing another illness, vaccinia or cowpox, Pasteur established the fundamental principle that attenuated cultures of an organism could confer protection (immunity) against the disease caused by that organism. Lister had previously given recognition to Pasteur for his importance to antisepsis and now Pasteur recognised Jenner's work by calling his own method "vaccination". Pasteur went on to develop an extract from the spinal cords of rabbits infected with rabies and first successfully applied it to humans in 1885 when he treated a young boy named Joseph Meister, who had been bitten by a rabid dog.

Following Pasteur's work, bacteriology and immunology expanded exponentially. Roux, a pupil of Pasteur, reported a filterable virus and, together with Alexander Emile Yersin, he isolated the diphtheria bacillus and developed an antitoxin. Increasing numbers of bacteria were subsequently discovered and numbers of vaccines and antisera were produced. Thus mechanisms of disease prevention became established.

Robert Koch (1843–1910) devoted a tremendous amount of his spare time as a county physician, to studying microorganisms. He isolated the tubercle bacillus as the cause of tuberculosis and defined the essential steps (the Koch's postulates) required to prove that an organism is the cause of disease. Ferdinand Cohn (1828–1898) at the University of Breslau was important in establishing bacteriology as a science. He also helped classify bacteria and gave support to Koch, Davaine and others working in the field.

In the 1890s, two important discoveries were made, namely the development of antitoxins and the discovery of viruses. Microbiologists had now shown that bacteria were capable of producing toxins which could directly damage the human body. The word toxin is derived from the Greek toxikon, bow poison, referring to poisonous material placed on arrows by Greek warriors. The implication is therefore that the bacterium produces a molecule that it "releases" to affect host cells at a distance. The term was first used by Roux and Yersin to describe the factor released into the culture medium by *Corynebacterium diphtheriae* that caused the death of recipient animals. von Behring and Kitasato showed that the body produced compounds which acted against these toxins. von Behring was able to obtain neutralising antitoxin sera from the blood of animals which were injected with intermittent doses of these toxins. Paul Ehrlich (1854–1915) used antitoxic sera to treat diseases and termed this "passive immunisation". This was to distinguish the treatment from active immunisation whereby attenuated cultures of the bacteria or its toxin were inoculated into the patient to produce protecting antibodies. Pasteur's anti-rabies inoculation and Jenner's smallpox vaccination were examples of this latter process.

Scientists were able to then use the presence of circulating antibodies against a particular infectious agent or toxin to confirm a diagnosis. In addition to such humoral immunity, workers continued to study the cellular immune response and Elie Metchnikoff (1845–1916) showed that some cells were able to kill bacteria by phagocytosis as well as by elaborating antibodies against them. Subsequent work showed that humans could be carriers for pathogenic organisms such as cholera, diphtheria, typhoid fever and meningitis without actually being ill themselves.

Although the majority of organisms responsible for human infectious disease were bacteria which could at that stage be seen under the microscope, some organisms, e.g. those for smallpox and rabies, could not be seen and were so small that even fine filters allowed them to pass through. Thus, following such observations on smallpox and rabies by Loeffler and Roux, the field of virology commenced. It subsequently became clear that there were intermediate size microbes between bacteria and viruses. These were named Rickettsia, after Howard Ricketts (1871–1910) who identified the cause of Rocky Mountain spotted fever. Identification of protozoal organisms followed, e.g. malaria. Further investigation revealed that some disease producing organisms underwent lifecycles in animal hosts in the course of which they could infect humans. Examples included lice in typhus, tsetse fly in sleeping sickness and ticks in filariasis. The mosquito was identified as the insect vector in malaria and yellow fever.

Scientists had now shown that microorganisms existed in their own right and that these agents could cause human disease and that spread between humans was possible. Furthermore, it had been shown that preventative measures using asepsis and antiseptic techniques were important and that modified forms of microorganisms or their toxins could be used either passively or actively to confer a degree of immunity. It was known that other vectors could play a significant role in infectious disease and that these vectors could be targets as part of a broader plan of disease eradication. In spite of these advances, agents were urgently needed to treat patients with established infectious disease. The work of Paul Ehrlich contributed enormously to this area of medical therapeutics. As a student, he was interested in the methodology of making cells visible under the microscope and his doctoral thesis was on histological staining. His work on the theory of specific affinity of cells for dyes was based on the work of George Hayems's techniques of colouring living cells (vital staining), together with Hermann Hoffmann's stains of bacteria and Carl Weigert's visualisation of bacterial cocci in tissues. He also demonstrated the staining characteristics of white blood cells with aniline dyes, which permitted others to understand abnormalities of blood cells and thus significantly aided the foundations of haematology. He was also an important contributor to developing the methodology for identification of the tubercle bacillus under the microscope. His work also encompassed techniques of neutralising toxins with antitoxins *in vitro* and he also demonstrated that some antibodies were heat-resistant and others heat-susceptible. This led to his "side-chain theory" which, although abandoned for many years, has now recently been revisited. In 1905, Fritz Schaudinn identified *treponena pallidum* as the causative organism of syphilis. After trying 605 compounds, Ehrlich synthesised an arsenical known as salvarsan in 1910, which was therapeutically effective in the management of patients with syphilis. Subsequently, less toxic arsenic compounds, e.g. neoarsphenamine followed from his work and thus treatment for patients with syphilis advanced significantly. Approximately 30 years later, sulphonomides were introduced into clinical practice for the treatment of bacterial infections. This was a direct follow-on from Ehrlich's work which demonstrated that dyes could, in themselves, be anti-bacterial agents. Although penicillin subsequently led to Ehrlich's drugs against syphilis being abandoned, he had nevertheless lit the fuse paper on the science of anti-microbial chemotherapy, which was to explode in the 20th century.

For the first 30 years of the 20th century, viruses were too small to be directly visualised and they could only be studied by their pathogenic effects on those animals they infected. When the electron microscope was developed, viral structure could be seen and appreciated in detail and the mechanism by which viruses infected cells could be clarified. Science then provided measures for growing cells in culture and for making monolayer cultures, which aided exact quantitative studies of viral multiplication and the action of specific antiviral antibodies. For the first time, it was possible to directly observe the factors involved in viral genetics and mutation. It was therefore possible to target viruses at particular stages in their development and life cycle and therefore antiviral agents and vaccines could be developed. For example, the poliomyelitis virus was first isolated by Landsteiner and Popper in 1909. In 1954, a non-infectious vaccine for this virus was devised by Salk and co-workers. Three years after introduction of this vaccine, the annual incidence of poliomyelitis had fallen significantly in the United States. Albert Sabin subsequently developed a live attenuated polio vaccine which could be taken orally. Vaccines have now been made against many viruses, e.g. measles, mumps and hepatitis A and B.

Paul Ehrlich wished to find a particular agent ("a magic bullet") which would seek out and destroy germs in the body without destroying host cells. The arsenical compounds which he successfully used in the treatment of syphilis was a tremendous advance in achieving this and his work led to Charles Heidelberger and W. A. Jacobs introducing sulfanilamide, an azo dye capable of destroying bacteria. Gerhard Domagk reported that prontosil, a textile dye stuff synthesised in 1908, acted against Streptococci in mice and this chemical dramatically improved the outcome of many human infections including septicaemia. Trefouel showed that prontosil caused the body to produce sulfanilamide, the active anti-bacterial substance in the dye. Further sulpha drugs were developed and introduced into clinical practice.

Pasteur and many workers who followed him noted antagonism between certain bacteria. It was subsequently shown that the higher bacteria, molds and fungi also destroyed certain bacteria. This process had earlier been named antibiosis by Vuillemin. In 1912, Westling named one such effective mould, *penicillium notatum*. Approximately ten years later, Lieske and Gratia showed that a species of penicillium dissolved anthrax bacilli and impeded Staphylococci. In 1929, Alexander Fleming reported in the *British Journal of Experimental Pathology*, his observations on the anti-bacterial action of penicillium (which he called penicillin). He suggested that it could be used to inhibit bacteria and thus help in obtaining their cultural isolation. In 1941, Howard Florey and Ernst Chain demonstrated that penicillin had tremendous therapeutic potential. Other types of antibiotics now followed rapidly. Selman Waksman obtained streptomycin from *streptomyces griseus* in 1944 and this was particularly useful against tuberculosis. Other strains of streptomyces yielded substances such as chloromycetin and aureomycin. Numerous classes of agents (many of which are synthetic) have subsequently followed, e.g. cephalosporins, aminoglycosides, lincosamides, quinolones, imipenem with cilastin, anti-fungal agents, anti-viral agents, isoniazid and anti-tuberculous agents, and many more. However, as new substances have been developed, bacteria have become resistant and hence there is a continuous ongoing struggle to identify new specific anti-microbial agents with a high therapeutic index.

Immunocompromise and Opportunistic Infection in Organ Transplantation

It follows that we live in a potentially hostile world filled with a bewildering array of infectious agents which are capable of causing disease. Fortunately, a variety of defense mechanisms can establish a state of immunity (from the Latin immunitas, meaning freedom from) against infections and it is the operation of this system which provides the basis for the study of immunology. We have established that the simplest way to avoid infection is to prevent the microorganisms from gaining access to the body. Lister appreciated that the major line of defense is the skin, which when intact is impermeable to most infectious agents. Mucus, which is secreted by membranes lining the inner surfaces of the body, acts as a protective barrier to block the adherence of bacteria to epithelial cells. Many of the secreted body fluids contain bactericidal compounds such as acid, zinc, lactoperoxidase and lysozyme. There is also microbial antagonism between the normal bacterial flora of the body, which can suppress the growth of potentially pathogenic bacteria and fungi at superficial sites by competition for essential nutrients or by production of inhibitory substances such as colicins or acid. The work of Metchnikoff and others has elucidated the role of phagocytic cells which kill microorganisms and the mechanisms by which phagocytic cells operate have been clarified, in particular, much work has focused on the polymorphonuclear neutrophil and macrophage. We have learnt that these cells engulf microorganisms following adherence of the microorganism to the phagocytic cell surface and that killing can occur via oxygen dependent and independent mechanisms. Our knowledge has also expanded to appreciate the role of complement in this process and the interplay between complement and other aspects of humoral immunity. Cytokines and interferons have been described and the role of other cells such as natural killer cells and eosinophils in the human defense against infection has been defined.

The immune system has evolved to protect us from the numerous potential pathogens which are present in the environment. The basis of immunity is the immune system's ability to recognise foreign molecules (antigens) and react to them while at the same time tolerating molecules of the body's own tissues. Broadly speaking, an individual's reaction to a pathogen consists of a non-adaptive or innate response and an adaptive immune response. The innate response is not improved by repeated encounters with a particular pathogen. However, the adaptive response has the important characteristics of specificity and memory. It can therefore specifically recognise a pathogen on first encounter. Should re-infection with that pathogen occur, an augmented immune response will follow. Immune recognition and the immune reactions that ensue are central to the subject of immunology and the defense against microorganisms. When the system breaks down, for example as a consequence of congenital or acquired disease or immunosuppressive therapy, a particular type of infection known as opportunistic infection[4] can occur. This term refers to infections caused by organisms which are not usually recognised as pathogens but which invade the body as a result of specific immunodeficiency. The definition includes infections by known pathogens in unusual sites, which gained access as a consequence of physical immunological breeches in host defenses.

Our knowledge of the immune system has shown us that immunodeficiency can be inherited or acquired. Examples of inherited defects include severe combined immunodeficiency syndromes (reticular dysgenesis and Swiss-type immunodeficiency), thymic hypoplasia (di George syndrome) and immunoglobulin deficiency syndromes. Acquired immunodeficiency can occur following infection, general ill health or chemotherapy for malignant disease, or as a consequence of immunosuppression following organ transplantation. Immunosuppressed patients are more susceptible to endogenous and exogenous infections than normal individuals, and such infection is often more aggressive and invasive than that found in non-immunocompromised individuals. It should always be remembered that infection in immunosuppressed patients may present in an atypical fashion.

Table 1 Opportunistic organisms which may produce infection in immunocompromised patients

Protozoa

- *Pneumocystis carinii*
- *Toxoplasma gondii*

Viruses

- Cytomegalovirus
- Varicella-zoster virus
- Herpes simplex virus
- Epstein-Barr virus

Fungi

- *spergillus fumigatus*
- *Candida albicans*
- *Mucor*
- *Cryptococcus neoformans*
- *Histoplasma capsulatum*
- *Coccidioides immitis*

Bacteria

- *Streptococcus pneumoniae*
- *Staphylococcus epidermidis*
- *Pseudomonas aeruginosa*
- Enterococci
- Gram-negative organisms
- *Nocardia asteroides*
- *Mycobacterium*
- Clostridium difficile

Common organisms producing opportunistic infection are listed in Table 1. It should be appreciated that this list is not exhaustive and that infections with other organisms can occur.

Protozoa

Pneumocystis carinii

Chagas discovered *Pneumocystis carinii* in 1909 when he mistakenly interpreted the organism for a trypanosome. Several years later, Delanoes identified *Pneumocystis carinii* as a separate genus and species and named the organism after Dr Carini who was an early worker in the field. Pneumocystis first came to medical attention after World War II when it was found to be the cause of interstitial plasma cell pneumonia, a disorder of institutionalised debilitated infants of Central and Eastern Europe. It was subsequently appreciated to be an important cause of immunocompromised pneumonia in the 1960s. Interest faded with the development of safe and effective chemoprophylaxis until the 1980s when a dramatic increase in the incidence of pneumocystis pneumonia was observed in patients who were subsequently shown to have the acquired immunodeficiency syndrome.

The taxonomy of *Pneumocystis carinii* has long been a matter of controversy. Many workers have classified the organism as a protozoan based on morphological structural appearances (e.g. ameboid appearance and presence of filopodia), lack of growth on fungal media and sensitivity to anti-protozoal drugs. However, other investigators favour a fungal classification for the organism. This opinion is based on similarities to the ascospore formation in yeasts, the poorly developed organelle system and the staining with methenamine silver (Fig. 1) and other reagents that stain fungi. Analysis of ribosomal RNA sequences suggest a closer relationship to fungi than to protozoa.

Studies in immunosuppressed animals have shown that *Pneumocystis carinii* is communicable via the airborne route. That person-to-person transmission occurs is suggested by outbreaks and clusters of cases in orphanages and hospitals and by the occurrence of the disease

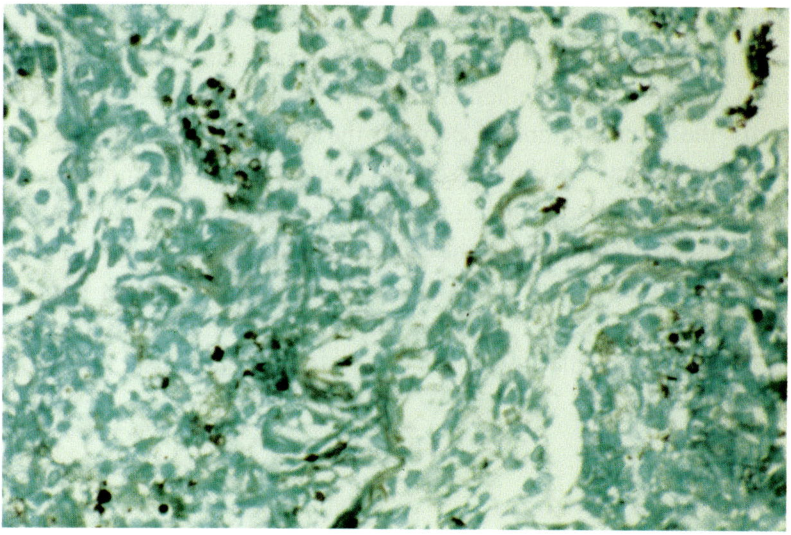

Fig. 1 Transbronchial biopsy from a renal transplant patient with *Pneumocystis carinii* pneumonia. The organisms are stained black (Grocotts silver methenamine).

in immunosuppressed patients who have prolonged contact with each other. The incubation period is between four and eight weeks. There is no evidence that *Pneumocystis carinii* infection is a zoonosis. Exposure to the organism does not result in protective immunity. It is now widely believed that impaired cellular immunity is more important than impaired humoral immunity in predisposing to *Pneumocystis carinii* infection, although the specific host immune defects involved in the process are poorly understood. Patients particularly at risk include premature malnourished infants, children with primary immunodeficiency diseases (particularly severe combined immunodeficiency disease), patients receiving immunosuppressive therapy and those with the Acquired Immune Deficiency Syndrome or AIDS.

Presenting features of infection may be non-specific and chest radiography is often normal. This may contrast markedly with the clinical presentation of profound respiratory failure and arterial hypoxaemia. Since there are no consistently reliable non-invasive methods of making the diagnosis, fibreoptic bronchoscopy with analysis of bronchoalveolar lavage fluid and transbronchial lung biopsy specimens using Grocotts silver methenamine preparation (Fig. 1) is necessary to identify the organism and to confirm the diagnosis. Treatment consists of high dose oral co-trimoxazole or intravenous pentamidine isethionate in patients allergic to co-trimoxazole. Alternative treatment comprises combination primaquine and clindamycin. Corticosteroids are prescribed when there is evidence of pulmonary fibrosis or severe arterial hypoxaemia. Effective prophylactic therapy consists of interrupted co-trimoxazole therapy or fortnightly nebulised pentamidine isethionate therapy.

Toxoplasma gondii

Toxoplasma gondii is an obligate intercellular protozoan that is ubiquitous in nature. It is an important cause of infection (toxoplasma infection) and disease (toxoplasmosis) in humans and in domestic animals. Toxoplasma infection is a worldwide zoonosis. The presence of cats appears to be of primary importance in transmission of the infection in most (but not all) endemic areas of the world. In immunosuppressed patients, latent *Toxoplasma gondii* infection may be re-activated or primary infection may occur from seropositive donor organs following transplantation. The clinical presentation is usually neurological although myocarditis, pneumonitis or disseminated lesions can occur. Diagnosis is made by a combination of clinical presentation and serological tests for antibodies to *Toxoplasma gondii* with histopathological identification of the organism in biopsy material (Fig. 2). Treatment is with pyrimethamine and sulphadiazine with folinic acid supplementation.

Viruses

The history of virology extends throughout recorded history. Indeed, viral diseases are among the oldest (e.g. rabies and polio) and most

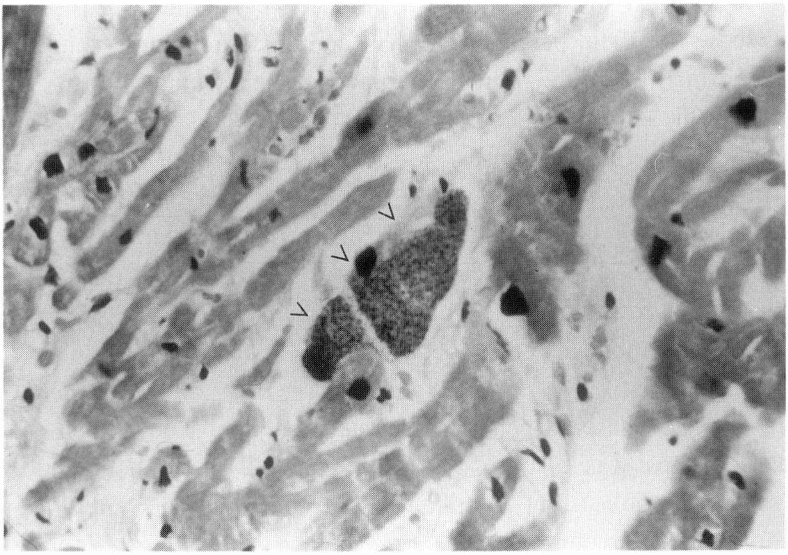

Fig. 2 Endomyocardial biopsy specimen from a cardiac transplant recipient with toxoplasma myocarditis (arrows). Haematoxylin and eosin, medium power.

recently described human diseases (e.g. Acquired Immune Deficiency Syndrome). During the 19th century, the rate of progress in the field of virology began to accelerate and led to the identification and differentiation of a number of viral illnesses (e.g. smallpox from chickenpox and measles from rubella). The work of Virchow and others allowed the pathological substrate of many viral diseases to be identified, and Pasteur introduced the systematic use of laboratory animals to study disease pathogenesis.

The first viruses were identified at the end of the 19th century. Beijerinck identified tobacco mosaic virus and Loeffler and Frosch discovered foot and mouth disease virus. The discovery of the first human disease-causing virus (the yellow fever virus) followed. Important work on its pathogenesis was performed by Walter Reed and the Army Yellow Fever Commission. Towards the end of the 1930s, tumour viruses, bacteriophages, influenza viruses, mumps and many arboviruses had been identified. The discovery of new viruses and knowledge of their pathogenesis has continued relentlessly until the present time, with the human immunodeficiency and human T-cell lymphotropic viruses being the most recent additions among the human disease-causing viruses.

In the 1940s, Delbruck, Luria and others used bacteriophages as a model and were able to establish many of the basic principles of microbial genetics and molecular biology. They identified major events in the viral growth cycle. Seminal work by Avery and associates on the transformation of pneumococcal types established DNA as the genetic material. Thus, the stage was set for the experiments of Hershey and Chase, which demonstrated that the genetic material of bacteriophages was also DNA. In the late 1940s, Enders and colleagues grew polio virus in tissue culture and this work subsequently led to the development of both the formalin inactivated (Salk) and live attenuated (Sabin) vaccines for polio. Thus, the modern era of virology was born. It was allowed to flourish by important technical advances such as the development of electron microscopy, ultracentrifugation and techniques for the electrophoresis of nucleic acids and proteins. It was therefore possible to improve our understanding of viral structure and to analyse purified virion components. X-ray crystallography has allowed structural definition of viruses and their components at an almost atomic level. Techniques of molecular biology have allowed the entire genomes of many viruses to be sequenced and the use of monoclonal antibodies has enabled specific domains on viral proteins to be precisely defined. These and newer techniques are continuously being applied to improve the diagnosis and treatment of viral illness. An example is the polymerase chain reaction (PCR) which allows small amounts of nucleic acid, e.g. the integrated DNA of a retrovirus such as the human immunodeficiency virus that is present in some circulating host cells to be dramatically amplified. This greatly facilitates viral detection.

Viruses were first classified as a group distinct from other microorganisms by their ability to pass through filters of a small pore size.

They were often referred to as "filterable agents". Initially, subdivisions were based primarily on pathological properties such as specific organ tropism (e.g. the enteroviruses) or on common epidemiological features, such as transmission by arthropod vectors (e.g. the arboviruses). In the 1950s, classification was based on morphological and physicochemical criteria, and more recently, along lines of genetic similarities. Currently, the key constituents of classification systems are:

1. The type and structure of the viral nucleic acid and technique used by the virus in its replication.
2. The type of symmetry of the virus capsid (helical versus icosa hedral).
3. The presence or absence of an envelope.

Cytomegalovirus

Cytomegalovirus (CMV) is the most common opportunistic viral pathogen and is an important cause of morbidity and mortality for transplant recipients. Like other herpes viruses, it has the unique capacity to remain latent in tissues after recovery of the host from an acute infection. Therefore, once the patient is infected they are always infected. This virus is therefore an opportunist par excellence because it is frequently already in hosts waiting to be activated when the host becomes immunosuppressed.

In 1881, Ribbert described "protozoan-like" cells in the kidney of a stillborn. By 1932, 25 cases of cytomegalic inclusion disease had been described. Farber and Wolbach carried out a post-mortem examination on 183 infants who died of various diseases and found that 12% of the submaxillary glands of these children had intranuclear and cytoplasmic inclusions. Murine CMV was identified in mouse culture by Smith and later human CMV was independently isolated by Smith, Rowe and Weller. It was Weller who coined the term cytomegalovirus to replace salivary gland virus or cytomegalic inclusion disease virus, and Klemola and Kaariainen first described recognisable CMV infection and disease in a healthy adult. It was subsequently found that CMV mononucleosis can occur sporadically and also after blood and leukocyte transfusions. The site of latency of CMV is not known but probably includes circulating peripheral mononuclear leukocytes. The virus can be spread by sexual contact, by blood transfusion, during birth to the fetus and following organ transplantation. In the transplant setting, post-operative viral infection is minimised by ensuring that CMV IgG antibody negative recipients receive CMV IgG negative donor organs and blood products whenever possible. If a CMV IgG negative recipient receives an organ from a positive donor, prophylactic post-operative ganciclovir and hyperimmune globulin to CMV is often effective in reducing the severity and delaying the onset of CMV infection. The virus has been implicated in the development of acute rejection and obliterative bronchiolitis following lung transplantation and in the development of transplant coronary artery disease in cardiac transplant recipients. More potent immunosuppressive drugs, e.g. antithymocyte globulin or OKT3 are associated with a higher incidence of post-transplant CMV infection. Furthermore, different incidences of CMV infection have been reported in different types of transplant recipients, e.g. a higher incidence is seen in heart/lung and lung transplant recipients when compared with renal transplant recipients.

The usual clinical manifestations of symptomatic CMV infection are fever, malaise, myalgia, abdominal discomfort, colitis, gastritis, leucopenia, thrombocytopenia, atypical lymphocytosis and hepatitis. Pneumonitis, encephalitis and retinitis can also occur. The diagnosis of CMV infection is made by a combination of the above clinical features, serological investigations and the detection of early antigen fluorescent foci in blood, saliva and urine. CMV pneumonitis must be excluded in an immunocompromised patient with an abnormal chest radiograph (Fig. 3). Fibreoptic bronchoscopy is necessary to reliably define pneumonitis in immunocompromised patients. The development of fibre-optic bronchoscopy and its application into clinical medicine

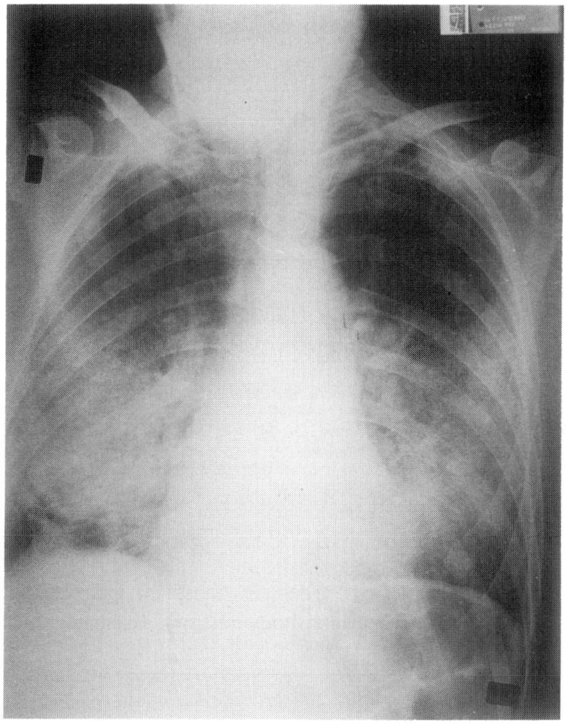

Fig. 3 Chest radiograph demonstrating interstitial pneumonitis in a cardiopulmonary transplant recipient with *Pneumocystis carinii* pneumonia.

has had tremendous impact in the diagnosis of opportunistic pneumonia. At bronchoscopy, bronchoalveolar lavage specimens are taken for culture and sensitivity, opportunistic pathogen screen and immunocytochemistry. Transbronchial lung biopsy specimens are sent for histopathological examination and culture. The diagnosis of CMV pneumonitis can be made by demonstrating the presence of CMV inclusion bodies in transbronchial lung biopsy specimens (Fig. 4). Treatment consists of intravenous ganciclovir which may be supplemented by hyperimmune globulin to CMV. Alternatively, foscarnet can be prescribed.

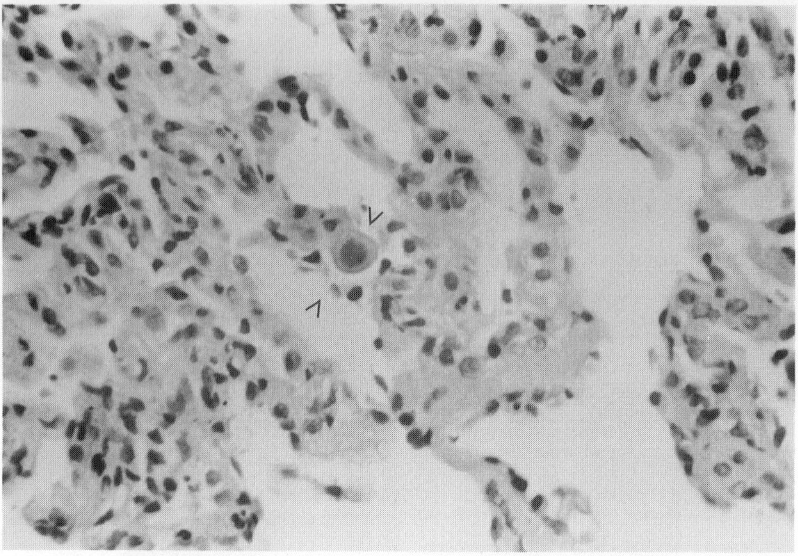

Fig. 4 Transbronchial biopsy from a hepatic transplant recipient with CMV pneumonitis demonstrating a CMV inclusion body (arrow). Haematoxylin and eosin, medium power.

Epstein-Barr virus

This is a ubiquitous member of the human herpes virus group. Filatov and Pfeiffer provided the initial description of the disease at the end of the 19th century when, almost simultaneously, they described an illness characterised by malaise, fever, hepatosplenomegaly, lymphadenopathy and abdominal discomfort. The illness came to be known as Drusenfieber (glandular fever) and was reported to occur in family outbreaks. Between 1910 and 1920, a number of authors reported cases of apparent spontaneous remission of leukaemia with a similar clinical course to glandular fever. In 1921, Sprunt and Evans described six cases of fever, lymphadenopathy and malaise occurring in previously healthy young adults. The authors demonstrated the presence of atypical mononuclear lymphocytosis that developed in each of the patients and contrasted the appearance of these lymphocytes to the uniform lymphocyte morphology observed in children with other infections. In 1923, Downey and McKinlay provided a more detailed morphological description of the atypical lymphocyte. In 1932, Paul and Bunnell were investigating immunological mechanisms in serum sickness and unexpectedly found high titres of spontaneously occurring sheep red blood cell agglutinins in the sera of patients with infectious mononucleosis. Davidsohn later enhanced the specificity of detection of this heterophile antibody by differential absorption of serum with guinea pig kidney and with beef erythrocytes. In 1958, Burkitt described an unusual lymphoma which involved the head and neck and six years later, in 1964, Epstein and colleagues described the presence of particles which closely resembled herpes viruses in biopsy specimens from patients with this condition. An indirect immunofluorescent antibody technique to the now called Epstein-Barr virus (EBV) was developed by Werner and Henle and titres of this antibody were detected in patients with Burkitt's lymphoma.

It has subsequently been found that infection with EBV is common, has a worldwide distribution and is largely sub-clinical in early childhood. It has been established as the aetiologic agent of heterophile-positive infectious mononucleosis, which occurs most frequently in late adolescence or early childhood. In addition to the association between EBV and African Burkitt's lymphoma, there is an association between EBV and nasopharyngeal carcinoma.

As experience with organ transplantation has progressed, it has become clear that the Epstein-Barr virus is an important aetiological factor in the development of lymphoproliferative disorders. These disorders occur in up to 2% of adults and 4% of paediatric transplant recipients, although some authors have reported higher incidences. They develop in patients whose T-cell mediated responses are impaired by cyclosporin-A, tacrolimus, OKT3 or antithymocyte globulin. Both B and T-cell types occur but T-cell types are more common and are frequently associated with EBV. They may be reversible after reduction in immunosuppression. T-cell disorders usually occur independently of diagnosed viral infection. Both types of abnormality have

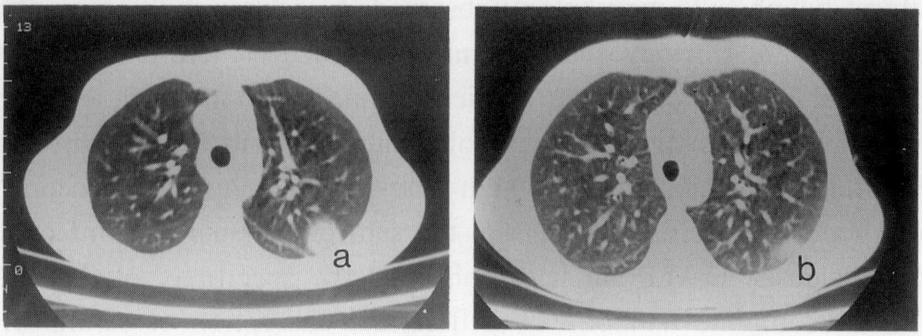

Fig. 5 Thoracic CT scan from a cardiopulmonary transplant recipient with a B-cell (Epstein-Barr virus associated) pulmonary lymphoproliferative disorder. Scan A shows the tumour at the time of diagnosis and scan B two weeks after commencing aciclovir therapy and reduction in immunosuppression. The patient recovered completely and has had no recurrence in the six years following successful treatment.

a variable clinical presentation, which includes an infectious mononucleosis-like illness, generalised lymphadenopathy and tonsillar enlargement or extra-nodal masses. No one single investigation is reliable and therefore the diagnosis is reached by a combination of clinical features, serological investigations (EBV antibody titres, clonal antibody activation to EBV), chest X-ray, computerised tomography scan of affected areas (Fig. 5) and biopsy of lesions. Due to the wide variety of histopathological appearances, ranging from changes consistent with viral lymphadenitis to variants of high grade non Hodgkin's lymphoma, accurate diagnosis depends on identifying EBV in biopsy material either by immunohistochemical staining or by *in situ* hybridisation. Unlike conventional lymphomas, lesions may be monoclonal or polyclonal for immunoglobulin light chain expression. Classification is based on the degree of cellular pleomorphism and clonality.

Initial treatment is conservative irrespective of the histopathological appearances and clonality and consists of reduction in immunosuppression and treatment with high dose acyclovir for B-cell disorders. T-cell disorders usually respond to reduction in immunosuppression alone. There is poor correlation between clinical behaviour and histopathological appearances and clonality, although up to 60% of patients respond to these conservative measures. These patients often include those with an infectious mononucleosis-like illness which may occur early after transplantation and are associated with a good prognosis. Conventional lymphoma chemotherapy may, however, be required in clinically aggressive cases that occur late after transplantation.

Herpes simplex and varicella zoster

Herpes simplex virus can cause orofacial lesions, keratitis, pneumonitis and encephalitis. Encephalitis is an important cause of morbidity and mortality after bone marrow transplantation. The development of acyclovir has had an important favourable benefit on the treatment of herpes viral infection following transplantation. Varicella zoster viruses causes varicella (chickenpox) as a primary infection and can cause recurrent disease later as zoster (shingles). In transplant patients, this has been shown to progress to a disseminated disease in some patients. Varicella zoster also responds to aciclovir therapy.

Fungi

Fungi can be classified as yeasts or molds. Yeast-like fungi are typically round or oval and reproduce by budding. Molds consist of tubular structures called hyphae and grow by branching and longitudinal extension. Not all pathogenic fungi can be categorised as yeasts or molds however. For example, *Coccidioides immitis* are round in tissue but do not bud. Instead, the cytoplasm divides up to form numerous internal spores that, upon rupture of the parent cell, are released to form new spherical structures. Some fungi can grow either yeast-like or as a mold. In patients with candida infection, the fungus is often seen in both tubular and rounded forms. The dimorphic fungi grow in the host as yeast-like forms but grow at room temperature *in vitro* as molds. These fungi include histoplasma and coccidioides. Most fungi reproduce by forming spores through mitosis.

In transplant recipients, fungal infection is usually due to candida and aspergillus organisms. Candida organisms are yeasts, i.e. fungi that exist predominantly in a unicellular form. Hippocrates and Galen described oral lesions that were probably thrush and, in 1839, Langenbeck found fungi in oral lesions of a patient. In 1841, Berg established the fungal aetiology of thrush by inoculating healthy babies with aphthous "membrane material". In 1843, Robin referred to the organism as *Oidium albicans* and over 100 synonyms for Candida albicans followed. The two which have persisted are *Monilia albicans*, which originated from Zopf in 1890 and *Candida albicans* first used by Berkhout in 1923. Zenker described the first documented case of invasive candida infection in 1861 and the first case of candida endocarditis was described in 1940. When antibiotics were introduced into clinical practice in the 1940s, previously undocumented manifestations of candida infections were increasingly described. Examples included arthritis, osteomyelitis, myocarditis, meningitis, peritonitis, myositis and endophthalmitis. In addition to the widespread use of antibiotics, immunosuppressive therapy for organ transplantation has been an important factor in the expanding incidence of observed candida infection. Indeed candida pneumonia is an important cause of opportunistic pneumonia in lung transplant recipients. *Candida glabrata* is sometimes also encountered in transplant recipients. Other yeast-like fungi which may cause infection include *Histoplasma capsulatum, Cryptococcus neoformans and Coccidioides immitis.*

Aspergillus species

Aspergillus is an ubiquitous mold found in the environment of most countries of the world. *Aspergillus fumigatus* is the usual cause of aspergillosis. *Aspergillus flavus* is the second most important species causing human disease in immunosuppressed patients, particularly in lesions which commence in the nose or paranasal sinus. *Aspergillus niger* and other species have been shown to cause infection in immunosuppressed patients.

Diagnosis and treatment

Fungal infection should always be considered in the differential diagnosis of opportunistic infection in the transplant patient. These infections are more commonly encountered in patients receiving high dose steroid therapy or antibiotics and in those who are neutropenic. The most common organisms are *Candida albicans* and *Aspergillus fumigatus*. There should be a high index of suspicion of fungal infection in patients who are failing to respond to antibiotic therapy or those with no obvious cause to explain their clinical presentation of infection. The diagnosis is made by a combination of clinical examination, microscopy and fungal culture of body fluids; detection of fungal antigen in blood; barium or endoscopic examination (GI endoscopy or bronchoscopy with biopsy of affected areas (Fig. 6) for histological

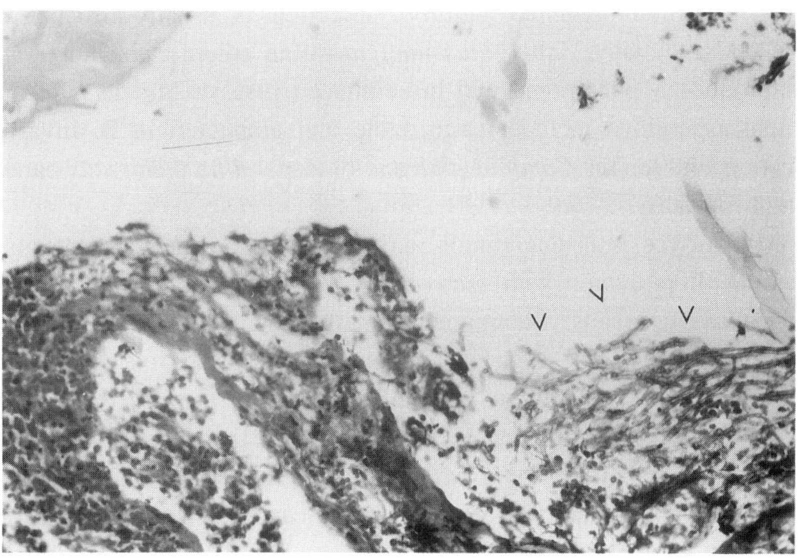

Fig. 6 Transbronchial biopsy from a bone-marrow transplant recipient with invasive aspergillosis. The fungal hyphae are clearly seen (arrow). Haematoxylin and eosin, medium power.

examination). The diagnosis of Cryptococcus meningitis is confirmed by detection of Cryptococci in cerebrospinal fluid by microscopy and culture and by detection of Cryptococcal capsular polysaccharides using a latex agglutination slide technique.

Cutaneous or oral candidiasis is treated with topical antifungal agents, e.g. nystatin or amphotericin lozenges. Oesophageal and pulmonary candidiasis is treated with fluconazole which may be supplemented with amphotericin, particularly as an aerosol treatment in patients with pulmonary candida infection. Disseminated candidiasis can be a particular problem following bone marrow transplantation, and is often preceded by stomatitis, oesophagitis or vulvovaginitis and may produce end ophthalmitis, osteomyelitis, arthritis, myocarditis, pneumonitis, meningitis and nodular skin lesions. Treatment consists of intravenous amphotericin B. The development of the less nephrotoxic liposomal formulation reduces the likelihood of synergism with cyclosporin A in promoting nephrotoxicity in transplant recipients with fungal infection. *Candida glabrata* infection is usually resistant to fluconazole therapy. *Aspergillus fumigatus* can colonise the lungs and can produce a pneumonia and disseminated disease. Agents effective against aspergillus include itraconazole and amphotericin B. Invasive infection with either *Candida albicans* or *Aspergillus fumigatus* carries a high mortality.

Mucormycosis is uncommon and can present with gastrointestinal bleeding. It is treated with intravenous amphotericin B. *Cryptococcus neoformans* can cause meningoencephalitis and if untreated is invariably fatal within a few weeks. Amphotericin B and five flucytosine are usually administered in combination as synergistic activity often results, allowing a reduction in the duration of therapy and total dose of amphotericin B. *Histoplasma capsulatum* can cause bronchopneumonia and pulmonary fibrosis with calcification. Disseminated histoplasmosis is fatal unless treated with amphotericin B. Coccidioides causes respiratory, skin, joint and cerebral disease and can also be treated with amphotericin B.

Bacterial Infection

In 1884, Christian Gram developed an important method of classifying bacteria when he observed that certain bacteria (Gram-positive) retain a crystal violet dye after decolourisation while others (Gram-negative) do not. This staining trait has been shown to be related to the differential structure of the cell wall of Gram-positive and Gram-negative bacteria. Both types of bacteria contain distinct important cell surface constituents but also share other constituents found on the cell surface of all bacteria.

Risk factors for the development of bacterial infection in transplant recipients include prolonged intubation and mechanical ventilation, drains, intravenous cannulas and bladder catheters.

Lung transplant recipients are at particular risk because the allograft is in constant communication with the external environment. Cystic fibrosis lung transplant recipients have been shown to have colonisation with pseudomonal species in their upper respiratory tract and this often leads to allograft infection. Bone marrow transplant recipients frequently develop bacteraemia on account of severe granulocytopenia, particularly in the first two or three weeks post-transplant. Cholangitis can occur following hepatic transplantation and urinary tract infection may develop following renal transplantation.

Important advances in microbiology and in intensive care medicine have shown that prophylactic perioperative antibiotic therapy, although often beneficial in reducing wound infection, may facilitate selection of antibiotic-resistant organisms colonising the skin, oro-pharynx or gastrointestinal tract and that these latter organisms may produce infection. We have observed that patients requiring inotropic support can develop gut ischaemia, which may facilitate translocation of bacterial pathogens into the blood stream leading to the development of multiple organ failure. Staphylococcal septicaemia can occur from the skin as a consequence of impaired local defense due to the presence of central lines or chest drains. Methicillin-resistant *Staphylococcus aureus* may be acquired in the hospital. The importance of aseptic and antiseptic techniques in limiting spread of infection has been mentioned previously.

It has also been shown that impaired gastric acid secretions, e.g. secondary to H2 receptor antagonists or vagal nerve injury during cardiopulmonary transplantation, may facilitate bacterial and fungal overgrowth in the stomach. These organisms can then be aspirated and cause pneumonia. Prolonged endotracheal intubation reduces local pulmonary defense mechanisms and favours the development of bacteria pneumonia. Urinary tract infection can complicate in-dwelling bladder catheters.

If antibiotic therapy is prescribed, infection with *Clostridium difficile* can occur and lead to the development of pseudomembranous colitis. T-cell immunosuppressive agents, such as cyclosporin A and tacrolimus, increase susceptibility to *Streptococcus pneumoniae* and to mycobacterial infection. Measures to prevent infection include prophylactic perioperative antibiotic therapy when central lines, chest or abdominal drains are *in situ*. Nebulised colistin sulphate is effective in reducing the incidence of allograft infection in cystic fibrosis lung transplant recipients. Partial gut decontamination using oral non-absorbable antibiotic regimes can be effective in reducing the rate of endogenous Gram-negative infection, although compliance can be a problem. H2 receptor antagonists should not be prescribed unless there is a definite clinical indication. Reduction of surface flora using chlorhexidine in baths may be beneficial in reducing bacterial colonisation. The majority of transplant units do not routinely employ reverse barrier nursing unless patients are neutropenic. Patients with a past history of tuberculosis receive prophylactic isoniazid.

It is important to have a high index of suspicion for bacterial infection and to commence appropriate treatment early. It should be remembered that signs of infection in immunocompromised patients may be different than in normal individuals. In addition to clinical findings, diagnosis is made by microbiological culture and examination of body fluids, mucosal and skin swabs and stool culture. Broad spectrum empirical antibiotic therapy is frequently employed while culture results are awaited. One should remember that aminoglycosides and vancomycin should be used with caution as these agents are potentially synergistic in producing nephrotoxicity when prescribed in conjunction with cyclosporin A. If *Clostridium difficile* infection is suspected, the organism or its toxin may be identified in stool specimens. Antibiotics effective against this agent include oral vancomycin or metronidazole. Pulmonary nocardiosis is rare and carries a high mortality. It can be treated with high dose co-trimoxazole or ampicillin. For *Nocardia asteroides*, there is synergism between co-trimoxazole and erythromycin. If patients with signs of infection fail to respond to broad spectrum antibiotic therapy, one should consider the following:

1. Inadequate initial therapy suggesting the need for additional or alternative agents.
2. Resistant bacterial super infection.
3. The presence of a focus requiring surgical drainage.
4. Infection with viruses of protozoa.
5. Fungal infection.

Bronchoscopy

It will now be apparent that bronchoscopy[5] is an important clinical adjunct in the diagnosis and management of opportunistic pneumonia. It was perhaps prompted by quotations like the following, "It adds a precious seeing to the eye" (William Shakespeare — *Love's Labour Lost*, 4.iii, 329) or "More is missed by not looking than by not knowing" (Thomas McCrae 1870–1935) or "If a man will begin with certainties, he will end with doubts; but if he will be content to begin with doubts, he will end in certainties" (Francis Bacon — *Advancement of Learning*), that independently and almost simultaneously, two pioneers, Gustav Killian in Europe and Chevalier Jackson in America, developed new medical instruments from which the new field of broncho-oesophagology emerged. In fact, the idea of bronchoscopy can be traced back as far as Hippocrates; however, the first usable rigid bronchoscopes were not developed until the turn of the century by Gustav Killian in Germany (1897) and Chevalier Jackson in the United States (1907). Rigid bronchoscopy initially remained within the remit of the thoracic surgeon in most hospitals until the introduction of the fibre-optic bronchoscope by Ikeda in Japan in 1966. Since then, rigid and flexible bronchoscopies are performed by both thoracic surgeons and physicians. The indications for bronchoscopy are many and varied and the diagnostic yield is high. Since the late 1970s, there has been a dramatic expansion in the role of the bronchoscope as a therapeutic instrument.

Almost any puzzling pulmonary condition should be considered for bronchoscopy. Modern techniques of anaesthesia and ventilation ensure the procedure's safety; patients must be seriously ill or have gross reduction is respiratory reserve before bronchoscopy becomes dangerous. In practice, there are now few contra-indications, although some patients who have gross reduction in respiratory reserve, those with advanced superior vena caval obstruction requiring urgent radiotherapy, and those in whom an aortic aneurysm is suspected, are usually contra-indicated from bronchoscopic examination. Bronchoscopy can be used to diagnose respiratory disease by macroscopic examination of the vocal cords and endobronchial tree, by taking trap sputum and bronchoalveolar lavage for cytological, microbiological and immunocytochemical analysis, by examining protected bronchial brush specimens for cytological examination and culture and by performing endobronchial and transbronchial biopsies to study tumour histology, diagnose interstitial lung disease, document ciliary motility studies or confirm opportunistic infection or allograft rejection (in lung transplant recipients). Since the 1970s, there has been a tremendous increase in the use of the bronchoscopy to treat patients with respiratory disease. Examples include removal of foreign body or secretions, palliation for bronchial carcinoma, stricture dilatation, endobronchial stent insertion, treatment of airway complications after lung transplantation and bronchoalveolar lavage for alveolar proteinosis.

The use of endobronchial laser therapy facilitates removal of granulation tissue, e.g. post-tracheostomy or at the level of the airway anastomosis following lung transplantation, and can also be used to debulk tumours. The word laser is the acronym of Light Amplification by Stimulated Emission of Radiation. The concept of an intense pencil beam of light as produced by a laser was probably first suggested in science fiction novels, for example H. G. Wells wrote in *War of the Worlds*, of alien invaders who attacked the earth using powerful beams of optical energy and, comic book heroes such as *Dan Dare* and *Buck Rogers* used death rays and ray guns. It was not until 1960 that the first laser was constructed. It has had an enormous impact on all areas of science, technology and medicine.

Consequences of Antimicrobial Therapy in Transplant Recipients

With increasing experience and the development of transplant medicine as a medical specialty in its own right, tremendous knowledge has been gained in post-transplant medical care over the past 20 years. No only have we improved our diagnostic modalities for allograft rejection and opportunistic infection but there has been ongoing development of newer more selective immunosuppressive therapy with a better therapeutic index, together with more potent anti-microbial agents. It should always be remembered, however, that drugs prescribed to treat infection may interact with immunosuppressive therapy. For example, erythromycin, itraconazole, ketoconazole and fluconazole may increase plasma cyclosporin and tacrolimus concentrations, whereas anti-tuberculous drugs may reduce plasma levels of both of these drugs as a consequence of liver enzyme induction. Thus, if agents known to interfere with cyclosporin A or tacrolimus are prescribed, the dose of the immunosuppressive drug must be adjusted appropriately and blood concentrations frequently monitored. In this way, the risks of toxicity or of rejection caused by sub-therapeutic concentrations can be avoided. It has also been shown that aminoglycosides, lincosamides and amphotericin may potentiate renal impairment when prescribed in conjunction with cyclosporin A or with tacrolimus. Ganciclovir and acyclovir can also impair renal function and ganciclovir may also cause bone marrow suppression. If renal impairment occurs, it may be necessary to reduce the dose of the calcineurin antagonist, thereby increasing the risk of allograft rejection. Should bone marrow depression occur as a consequence of co-trimoxazole or ganciclovir therapy, it may be necessary to reduce the dose of or stop azathioprine therapy. The combination of antibiotics and corticosteroids can facilitate fungal infection. If patients develop pseudomembranous colitis, absorption of immunosuppressive drugs may be impaired thus increasing susceptibility to allograft rejection. In this situation, intravenous immunosuppressive therapy is necessary until normal bowel function returns.

We have also seen that many transplant recipients wish to resume as normal a quality of life as possible, and in this regard, many wish to travel overseas. They are advised of the fact that diarrhoea and vomiting can lead to impaired absorption of immunosuppressive therapy and therefore increased likelihood of rejection. Medical attention should be sought urgently should diarrhoea and vomiting occur. Live viral (smallpox, yellow fever, measles, rubella and oral polio) and bacterial (BCG) vaccines are not permitted on account of the risk of producing infection in an immunocompromised patient. Inactivated viral (influenza, typhus, rabies and polio [Salk]) and bacterial (typhoid, paratyphoid, cholera and pertussis) vaccines are permitted as are toxoids (tetanus and diphtheria). The antibody response to genetically engineered vaccines (hepatitis A and B) is unpredictable on account of immunosuppression.

Conclusion

The aim of transplantation is improvement in survival and quality of life. Excellence in preservation and surgical techniques, together with post-operative medical care, has led to improved survival and quality of life for the successfully transplanted patient. We are indebted to those pioneers who throughout the centuries have tirelessly and relentlessly pursued their clinical observations and therefore have provided us with a tremendous understanding of infectious disease, diagnostic modalities and therapeutic options. They have contributed so much to transplant medicine of today. We look forward to ongoing research into this area which will undoubtedly make post-operative management of the transplant recipient even better in the future.

References

1. Lyons AS, Petrucelli RJ (1987). *Medicine: An Illustrated History.* New York, NY: Abradale Press, Harry N Abrams, Inc.
2. Mandell GL, Douglas RG, Bennett JE (1990). *Principles and Practice of Infectious Diseases, 3rd Ed.* New York, NY: Churchill Livingstone Inc.
3. Hakim S (1997). *Introduction to Organ Transplantation.* London: Imperial College Press.
4. Madden BP (1998). Immunocompromise and opportunistic infection in organ transplantation. *Surgery* **16**(2), 37–40.
5. Madden BP, Murday AJ (1996). Bronchoscopy in the diagnosis and management of respiratory disease. *Surgery* **14**(1), 6–8.

XIX

THE HISTORY OF ETHICAL ISSUES IN TRANSPLANTATION

R Randal Bollinger
Jeremy Sugarman

Introduction

Transplantation, the transfer of tissues from one individual to another, finds mention in Greek mythology and early Christian legends. In the Iliad, Book VI, Homer described the mythical chimera, part lion, part goat and part serpent. In the early centuries of Christianity, there appeared the legend of Damian and Cosmas who posthumously attached the severed leg of a recently dead Moor to the amputation stump of a parishioner.

In the 18th century, John Hunter experimented with auto-, allo- and xenotransplantation in chickens, yet clinical transplantation is a product of the 20th century. As the barriers to transplantation have been overcome, ethical issues have arisen at each juncture and have been addressed by the profession, the public, legislative bodies and the courts. Indeed during the second half of the 20th century, advances in transplantation have produced a nearly continuous challenge to medical ethics. For example with increasing patient and graft survivals for kidney, liver, heart, pancreas and lung transplantation, procedures that had been unethical to apply to patients became ethical to attempt and in some cases unethical not to try.

Nations and regions vary in their state of medical resources and capabilities that affect their ability to apply the expanding knowledge of transplantation. In areas with limited resources to meet the health care needs of their populations, the ethical principle of justice suggests that priority be given to general medical and public health programs, such as immunization and treatment of infectious diseases. Such programs benefit everyone in contrast to transplantation which at best can help only a few individuals. Although organ transplantation may be made less costly, and is already more cost-effective than some alternative treatments of end stage organ failure (e.g. renal transplantation versus hemodialysis), it would still be ethically inappropriate to apply it in situations where its still formidable cost precludes meeting higher priority health care needs of the population. Nations vary also in their legal, political, religious and social readiness to undertake human organ transplantation. What is morally appropriate in one society with well established criteria for neurological or "brain" death codified into its system of laws may be morally inappropriate in another society that does not recognize the neurological diagnosis of death. Japan, for example has been for decades among the nations with the most developed social and economic systems in the world. Japan, however, did not until 1997 accept the neurological diagnosis of death, so harvesting of vital organs from cadavers was largely precluded. Consequently, prior to this acceptance of such criteria, it was unethical to attempt heart or lung transplantation in Japan even though the medical knowledge and resources to perform the procedure successfully were available.

Many other ethical issues associated with transplantation are also important to consider with a global perspective. For instance, there is no universally accepted agreement on use of living donors, use of animal donors, payment for organs, indications for recipient listing, donor suitability and allocation of organs. Such disagreements call out for ethical analysis, even if no single, ethically "correct" answer emerges that will serve all people in all places for all times.

Advances in transplantation technology also continually challenge prior conclusions about what is appropriate. For example, the artificial heart as a possible bridge to transplant has pitted its developers against the society that must contend with its consequences. Thomas A. Preston states: "It is a classic conflict between the economic independence of individuals and the determination of the public good in an increasingly interdependent society."[1] Under these circumstances of continual change, persons involved in transplantation in one place may learn and benefit from the experiences and ethical deliberations of others who have faced similar problems in the past. Although much

Table 1 Resolved and Unresolved Ethical Issues in Transplantation

Issue	Current Concensus	Controversial Special Cases
Living donors	Resolved in favor	–Unrelated living donors –Segmental liver, lung, pancreas –Adolescent donors
Brain death	Resolved in favor	–Anencephalic –Vegetative states –Ventilation for donation
Payments for organs	Resolved against	–Rewarded gifting –In kind incentives –Preferential status
Informed consent	Resolved in favor	–Expanded donor pool –Presumed consent –Binding advanced directives
Allocation of organs	Controversial	–Sickest first –Equalize waiting time –Local vs national vs international use
Suitability of organs	Controversial	–Split organs –Non-heart beating donors –Aged or diseased donors
Xenotransplantation	Controversial	–Bridge to transplant –Non-human primates –Genetic engineering

can be gained from understanding the evolution of transplantation in societies with beliefs different from one another, this chapter focuses on the history of transplantation ethics in developed, especially Western European, North American and Pacific rim countries because most solid organ transplants have been performed there. Selected experiences of South American, African, Middle Eastern and Asian countries are recounted as well as the historical ethical issues in transplantation vary in their degree of resolution. Even where consensus was reached, there are new or special cases which arouse controversy (Table 1). Of course, the ethics of transplantation are evolving even in countries where transplantation is prevalent, in part because outcomes continue to improve and medical and scientific capabilities increase. This chapter will deal sequentially with the historical development of the ethics of taking organs from living donors, brain death, payment or other rewards for organs, allocation of human organs as a scarce resource, suitability of organs for transplantation and xenotransplantation. Related issues such as informed consent, confidentiality, publicity, scientific trials and proof, the roles of government, professional and other organizations, the laws of biology and society in general in the historical development of the ethics of transplantation are addressed.

The Use of Living Donors

Until successful outcomes could be achieved, transplantation remained theoretical and experimental. Emerich Ullmann's first functioning renal transplant in a dog reported in 1902[2] led to numerous attempts to achieve long term survival using animal organs from living donors. These were uniformly unsuccessful and transplantation remained a theoretical possibility for human patients for the next half century. The coincidence of several developments changed the situation in the middle of the 20th century. Wilhelm Kolff developed an artificial kidney, which for the first time kept patients with acute kidney failure alive for extended periods. Drs Jean Hamburger[3] in Paris and David

Hume in Boston attempted to place kidneys into some of these patients who had chronic renal failure. Because their success with cadaver donors was limited in part by organ viability, the ethics of removing healthy kidneys from living donors was among the first issues to be addressed when transplantation became possible.

Dr Joseph Murray[4] who received the 1990 Nobel Prize in Medicine for the first successful transplant in man faced the issue squarely before undertaking renal transplantation between identical twins. The issue was brought before a court in Massachusetts in 1954. The presiding judge ruled that the possible harm to the donor from the loss of a kidney was more than matched by the psychiatric effects if the potential donor were to lose his identical twin brother. The transplant was allowed to proceed and living related donation was established as a means of providing kidneys for end stage renal disease patients.

The concept of living donors was extended to non-identical twins in 1959 and subsequently to biologically related, then to non-genetically related family members. Use of living donors was initially based upon the improved outcomes associated with high degrees of HLA match. Kidney graft survivals in some programs were 40% higher at one year with organs from living donors in comparison to cadaver donors. New immunosuppresive agents that became available in the last decades of the 20th century have greatly reduced the relative survival advantage of living donor transplants. The new immunosupression has also improved early survival of living unrelated donor transplants to nearly equal that of related donor.

The small advantage of living donor graft survival must be weighed against the physical and psychological risks to the donor, albeit the risk is small for kidney transplantation. Mortality is statistically insignificant, but morbidity, including potentially life-threatening problems like hemorrhage and pulmonary embolus remain relevant. The physical risks are greater for living donors of segmental liver,[5] segmental lung,[6] and segmental pancreas[7] transplants.

The very act of asking a relative to serve as a living donor also raises ethical issues related to privacy and perceptions of coercion.

As has been observed, the person being asked to donate may have some reason to object to being a donor, whether a personal fear or one that relates to the safety of the organs. For example, the prospective donor may knowingly be infected by a potentially transmittable agent and be reluctant to tell family members. Further, despite a prospective donor's desire not to be a donor, there may be intrinsic family pressure to do so. In some cases, transplant teams have been asked by prospective donors to lie to family members as a means of protection. Accordingly, the decision to deem a donor "acceptable" must be sensitive to these issues.

In addition, even if the outcome advantage of living donors has narrowed relative to that of cadaver donor grafts, the elements of availability and immediacy remain as distinct advantages of live organ donation. Human organs are scarce and waiting times are rising. Long waits and gradual deterioration on support systems are bypassed if a living donor is available. Immediacy is particularly advantageous for children in urgent need of lung or liver transplants. Segmental grafts removed from their parents may be life saving. Parental instincts for preservation of their children create irresistible pressures to donate organs or parts of organs to their children.

In summary, the process of live organ donation has resulted in good biological outcomes (Table 2). Despite improved results with cadaver donor transplants which reduce the outcome advantage of live donor organs, the scarcity of human organs and the difficulty of providing cadaver organs urgently will preserve the role of live organ donation as an alternative for the near future. A contrary view is expressed by Dr Thomas E. Starzl, the founding father and driving force behind liver transplantation, who stated: "soon, if current trends continue, it may be hard to justify using living donors." He asked pointedly whether "...knowing what we know now should we go on encouraging living relatives to provide organs — taking whatever risks that entails — when families of brain-dead people may never be asked whether they wish to donate their dead relatives' organs? Is it ethical to harm the living before harvesting the dead? Such questions need to be debated in the light of changing clinical information."[8]

Table 2 The historical evolution of living donor transplantation

1951	Transplantation of allogeneic kidneys removed for therapeutic reasons (Kuss, 1951).
1954	Successful renal transplantation between identical twins (Murray, 1955).
1962	Successful renal transplantation from non-identical twin living donors (Hamburger, 1991).
1978	Segmental pancreas transplantation from living donors (Sutherland, 1984).
1989	Segmental liver transplantation from living donors (Raia, 1989).
1990	Segmental lung transplantation from living donors (Starns, 1994).

The improvements in graft survival are not limited to cadaver donor grafts, however. Living unrelated donor transplants perform exceeding well, which fact has prompted Martyn Evans to espouse the view that organ donations should not be restricted to relatives but rather "should be welcomed when clinically appropriate and truly voluntary."[9] In the cases of live donor transplants, the risks and benefits are not equal since the donor bears most of the risk but the recipient receives most of the benefit. As discussed by Shaw *et al.* in the case of lung transplantation from live donors, physicians, surgeons and institutional review boards or research ethics committees have a special obligation under these circumstances to minimize risks and maximize benefits through the selection process and to promote optimal decision-making by donors and recipients through the best possible consent process.[10] They point out from an international perspective that live donors may be the sole source of lungs for transplantation in countries such as Japan or India, where cadaveric donation is limited by religious and cultural prohibitions against the concept of brain death. The same may be said for kidney, liver and pancreas transplantation in such areas. In contrast, live donor heart transplantation is universally condemned because human sacrifice is not condoned.

Living donation is an exception to the rule that procedures that risk harm to patients should be intended to serve their welfare in that the donors are required to take risks and perhaps be harmed for the sake of the recipients. When competent people give consent to such procedures, a conflict is created between the principles of beneficence and autonomy as pointed out by Elliot. He has drawn a distinction between allowing a person to risk harm to himself and encouraging it: "Substantial payment to organ donors or volunteers for dangerous research arguably crosses the line between allowing and encouraging." Procurement of organs from cadaveric sources eliminates this conflict and is a strong argument to develop better programs to identify, procure and preserve cadaveric organs.

Brain Death

Death has historically been declared when the heart has stopped beating, breathing has ceased, and neither returns. However, modern intensive care techniques with respiratory, cardiac and hemodynamic support allow cardiopulmonary function to be maintained long after all cognitive and other brain functions of the individual are irrevocably lost. The question of deciding when to stop artificial support of such patients led to intense deliberations and ultimately to the concept of neurological or "brain" death.

Prior to the acceptance of the brain death statutes (Table 3), the law assumed that death was declared when all vital signs were absent. Death was defined as "the cessation of life; the ceasing to exist; defined by physicians as a total stoppage of the circulation of the blood, and a cessation of the animal and vital functions consequent thereupon, such as respiration, pulsation, etc." The acceptance of brain death occurred in several stages. In 1968, the Ad Hoc Committee of the Harvard Medical School to Examine the Definition of Brain Death developed a definition of irreversible coma.[11] A collaborative study of 503 comatose and apneic patients in the United States reported criteria that would allow the diagnosis of a dead brain.[12] In 1981, the

Table 3 The acceptance of brain death in the United States

Year	Event (Reference)
1968	Definition of irreversible coma (Ad Hoc Committee of Harvard Medical School)
1977	Criteria that allow diagnosis of a dead brain (A collaborative study)
1981	Guidelines for the determination of death (President's Commission for the study of Ethical Problems)
1981–1990	Uniform Determination of Death Act in all 50 states and District of Columbia

President's Commission for the Study of Ethical Problems in Medicine and Biomedical and Behavioral Research received a report from their medical consultants on the diagnosis of death and published "Guidelines for the Determination of Death".[13] The President's Commission opinioned "when respiration and circulation have irreversibly ceased, there is no need to assess brain functions directly. When cardiopulmonary functions are artificially maintained, neurological criteria must be used to assess whether brain functions have ceased irreversibly" and proceeded to outline the criteria for determination of neurological death as well as to discuss complicating conditions. The President's Commission proposed a model statute intended for adoption in every jurisdiction in the United States:

Uniform Determination of Death Act

An individual who has sustained either

1. irreversible cessation of circulatory and respiratory functions or
2. irreversible cessation of all functions of the entire brain, including the entire brain stem, is dead.

A determination of death must be made in accordance with accepted medical standards. After years of additional debate in Europe and North America, the concept that death may be diagnosed by either traditional or neurological means has gained general acceptance and is no longer a topic of intense ethical debate. All 50 states and the District of Columbia now accept this standard. In fact, brain death is now well-established in law and medical practices throughout the Western world.[14]

Brain death has proved to be a most important concept for the progress of organ transplantation from cadaver donors. The long periods of warm ischemia which were associated with declaration of death by cessation of cardiac function precluded transplantation of the heart, liver and lungs which suffered irreversible damage after a few minutes of warm ischemia. Even the kidneys, which tolerated better the insult of warm ischemia, became unusable after an hour of warm ischemia. In those parts of the world where brain death was given legal standing and became standard practice, vital organ transplantation increased rapidly.

In other parts of the world where religious and social prohibitions prevented the declaration of brain death, transplantation was severely limited to occasional human living donors and to animal experimentation. In Japan, for example, no cardiac transplantation was performed between 1968 and September 1997 due primarily to "social disapproval".[15] In October 1997, a new Japanese law went into effect which permits organ harvesting from patients with brain death who had previously expressed their will to donate organs and expressed this desire in writing. Clinical transplantation limited previously to those procedures for which living donors were available has subsequently blossomed. Heart transplantation has begun in Japan and the ethical tensions created by Japanese patients traveling abroad to receive heart transplants has abated. That is, when patients traveled abroad to obtain organs from brain dead donors, they used scarce resources, depriving some members of that society from doing so.

Despite consensus regarding criteria for brain death, the persistent vegetative state in which all higher brain functions are lost but brain stem and hypothalamic activity persist, is a tragic one that still incites eated ethical discussions. Note that a large part of the debate about using organs from these persons is a debate about withdrawal of support. Although in many cases the organs from patients in persistent vegetative states would be usable for transplantation, the appropriateness of harvesting them is not settled and will likely remain controversial for some time to come.

Similarly, anencephalic infants, which are born brain absent may survive for short periods with brain stem function only. They are born with total and permanent loss of consciousness but are not dead by the criteria of the Act. Some have argued that anencephalic infants are "so very different from other living infants — and their future so radically limited — that it is permissible with the fully informed and freely-given consent of the parents to remove their organs for transplantation."[16] This approach would suggest a special category, neither alive nor dead, for the infant born without a brain raising conceptual, ethical and legal issues.

Establishing new categories to include not only anencephaly but also persistent vegetative states and other types of coma, requires consideration of these issues. Additional challenges relate to the fact that total and permanent loss of consciousness may be difficult to prove and starts down the ethical "slippery slope" of considering other neurologically damaged persons as potential organ donors. Further, there seems to be marked variation among different members of society regarding those conditions that will warrant substantial attention and respect as efforts are made to develop practical procedures.

Payments of Rewards for Organ Donation

Are human organs for transplantation a gift given altruistically to help one's fellow man or are they a commodity to exchange through a commercial market by payment or barter? Whether human organ donation and distribution should be commercialized has evoked widespread debate among clinicians, economists, entrepreneurs, philosophers, ethicists, lawyers, legislators and the public. In their 1988 symposium on Organ Transplantation Policy: Issues and Prospects, Drs Blumenstein and Sloan discuss development of a market economy for human body parts. Organs are viewed as products and transplantation as a commercial industry that would be improved by closer conformity to a market model.[17] Market considerations are prompted by the disappointing rate of altruism in the population, even after public education in medically sophisticated areas such as the Eurotransplant region.[18] On the one hand, proponents of a free market system point out the large advantages in terms of organ supply that might follow from providing a monetary reward system for persons or their families who would consent to donation of cadaver organs. On the other hand, there are strong objections to making humans or their organs and tissues a commodity in... "a wholly secularized marketplace that permits one to reduce any and all things to assets for sale. Nothing is sacred; neither the body as a whole nor any of its parts. A thing is worth only what someone is willing to pay for it."[19] William F. May summarizes "in matters so fundamental as the donation of human organs, it is argued by Christian theologians (and others), *giving and receiving* are better than routine *taking and getting* and certainly to be preferred to *buying and selling*."[19]

The consideration of permitting, even encouraging, commercial transactions in human organs and tissues was discussed intensively at the first joint meeting of the European Society for Organ Transplantation and the European Dialysis and Transplant Association in December 1990.[20] It is clear from review of the 12 papers presented on ethical issues that no universal consensus exists concerning the possible role of commerce and transplantation. Indeed, practices and policies differ across the globe. All North American and most South American and European countries have legal prohibitions on commerce in human organs. Many African, Middle Eastern and Asian countries do not, and proscription of the sale of human organs has not been accepted universally by individuals of different cultures.[21]

K. C. Reddy points out that the poorer the society, the greater the social acceptability of less optimal forms of end stage renal disease management.[22] In Third World countries which lack maintenance hemodialysis and cadaver transplantation programs, living donor transplantation, including organs from paid donors is often the only available alternative to death. India has no laws which make the "sale" of organs illegal, has a large population of poor people willing to sell their organs, and accepts sacrifice as a way of life in the Hindu culture. The concept of "rewarded gifting" is widely accepted and practiced. More than half of the end stage renal disease (ESRD) patients in India have no suitable living related donors so, in the absence of hemodialysis or cadaveric donor transplantation programs, will die if not permitted to obtain organs from living donors. Regarding the risk involved with living donor donation, Reddy states "it is illogical to suggest that the risk involved is all right if it is taken in the spirit of love, but not when money exchanges hands between the parties involved for mutual benefit. When money is involved to achieve a good end with harm to none, the use of that money does not make the transaction immoral."

Accordingly, Reddy and his colleagues in Madras, India state "we choose to take a stand to buy rather than to let die, to attempt to return to useful lives some of the patients afflicted with ESRD, and at the same time give an opportunity to the donor and their families to improve their standard of living."[22] A. S. Daar of Oman differentiates rewarded gifting from commercialism. He believes that donations with appropriate and governmentally controlled incentives can be practiced without rampant commercialism.[23]

Objections, coming primarily from the developed Western world, regard paid human organ donation variously as a type of slavery, a means of victimizing the poor, an impediment to the development of cadaveric donation and/or a form of corporeal prostitution. Moreover, there is a high mortality rate among recipients of bought living-unrelated donor kidneys.[24]

Furthermore, the bad experiences in Greece with a private transplantation trading program are outlined by representatives of the Hellenic Transplantation Society.[25] Rizvi and Naqvi point out the negative fall-out in Pakistan of organ donation commercialism in neighboring India and urge that transplant teams resist the temptation of offering "rewarded gifting" in the developing world to avoid exploitation of the poor where regulation and supervision are impossible.[26] Abouna *et al.* based upon their experience in Kuwait considered carefully the numerous negative impacts of paid organ donation on donors, recipients, local transplant programs, the medical profession, local society and the international community. They concluded that "in view of the profound negative feedback effect of organ sales on voluntary and legitimate organ donation, it is imperative that all forms of paid organ donation be made illegal in all countries of the world if the efforts to create local cadaver procurement programs are to succeed."[27] As a result of experiences and arguments such as these, the 1990 European Congress accepted resolutions that prohibited commerce to obtain organs or tissues for allotransplantation. Pointedly, the Congress did *not* accept "rewarded gifts" as tolerable or permissible even in cultures where they are distinguishable from commerce.

Nickerson *et al.* conclude that "nothing short of a market test can demonstrate conclusively the impact that financial incentives would have on requests for organ donation...."[28] Neither the well designed empirical research suggested by them nor a controlled market test have yet been performed. The destruction of altruism and voluntary gifting, which it is feared would result from large monetary payments to donors or their families, is believed to be less likely when gifts in kind or non-monetary incentives are used. For example, a modest contribution to the funeral expenses of an organ donor may be helpful to the family without converting their altruistic donation to a commercial transaction. Similarly, giving priority to patients who had previously signed organ donor cards for organs if they themselves require transplantation would create the incentive of "preferential status."

Controversies clearly exist over the moral justification for regulating the gifting, trading or selling of organs. Previously used bases for resolving controversies over moral authority are: (1) force, (2) conversion to a common point of view, and (3) appeal to sound rational arguments.[29] The first two possibilities are inappropriate or impossible and the last is unlikely given the very different sources of moral authority invoked today by different societies and cultures of the world. However, mutual consent and mutual respect can, as pointed out by Englehardt, provide a source of common moral authority.

Are mutual consent and mutual respect leading to a consensus on commercial transactions in human organs and tissues? This appears to be the case as more than 50 nations from all continents have established laws prohibiting profit but permitting compensation for donor expenses and losses.[21] The Transplantation Society which represents the profession worldwide has stated clearly and repeatedly "no transplant surgeon/team shall be involved directly or indirectly in the buying or selling of organs/tissues or in any transplant activity aimed at commercial gain for himself/herself or any associated hospital or institute. Violation of these guidelines by any member of The Transplantation Society may be cause for expulsion from the Society." The World Medical Association has condemned the purchase and sale of human organs for transplantation and called on governments to prevent the commercial use of human organs. The Council of Europe has stated that "a human organ must not be offered for profit by any organ exchange organization, organ banking centre or by any other organization or individual whatsoever. However, this does not prevent the compensation of living donors for loss of earning and any expense caused by the removal or preceding examination." The Council of Europe also prohibits advertising outside the national territory either for donation or transplantation. The League of Arab States has stated that "the sale, purchase or remunerated donation of organs is prohibited, and no specialist may perform a transplant operation if he knows the organ to have been provided through such means." The World Health Assembly which governs the World Health Organization

(WHO) in regard to the trade for profit in human organs among living human beings has stated that "such trade is inconsistent with the most basic human values and contravenes the Universal Declaration of Human Rights and the spirit of the WHO constitution" in WHA resolution 40-13. Furthermore in WHA resolution 42-5, the Assembly called upon member states to take appropriate measures to prevent the purchase and sale of human organs for transplantation, to introduce legislation to prohibit trafficking in organs, and to discourage all practices which facilitate commercial trafficking in organs. In the United Kingdom, physicians have been barred for life from practicing medicine in England for their role in kidney transplants involving four paid Turkish donors. In the United States, the National Organ Transplant Act specifically prohibits the sale of human organs. The Bellagio Task Force Report on Transplantation, Body Integrity and the International Traffic in Organs in 1997 found "...no unarguable ethical principle that would justify a ban on the sale of organs under all circumstances."[30] However, they found that "...present abuses are so grave that the self-determination of would-be sellers of organs must be curtailed to protect the most vulnerable." They recommended a two-pronged policy with a ban on the sale of solid organs from live unrelated donors and experimentation with close evaluation of programs to reward families of donors of cadaveric organs. The Task Force rejected the use of organs from executed prisoners and from prisoners of war because neither are in a position to give voluntary informed consent to organ donation.

Thus, while consensus has evolved that commercial sale of human organs is unacceptable, there remains controversy about other alternatives, such as a paired-kidney-exchange program between two living donor-recipient pairs where the donors are ABO-incompatible with the intended recipients but are ABO-compatible with other recipients,[31] or the giving of preference to prior volunteers when allocating organs for transplantation.[32] Nevertheless, continued deliberation about these possible approaches, accompanied by empirical data regarding public attitudes concerning them would be welcome as attempts are made to resolve problems related to organ scarcity.

Organ Suitability

The severe worldwide shortfall in the number of donor organs for transplantation has stimulated re-examination of past donor practices in an effort to increase the number of organ failure patients who can be transplanted. Segments of liver, lung and pancreas are being taken from living donors, and unrelated living donors and are being used. Moreover, the definition of a suitable cadaver donor has undergone relentless expansion. Ethical issues arise continuously at the limits of current medical practice regarding the use of split organs, non-heart beating donors, and aged or diseased donors. The success in expanding the liver donor pool by using segments of living donors[33] has raised the possibility that more than one potential recipient could be treated if cadaver livers were also split into two grafts. The techniques of split liver grafting are now well established, especially to meet simultaneously the needs of a child and an adult.[34] The results with split liver transplants have been good, but not as good as with whole organ grafts, particularly in urgent situations, raising the ethical issues of when this technical advance should be applied and how to balance the risks and benefits of the resulting incremental increase in donor supply.

The debate over non-heart-beating cadaver donors in North America and Europe is summarized well in a 1995 monograph on the subject.[35] Patients who were declared dead by cardiopulmonary rather than neurological criteria were the original cadaver organ donors in the 1950s and 1960s. However, non-heart-beating cadaver donors were essentially not used once neurological criteria for death were accepted because of the organ damage from warm ischemia that occurred with the former but not the latter type of organ donor. The increasingly severe organ shortfall has led to a re-awakening of interest in obtaining organs from non-heart-beating cadavers by limiting the period of warm ischemia, hence preservation damage, associated with their procurement. Both *in situ* preservation with cold solutions and rapid retrieval of organs after therapy is withdrawn under controlled conditions have been used to limit warm ischemia. Each of these approaches raises ethical tensions between therapeutic and non-therapeutic treatment of the donor, informed consent for treatments and donation, conflicts between the interests of transplant centers and the potential donor and donor's family, and the timing of the determination of death. While some have seen non-heart-beating protocols as an opportunity for altruism by preserving the option of donation for family members,[36] others have viewed it as "an ignoble form of cannibalism."[37] Arnold and Younger note society's growing acceptance of circumstances where the timing of the patient's death is effected by medical intervention.[38] They suggest that rather than continually revising the line between life and death, it would be preferable to abandon the dead donor rule, which states that persons must be dead before their organs are taken, and rely instead entirely on informed consent, by the patient or family before organ retrieval as a primary safeguard against abuse.

International differences in views of euthanasia add to the controversy. Patients may be allowed to die in the United States, but must not be actively killed. In contrast, under certain circumstances active euthanasia is permitted in The Netherlands. Moreover, one state, Oregon, has outlined a mechanism permitting physician-assisted suicide. Regardless of increased attention to treatment at the end of life, it is unlikely that euthanasia for purposes of organ donation will be adopted.

The Institute of Medicine of the United States has examined the medical and ethical issues in recovering organs from non-heart-beating donors who do not meet the standard of brain death.[39] The study report concludes that "the recovery of organs from non-heart-beating donors is an important, medically effective, and ethically acceptable approach to reducing the gap that exists now and will exist in the future between the demand for and the available supply of organs for transplantation." The report outlines guiding principles and makes recommendations for a national policy on the topic to include written, locally approved protocols, public openness, case by case decisions on anticoagulants and vasodilators for organ preservation, family consent for pre-mortem cannulation, conflict of interest safeguards, determination of death, preservation of the family's option to be in attendance at life support withdrawal, and financial protection for the donor and the donor's family. Regarding the cessation of cardiopulmonary function, the report recommends "that not less than a five-minute interval, determined accurately by electronic and arterial pulse pressure monitoring, be required to determine donor death in controlled, non-heart-beating donors."

Expansion of the donor pool by the use of organs from the very young and very old or from those with known diseases like diabetes or infections have raised serious ethical issues. Kidneys from patients at the extremes of age have poorer function and shorter survival than those from patients between the ages of 5 and 65. Moreover, the percentage of older donors is increasing as the population ages. Is it ethical to transplant a kidney with statistically known inferior function such as reduced creatinine clearance, histological glomerular sclerosis or intra-renal vascular changes into a patient with end stage renal disease? An appropriate alternative may be to place both kidneys from a very young or a very old donor into the same recipient. Double adult renal allografts from donors greater than 60 years of age with abnormally low creatinine clearances can afford recipients renal function comparable to that achieved with younger kidneys.[40] Additional ethical concerns about organ and recipient suitability arise when the potential donor[16] and the potential recipient[41] are fetuses identified *in utero* by prenatal ultrasound.

Utilization of organs from diseased donors remains controversial. Sound decision-making requires that the circumstances in each case be examined individually. Medically important features related to the disease found in the donor must be delineated and decisions made based upon the potential effects it might have upon a recipient. For example, certain diseases, such as diabetes cause changes in organs like diabetic nephropathy which may be reversed by transplantation of the diabetic kidney into a non-diabetic recipient. In contrast, patients with infections such as human immunodeficiency virus (HIV), hepatitis B (HVB) and hepatitis C (HVC) may well have infectious virus in one or many of their solid organs, organs which would almost certainly infect the immunosuppressed recipient. In fact, transmission of disease has been one of the reasons for high mortality among recipients of bought living-unrelated donor kidneys.[18, 27] Although it may be literally impossible to eliminate all risk of transmissible disease in organ transplantation, medical factors may determine the degree of risk in particular circumstances. Hepatitis B is more likely to be transmitted by liver transplantation than by heart transplantation. Some would use organs even if they came from donors with known HIV, HBV or HCV infections if used in recipients already infected with the corresponding virus while others would not. Further complicating matters are those at risk for infection. For instance immunodeficiency virus (HIV) is more likely to be present in donors who inject illicit drugs and share needles or who are sexually promiscuous than in donors who do not. In the absence of proven infection, should donors with such risk factors for being infected be used for life saving organ transplants when potential recipients will die without them? From the viewpoint of someone faced with imminent death, answering such a question might not seem difficult, despite the desire to avoid harm in attempting to treat his disease. This state of despair poses obvious risks wherefore ethical use of the organs depends upon informed consent by the recipients (as discussed later in this chapter). Obtaining consent can be challenging in such settings.

Organ Allocation

Vital human organs for transplantation are a life-saving resource in short supply. Under conditions of scarcity, the principles and rules used for allocation of solid organs arouse heated ethical debates. It is obvious that for patients and families who receive needed organs in time the allocation rules are good, but for those who die waiting they are bad. Since the allocation decisions are life and death ones for patients, and the donations hinge in large part on a trustworthy system, the ethical basis for them should be widely known among the donating public and defensible in public forums.

Many factors may be considered for inclusion in the allocation formula. As a matter of justice, there must be equity or fairness throughout the process as well as good outcomes of transplants. In determining allocations, time waiting and medical urgency are important factors. Factors of utility or outcome are also essential and may be measured by patient and graft survivals. As a practical matter then, histocompatibility match, preservation time, procurement cost and geographic distance, and other factors which are associated with enhanced survival become important considerations. Should a scarce resource go to the person who would benefit most by receiving it or should every candidate have an equal chance of getting that resource? Should the organ go to the potential recipient who has waited the longest or the one who is sickest or the one who will realize most benefit? The lack of universally agreed upon and uniformly applied listing criteria for transplant recipients make difficult the process of organ allocation in practice.

Veatch has argued that, in principle, even perfect knowledge of the medical facts about the patients who are candidates cannot determine which patient should receive an organ...."[42] He contends that the clinician should provide data for those who create and run the allocation system, they should not make the allocation itself." Physicians are compelled to give undivided loyalty to the patients to preserve their patient's lives, relieve their suffering and maintain their health. The social good of developing a just system of allocation is a task for the general public, which supplies the pool of cadaver organs. The balancing of maximum utility and maintaining fairness is a task of the entire society, not just the medical profession.

The need for a new partnership between society and the medical profession in coming up with appropriate hospitals to the issues at hand was recognized by Senator Albert Gore, Jr., later Vice-President of the United States. Mr. Gore wrote "cases such as Baby Fae involve more than questions of medical science, they involve societal questions of medical ethics, resource allocation, and health care policy, and for answers they require new decision-making process that has as its cornerstone public participation."[43] Gore saw the need to meld society's views with the facts of medical science: "while a physician's scientific expertise plays an essential role in the decision-making process, we must through our institutions and our laws, develop the means by which society can collectively provide the guidance needed to make these decisions." These views were placed into action by the creation of a transplantation task force, passage of a national organ transplant act in the United States, and establishment of a unique partnership between government and the private sector, both public and professional, known as the United Network for Organ Sharing (UNOS). UNOS created a forum in the United States for the discussion and resolution of complex ethical issues surrounding organ donation, allocation, outcomes and quality that arose from clinical transplantation.

In the United States, a point system developed by the United Network for Organ Sharing (UNOS) is used to assign by formula the relative weight given to many different factors, factors which differ according to type of organ transplant. The selected balance between factors of justice (like urgency and waiting time) and factors of utility (like histocompatibility, preservation time or cost) has varied drastically at different points in the history of United States organ allocation. The current effort of UNOS is directed towards achieving an approximate equal balance between factors of utility and justice in the allocation of all solid organs for transplantation. However, no matter what balance is chosen, the realities of current medical knowledge and biological and logistical characteristics limits its application. For example, even if justice demanded that a certain urgent patient who had waited the longest be given a newly donated heart, the transplant could not be done if long distance transport caused the preservation time to exceed the six-hour maximum for that organ after which viability is threatened. In the case of kidney transplantation where time is not as critical, there is often a trade-off between the utility factors of high HLA match and preservation time. Cadaver donor kidneys are often transported long distances to achieve transplants between phenotypically identical donors and recipients which improves long term graft survival. The price is long preservation time, which diminishes graft function, at least in the short term. A similar trade-off results when kidneys are transported long distances to the patients who are waiting the longest or who are the sickest.

A question has arisen in many parts of the world whether organs for transplantation are a *local* resource given by individuals to help members of their community or are a *national* resource belonging to the entire society that has organized the system of organ donation and transplantation. In an area such as Europe with numerous nations, should sharing occur internationally? Should organs be allocated to non-resident or legal aliens who happen to be in a territory when their medical need for transplantation arises or who travel from a society without cadaveric organ donation to one where cadaver donor organs are available? The answers are unclear but in most situations, geography seems to be a poor means for the allocation of scarce, life-saving resources. At a minimum, within a society if all members are expected to donate organs, all must be eligible to receive them in times of need. Creating a system of equal access is a great moral and ethical challenge, one which requires consideration of the ethical impact of other factors like self-interest and ability to pay, as well as nationality. However, creating a system of equal access should encourage cadaveric programs to grow and stimulate the creation of cadaveric programs where none now exist.

Xenotransplantation

Xenotransplantation, the transplantation of organs or tissues between members of different species, has aroused an enormous degree of emotional debate. Although the large and ever-increasing disparity between patients waiting for transplants and the number of organ donors available to meet their needs has led to the tragedy of many patients dying while waiting, it has also led to emotional pleas for using organs from non-human primates and lower animals. Such pleas have evoked an equally emotional response from animal rights activists and those who oppose genetic engineering.

The idea of transplanting organs from lower animals to man is an old one. In 1906, Jaboulay attempted a pig-to-man kidney transplant, and in 1910, Unger performed an ape-to-man kidney transplant, both without success. In 1963, Reemtsma proposed that "new exploration of this field now seems warranted because of success with immunosuppresive measures in homografts and because of difficulties in obtaining suitable human organs",[44] conditions which pertain equally well today. He performed several transplants from chimpanzees to humans, one of which survived nine months.

The modern re-awakening of interest in this field (Fig. 1), spawned by the increasing success of allogeneic transplantation with newer immunosuppressive agents (e.g. cyclosporine, tacrolimus and mycophenolate mofetil) and monoclonal antibodies together with the ever more severe shortage of organs for transplantation, has prompted renewed efforts which in turn have sparked vigorous ethical controversy. The transplantation of a baboon heart into "Baby Fae" on October 14, 1984, in Loma Linda, California allowed the baby born with a hypoplastic heart to survive until November 15, 1984.[45] Huge controversies erupted over animal use, informed consent and human experimentation in children.[46] Similar reactions occurred in June and July 1992 after baboon livers were transplanted to human patients in Pittsburgh, Pennsylvania.[47] Although these transplants failed soon after implantation, they resulted in efforts to create a more suitable porcine donor using transgenic technology. The antibody-mediated, complement-dependent hyperacute rejection which characterized discordant xenografts has been overcome by placing human genes for complement regulatory proteins into the germ line of pigs. In this way,

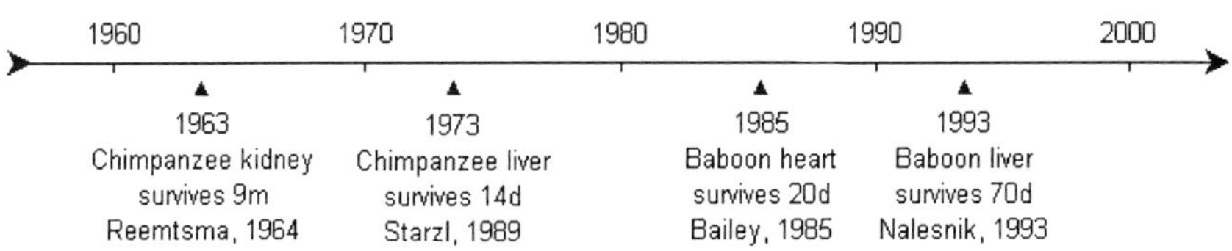

Fig. 1 Key points along the time-line of modern human xenotransplantation.

genetic engineering of porcine donors has solved at least partially the problems of acute xenograft rejection. At the same time, new and pressing ethical issues in xenotransplantation have been pushed to the forefront. Genetic engineering, cloning of individuals, animal rights, animal use, population-wide risk of zoonoses, human experimentation, informed consent and patient rights have all been the subject of ethical debates.

The barriers to xenotransplantation are formidable but the rewards of success in overcoming with respect to organ availability are huge. The donor supply would become essentially unlimited. Porcine donors have the advantages that they come in all sizes, are readily available, have few human pathogens, have frequent litters, and can be genetically engineered. In these regards, pigs differ significantly from non-human primate organ donors. Given these numerous potential advantages, it seems remarkable that so few human transplants of porcine organs have been attempted. The powerful restraints of medical ethics are partially responsible. Beneficence requires that transplant surgeons provide treatments that help or at least have the potential to help their patients. However, *ex vivo* perfusion of porcine livers has been used successfully to bridge liver failure patients to allotransplantation.[48]

The movement of xenotransplantation from the realm of clinical research to therapeutic intervention has been further restricted by the recent discovery of two sets of human-tropic pig endogenous retroviruses (PERVs). These viruses, present in all porcine cells, have envelope genes that determine their tropism. Both are capable of infecting human cells in culture. They exist in two forms, PERV-a and PERV-b which have 92% amino acid identity. Each porcine cell tested to date has ten to 23 copies of the PERV-a provirus and seven to ten copies of the PERV-b provirus.[49] Despite the fact that these viruses have never been demonstrated to infect humans, their existence and the fact that they can infect human cells *in vitro* have led to a moratorium in Europe and the United States on further human trials with porcine organ transplants. Means of monitoring PERVs and other potential xenograft infections and contingency plans for dealing with positive tests must be developed before further human trials of porcine organ transplants can be undertaken in the United States and similar restrictions have been applied in the United Kingdom.

The possibility of zoonoses raises a set of powerful questions about xenotransplantation and perhaps limits experimentation with it. Even if mechanisms are developed to monitor some forms of these agents, it is unclear whether it is permissible to experiment on individuals when it poses threats to society even though the individual might stand to benefit. Thus, continued deliberation about this possibility and establishing appropriate levels of review and monitoring for safety will be critical.

In addition, even if the immunological and infectious barriers to using porcine organs in humans are overcome, the concerns about animal rights will remain. A complete historical discussion of the ethics of animal use is beyond the scope of this chapter. The reader is referred to the seminal work of Peter Singer[50] for the philosophical underpinnings of the animal rights movement.

James Lindeman Nelson has pointed out that we have a fundamental problem with debates about such issues, namely "we don't really know what we are talking about. The distinction of humans and the rest of creation rests upon.... the range and power of the human intellect, the complexity and depth of our interpersonal relationships, our passions, both personal and esthetic, our sense of morality and of tragedy...." However, not only animals, but many humans, such as those who are brain injured, comatose and mentally ill, lack these strengths. Medical ethics preclude our taking vital organs from ill, injured or mentally impaired humans for the benefit of other persons. Nelson states "we are at a loss to say what it is about baboons that makes their livers fair game, when we wouldn't take vital organs from those of our own species whose ability to live rich full lives are no greater than those of the non-humans we seem so willing to prey upon. Unless we are able to isolate and defend the relative moral distinction, we should reject the seductive image of solving the problem of organ shortage by maintaining colonies of animals at the ready for transplantation on demand."[51] The ethical challenge is diminished but not eliminated by turning to farm animals, which are produced and used by humans in prodigious quantities for food and fiber. If, in addition to their agricultural use, such animals also produced life-saving organs for humans, the morality of their use by and for humans might actually be increased. The objections of those who believe all animals have rights comparable to those of man or who have religious beliefs that preclude any use of animals will continue to produce ethical and legal challenges to xenotransplantation.

Consent for Transplantation

Informed consent is now a familiar mechanism of providing protection to patients and research participants. Informed consent evolved separately but largely concurrently in medical practice and research.[52] Informed consent for medical practice grew out of a series of court cases, where procedures were done to patients without their permission. Gradually informed consent became established in medical practice, sometimes based on the rulings on particular cases and at other times based on governmental policies. For research, informed consent evolved following the revelation of scandals and the work of international and national commissions and became established in all codes of research ethics.

From the standpoint of ethics, informed consent involves giving those persons with adequate decision-making capacity who are positioned to make a voluntary choice, adequate information about a proposed intervention in a way that they can understand and then authorize their agreement to participate.

Transplantation has produced several particular challenges to informed consent as discussed in previous sections of this chapter. Does the urgency of organ failure allow time for the process of patient education, understanding and agreement that constitutes valid informed consent? Can commercial payment for human organs be made without corrupting voluntary, informed consent? What is the role of informed consent for living donors, especially parents who give parts of their lungs or livers to their children, when the risk is to the donor but the benefit is to the recipient?

In addition to these and other issues of consent related to the current practice of transplantation, there is a looming issue of informed consent for ever more borderline organs from the expanded donor pool. By expanding the donor pool to include persons with potentially transmissible infections, malignancies, extremes in donor age, anatomic anomalies and other known detrimental circumstances, there are challenges to the informed consent process to donation and transplantation. Sicker donors may have trouble in understanding what is asked and those with stigmatizing conditions may be reluctant to provide essential medical information for potential recipients. It is essential that they understand the trade-offs in using these organs. Worse outcomes and known biological risks may occur with organs from an expanded pool. At a minimum, the process of informed consent should give patients outcome information regarding the special risks of the organs they are about to receive.

Public trust in general and patient trust in particular depend on full and open discussion of the potential risks and benefits of medical treatments and informed consent requires such disclosure even if some elect to refuse transplantation from an expanded pool. The American Society of Transplant Surgeons Ethics Committee has debated this issue and concluded that transplant surgeons should disclose to patients information about any proven increased risks involved when using organs from the expanded donor pool of non-traditional donors.

Conclusion

The relative roles of government, society and the medical profession in the definition of transplant ethics varies around the world but each has had a significant historical part to play. Despite greater understanding about the ethical issues related to transplantation, a series of ethical concerns linger. Some have rejected the... "rescue-oriented and often zealous determination to maintain life at any cost; and a relentless hubris-ridden refusal to accept limits. It is disturbing to witness, over and over, the travail and distress to which this outlook can subject patients....."[53] They raise the ethical issue, seen also in modern medical management of malignancy and the practice of intensive care medicine, of "when to stop." Paul Ramsey argued in 1970 that we should discover the moral limits properly surrounding efforts to save life.[54] The problem was epitomized by surgeons who performed numerous re-transplants when the transplanted organs failed because of the physician's commitment to his or her patient and belief that life must be maintained at any cost and even if human suffering was the primary result. A more balanced approach would consider the implications of re-transplantation and the fair use of societal resources and the need to avoid hurting patients in the process of trying to help them. Important medical considerations include whether organ failure occurred earlier than programmed and if the quality, not just the quantity, of life were likely to be improved by the re-transplantation procedure. Another as yet unresolved set of ethical issues is the proper role of biology in decisions made in human transplantation. Tolerance, programmed death, the aging process and our ultimate mortality are appropriate topics of discourse. Whereas some lament the pervasive reluctance of humans to accept the biological limits imposed by the aging process and view transplantation as one manifestation of that reluctance,[55] others see in those same biological limitations a challenge for future research and understanding. Advances in our ability to control biological processes like the aging of human beings, and the analogous processes in their organ transplants which now lead over time to graft loss, could alter significantly our views of transplantation ethics. As the future of human beings and their organs is extended, the ethical virtues of a future market in cadaveric organs[56] may become apparent. Recent developments in the biology of cloning and embryonic stem cells forsage new and continuing challenges to the ethics of transplantation. Cells, tissues and possibly even organs for transplantation into humans may someday be created by applying to humans the somatic cell nuclear transfer cloning technique used to create Dolly, the cloned sheep,[57] or from pluripotent cell lines isolated directly from the inner cell mass of human embryos at the blastocyst stage.[58] In transplantation ethics, history is but prologue to a more challenging future.

Acknowledgements

The assistance of Ms. Mary Ann Rohrer in the preparation of this manuscript is gratefully acknowledged.

References

1. Preston TA (1985). Who benefits from the artificial heart? *Hastings Cent Rep* February 1985, pp. 5–7.
2. Ullmann E (1902). Experimentelle nierentransplantation. *Wiener klinische Wochenschrift* **15**(11), 1.
3. Hamburger J (1991). Memories of old times. In Terasaki PI (ed.) *History of Transplantation: Thirty-Five Recollections.* Los Angeles: UCLA Tissue Typing Laboratory, pp. 61–72.
4. Murray JE, Merrill JP, Harrison JH (1955). Renal homotransplantation in identical twins. *Surg Forum* **6**, 432.
5. Raia S, Nery JR, Mies S (1989). Liver transplantation from live donors. *Lancet* **ii**: 497–498.
6. Starnes VA, Barr ML, Cohen RG (1994). Lobar transplantation: indications, technique and outcome. *J. Thorac Cardiovasc Surg* **108**, 403–411.
7. Sutherland DE, Goetz FC, Najarian JS (1984). One hundred pancreas transplants at a single institution. *Ann Surg* **200**(4), 414–440.
8. Starzl TE (1985). Will live organ donations no longer be justified? *Hastings Cent Rep* April 1985, p. 5.
9. Evans M. (1989) Organ donations should not be restricted to relatives. *J Med Ethics* **15**, 17–20.

10. Shaw LR, Miller JD, Slutsky AS, *et al.* (1991). Ethics of lung transplantation with live donors. *Lancet* **338**, 678–680.
11. Ad Hoc Committee of Harvard Medical School (1968). A definition of irreversible coma. *JAMA* **205**(6), 85–88.
12. A Collaborative Study (1977). An appraisal of the criteria of cerebral death. *JAMA* **237**(10), 982–986.
13. President's Commission for Study of Ethical Problems in Medicine (1981). Guidelines for the determination of death. *JAMA* **246**(19), 2184–2186.
14. Pearson IJ(1997). Brain death. In Chapman JR, Deierhoi M, Wight C (eds.) *Organ and Tissue Donation for Transplantation*. London: Arnold, pp. 69–90.
15. Koretsune Y, Hori M, Sato N, *et al.* (1998). Natural history of patients referred to the Osaka University cardiac transplant program in relation to their medication. *Transpl Proc* **30**, 94–95.
16. Ethics and Social Impact Committee, Transplant Policy Center, Ann Arbor, MI (1988). Anencephalic infants as sources of transplantable organs. *Hastings Cent Rep* October/November 1988, pp. 28–30.
17. Blumenstein JF, Sloan FA (eds.) (1989). *Organ Transplantation Policy: Issues and Prospects*. Durham, N.C.: Duke University Press.
18. Schutt G, Duncker G (1995). Disappointing rate of altruism in the population. In Touraine JL, Traeger J, Betuel H, *et al.* (eds.) *Organ Shortage: The Solutions*. Dordrecht, The Netherlands: Kluwer Academic Publishers, pp. 49–52.
19. May WF (1985). Religious justifications for donating body parts. *Hastings Cent Rep*, February 1985, pp. 38–42.
20. Land W, Dossetor JB (eds.) (1991). *Organ Replacement Therapy: Ethics. Justice. Commerce*. Berlin: Springer-Verlag.
21. Fluss SS (1991). Preventing commercial transactions in human organs and tissues: an international overview of regulatory and administrative measures. In Land W, Dossetor JB (eds.) *Organ Replacement Therapy: Ethics. Justice. Commerce*. Berlin: Springer-Verlag, pp. 154–163.

22. Reddy KC (1991). Organ donation for consideration: an Indian view point. In Land W, Dossetor JB (eds.) *Organ Replacement Therapy: Ethics. Justice. Commerce.* Berlin: Springer-Verlag, pp. 173–180.
23. Daar AS (1991). Rewarded gifting and rampant commercialism in perspective: is there a difference? In Land W, Dossetor JB (eds.) *Organ Replacement Therapy: Ethics. Justice. Commerce.* Berlin: Springer-Verlag, pp. 181–189.
24. Salhudeen AK, Woods HF, Pingle A, *et al.* (1990). High mortality among recipients of bought living-unrelated donor kidneys. *Lancet,* **336**, 725–728.
25. Koniavitou-Hadjiyannaki K, Protogerou D, Drakopoulos S, *et al.* (1991). The ugly head of commercialism in organ transplantation in Greece. In Land W, Dossetor JB (eds.) *Organ Replacement Therapy: Ethics. Justice. Commerce.* Berlin: Springer-Verlag, pp. 200–202.
26. Rizvi SA, Naqvi SA (1991). Fallouts of commercialism in organ donation as seen in Pakistan. In Land W, Dossetor JB (eds.) *Organ Replacement Therapy: Ethics. Justice. Commerce.* Berlin: Springer-Verlag, pp. 203–205.
27. Abouna GM, Sabawi MM, Kumar MSA, Samhan M (1991). The negative impact of paid organ donation. In Land W, Dossetor JB (eds.) *Organ Replacement Therapy: Ethics. Justice. Commerce.* Berlin: Springer-Verlag, pp. 164–171.
28. Nickerson CAE, Jasper JD, Asch DA (1998). Comfort level, financial incentives, and consent for organ donation. *Transpl Proc* **30**, 155–159.
29. Englehardt Jr HT (1991). Is there a universal system of ethics or are ethics culture-specific. In Land W, Dossetor JB (eds.) *Organ Replacement Therapy: Ethics. Justice. Commerce.* Berlin: Springer-Verlag, pp. 147–153.
30. Rothman DJ, Rose E, Awaya T, *et al.* (1997). The Bellagio task force report on transplantation, bodily integrity, and the international traffic in organs. *Transpl Proc* **29**, 2739–2745.

31. Ross LF, Rubin DT, Siegler M, *et al.* (1997). Ethics of a paired-kidney-exchange program. *New Engl J Med* **336**, 1752–1755.
32. Jarvis R (1995). Join the club: a modest proposal to increase availability of donor organs. *J Med Ethics* **21**, 199–204.
33. Emond J, Heffron T, Kortz E, *et al.* (1993). Improved results of living-related liver transplantation with routine application in a pediatric program. *Transplantation* **55**(4), 835–840.
34. Fawcett J, Balderson G, Lynch SV, Strong RW (1998). Split liver transplantation: two grafts from one donor is the optimal use of a scarce resource. *Transp Rev* **12**(2), 64–73.
35. Arnold RM, Youngner SJ, Schapiro R, Spicer CM (eds.) (1995). *Procuring Organs for Transplant.* Baltimore: Johns Hopkins University Press.
36. Orlowski J (1995). The opportunity for altruism: preserving options for family members. In Arnold RM, Youngner SJ, Schapiro R, Spicer CM (eds.) *Procuring Organs for Transplant.* Baltimore: Johns Hopkins University Press, pp. 187–194.
37. Fox RC (1995). "An ignoble form of cannibalism": reflections on the Pittsburgh protocol for procuring organs from non-heart-beating cadavers. In Arnold RM, Youngner SJ (eds.) *Procuring Organs for Transplant.* Baltimore: Johns Hopkins University Press.
38. Arnold RM, Youngner SJ (1995). The dead donor rule: should we stretch it, bend it, or abandon it. In Arnold RM, Youngner SJ, Schapiro R, Spicer CM (eds.) *Procuring Organs for Transplant.* Baltimore: Johns Hopkins University Press, pp. 219–234.
39. Herdman R, Potts JT (1997). *Non-Heart-Beating Organ Transplantation. Medical and Ethical Issues in Procurement.* Washington, D.C.: Institute of Medicine, National Academy Press.
40. Johnson L, Kuo PC, Schweitzer J, *et al.* (1998). Double adult renal allografts. *Transpl Rev* **12**(2), 59–63.
41. Michaels MG, Frader J, Armitage J (1993). Ethical considerations in listing fetuses as candidates for neonatal heart transplantation. *JAMA* **269**(3), 401–402.
42. Veatch RM (1991). Who empowers medical doctors to make allocative decisions for dialysis and organ transplantation? In

Land W, Dossetor JB (eds.) *Organ Replacement Therapy: Ethics. Justice. Commerce.* Berlin: Springer-Verlag, pp. 331–336.

43. Gore Jr A (1985). The need for a new partnership. *Hastings Cent Rep* February, 1985, p. 13.
44. Reemtsma K, McCracken BH, Schlegel J, Pearl MA, DeWitt CW, Smith PE, Hewitt RL, Finner RL, Creech O (1964). Renal heterotransplantation in man. *Ann Surg* **160**(3), 384–410.
45. Bailey LL, Nehlsen-Cannavella SL, Concepcion W, Jolley WB (1985). Baboon to human cardiac xenotransplantation in a neonate. *JAMA* **254**, 3321–3327.
46. The Subject is Baby Fae (1985). *Hastings Cent Rep* February 1985, pp. 8–17.
47. Nalesnik MA, Fung JJ, Starzl TE, *et al.* (1993). Pathology studies in two baboon to human liver xenograft cases. In *Proceedings Second International Congress on Xenotransplantation, Cambridge, UK September 26, 1993.*
48. Harland RC, Bollinger RR (1994). Extracorporeal hepatic perfusion in the treatment of patients with hepatic failure. *Transpl Rev* **8**, 73–79.
49. LeTissier P, Stoye JP, Yasuhiro T, *et al.* (1997). Two sets of human tropic pig retroviruses. *Nature* **389**, 681–682.
50. Singer P (1990). *Animal Liberation,* 2nd Ed., New York: Random House.
51. Nelson JL (1992). Transplantation through a glass darkly. *Hastings Cent Rep* September–October 1992, pp. 6–8.
52. Faden RR, Beauchamp TL (1986). *A History and Theory of Informed Consent.* New York: Oxford University Press.
53. Fox RC, Swazey JP (1992). The participant observers: final journeys. In *Spare Parts.* New York: Oxford University Press, p. 204.
54. Ramsey P (1970). *The Patient as a Person: Explorations in Medical Ethics.* New Haven: Yale University Press.
55. Fox RC, Swazey JP (1992). Leaving the field. *Hastings Cent Rep* September-October 1992, pp. 9–15.
56. Cohen LR (1991). The ethical virtues of a future market in cadaveric organs. In Land W, Dossetor JB (eds.) *Organ Replace-*

ment Therapy: *Ethics. Justice. Commerce.* Berlin: Springer-Verlag, pp. 302–310.

57. Klotzko AJ (1998). Voices from Roslin: the creators of Dolly discuss science, ethics and responsibility. *Cambridge Quart Healthcare Ethics* **7**, 121–140.
58. Thompson J, *et al.* (1998). Embryonic stem cell lines derived from human blastocysts. *Science* **282**, 1145–1147.

INDEX